# Fundamentals of EMS, NMS, and OSS/BSS

## OTHER TELECOMMUNICATIONS BOOKS FROM AUERBACH

**3D Television (3DTV) Technology, Systems, and Deployment**
Daniel Minoli
ISBN 978-1-4398-4066-5

**Applied Aspects of Optical Communication and LIDAR**
Nathan Blaunstein, Shlomi Arnon, Natan Kopeika, and Arkadi Zilberman
ISBN 978-1-4200-9040-6

**Cloud Computing and Software Services: Theory and Techniques**
Edited by Syed A. Ahson and Mohammad Ilyas
ISBN 978-1-4398-0315-8

**Cognitive Radio Networks: Architectures, Protocols, and Standards**
Edited by Yan Zhang, Jun Zheng, and Hsiao-Hwa Chen
ISBN 978-1-4200-7775-9

**Cooperative Wireless Communications**
Edited by Yan Zhang, Hsiao-Hwa Chen, and Mohsen Guizani
ISBN 978-14200-6469-8

**Evolved Cellular Network Planning and Optimization for UMTS and LTE**
Lingyang Song and Jia Shen
ISBN 978-1-4398-0649-4

**Fundamentals of EMS, NMS and OSS/BSS**
Jithesh Sathyan
ISBN 978-1-4200-8573-0

**Handbook on Mobile Ad Hoc and Pervasive Communications**
Edited by Mieso K. Denko and Laurence T. Yang
ISBN 978-1-4398-4616-2

**HSDPA/HSUPA Handbook**
Edited by Borko Furht and Syed A. Ahson
ISBN 978-1-4200-7863-3

**IP Communications and Services for NGN**
Johnson I. Agbinya
ISBN 978-1-4200-7090-3

**Mobile Device Security: A Comprehensive Guide to Securing Your Information in a Moving World**
Stephen Fried
ISBN 978-1-4398-2016-2

**Mobile Opportunistic Networks: Architectures, Protocols and Applications**
Edited by Mieso K. Denko
ISBN 978-1-4200-8812-0

**Orthogonal Frequency Division Multiple Access Fundamentals and Applications**
Edited by Tao Jiang, Lingyang Song, and Yan Zhang,
ISBN 978-1-4200-8824-3

**Overlay Networks: Toward Information Networking**
Sasu Tarkoma
ISBN 978-1-4398-1371-3

**RFID and Sensor Networks: Architectures, Protocols, Security, and Integrations**
Edited by Yan Zhang, Laurence T. Yang, and Jiming Chen
ISBN 978-1-4200-7777-3

**Security of Self-Organizing Networks: MANET, WSN, WMN, VANET**
Edited by Al-Sakib Khan Pathan
ISBN 978-1-4398-1919-7

**Security of Mobile Communications**
Noureddine Boudriga
ISBN 978-0-8493-7941-3

**Socio-Technical Networks: Science and Engineering Design**
Edited by Fei Hu, Ali Mostashari, and Jiang Xie
ISBN 978-1-4398-0980-8

**Transmission Techniques for Emergent Multicast and Broadcast Systems**
Mario Marques da Silva, Americo Correia, Rui Dinis, Nuno Suoto, and Joao Carlos Silva
ISBN 978-1-4398-1593-9

**Underwater Acoustic Sensor Networks**
Edited by Yang Xiao
ISBN 978-1-4200-6711-8

**Wireless Sensor Networks: Principles and Practice**
Fei Hu and Xiaojun Cao
ISBN 978-1-4200-9215-8

**ZigBee Network Protocols and Applications**
Chonggang Wang, Tao Jiang, and Qian Zhang
ISBN 978-1-4398-1601-1


**AUERBACH PUBLICATIONS**
www.auerbach-publications.com
To Order Call: 1-800-272-7737 • Fax: 1-800-374-3401
E-mail: orders@crcpress.com

# Fundamentals of EMS, NMS, and OSS/BSS

JITHESH SATHYAN

CRC Press is an imprint of the
Taylor & Francis Group, an **informa** business
AN AUERBACH BOOK

Auerbach Publications
Taylor & Francis Group
6000 Broken Sound Parkway NW, Suite 300
Boca Raton, FL 33487-2742



Printed in the United States of America on acid-free paper
10 9 8 7 6 5 4 3 2 1

International Standard Book Number: 978-1-4200-8573-0 (Hardback)



**Library of Congress Cataloging-in-Publication Data**

Sathyan, Jithesh.
Fundamentals of EMS, NMS, and OSS/BSS / author, Jithesh Sathyan.
p. cm.
Includes bibliographical references and index.
ISBN 978-1-4200-8573-0 (hardcover : alk. paper)
1. Computer networks--Management. 2. Support services (Management)--Data processing. I. Title.

TK5105.5.S14 2010
004.6--dc22 2010017682

**Visit the Taylor & Francis Web site at**
**http://www.taylorandfrancis.com**

**and the Auerbach Web site at**
**http://www.auerbach-publications.com**

# Contents

# Foreword

In the telecommunications industry, the rapidly evolving technologies and their changes make it very difficult to develop efficient support systems. In the past, different vertical networks addressing different needs and distinctive networks with definite management strategy were the easiest approaches taken by operators. However, in a world where operators need to find ways to differentiate their offerings and keep the margins healthy, the above "stovepipe" approach is no longer sufficient and could hamper good business practice.

With advances in technology, the communications network and its management are becoming more complex as well as more dynamic and fluid in nature. As with any other field of study involving the term *management*, the world of telecommunications management also tends to be opaque and filled with jargon and terminology. It is a field that would benefit immensely from literature that demystifies and simplifies all its various aspects.

*Fundamentals of EMS, NMS, and OSS/BSS* has excellent coverage of the basics of telecom management, from legacy management systems to the next generation systems. In this book, Jithesh introduces the principles of EMS, NMS, and OSS/BSS and goes in-depth on implementation guidelines including the most common design patterns in use. This book makes two important contributions. First, it shows the roles that EMS, NMS, and OSS/BSS can play in the telecom environment with complex real-time systems. Second, it provides a very realistic view of use cases and implementation examples illustrating how the aspects covered in the first three sections of the book can be put into practice.

With his experience in delivering telecom projects, Jithesh has explained element management systems (EMS), network management systems (NMS), and operation/business support systems (OSS/BSS) through concepts, functionality descriptions, applicable protocols, and implementation guidelines. Each section touches upon definitions and models, explaining the complex concepts without losing simplicity or introducing technical jargon, and finally directing the user to further reading material to enable a better, more detailed understanding.

I am honored to have had the opportunity to work directly with Jithesh in architectural design efforts for the telecom support systems and I am sure this book will be a valuable asset for academicians and professionals. It is with great pleasure that I introduce this book. I wish Jithesh all the success in this and future endeavors.

**Manesh Sadasivan**
*Principal Architect, Product Engineering*
*Infosys Technologies Limited (www.infosys.com)*

# Preface

The communication service industry is moving toward a converged world where services like data, voice, and value-added services are available anytime and anywhere. The services are "always on," without the hassle of waiting or moving between locations to be connected or for continuity in services. The facilities offer easy and effortless communications, based on mobility and personalized services that increase quality of life and lead to more customer satisfaction. Service providers will have much more effective channels to reach the customer base with new services and applications. With changes in services and underlying networks, the ease in managing the operations becomes critically important. The major challenge is changing business logic and appropriate support systems for service delivery, assurance, and billing. Even after much maturity of IMS technology, there has been a lag in the adoption of this service because of the absence of a well-defined OSS/BSS stack that could manage the IMS and proper billing system to get business value to service providers to move to IMS. This makes operations support systems a topic of considerable interest in today's ICT industry.

Operations support systems (OSS) as a whole includes the systems used to support the daily operations of the service provider. These include business support systems (BSS) such as billing and customer management, service operations such as service provisioning and management, element management, and network management applications. In the layered management approach, BSS corresponds to business management and OSS corresponds to service management, while the network level operation support is done using a network management system and individual resources are managed with an element management system. The same terminology is used in the title of the book to explain the different layers of telecom management that are covered in this book.

*Fundamentals of EMS, NMS, and OSS/BSS* completely covers on the basics of telecom resource and service management. This book is designed to teach all you need to understand to get a good theoretical base on telecom management. The book has four sections: Element Management System, Network Management System, Operation/Business Support Systems, and Implementation Guidelines. The first section covers element management system in detail. Initial efforts in managing elements to the latest management standards are covered in this part. The second section deals with the basics of network management; legacy systems, management protocols, standards, and popular products are all handled in this part. The third section deals with OSS/BSS, covering the process, applications, and interfaces in the service/business management layers in detail. The final section gives the reader implementation guidelines to start developing telecom management solutions.

This book is dedicated to my beloved wife for her patience and consistent support that helped me a lot in completing this book. I would like to thank the product engineering team at Infosys for giving me numerous engagements in EMS, NMS, and OSS/BSS space with a variety of telecom clients in multiple countries that went a long way in giving me the knowledge to write this book. Writing this book has been a very good experience and I hope you enjoy reading it as much as I enjoyed writing it.

**Jithesh Sathyan**
*Technical Architect*
*Infosys Technologies Limited*

# About the Author

**Jithesh Sathyan** is a technical architect at Infosys Technologies Limited (NASDAQ: INFY). He is the sole inventor of the first granted patent of Infosys at the U.S. Patent and Trademark Office. He has to his credit multiple international papers in the telecom domain, specifically on EMS, NMS, and OSS/BSS. He is a gold medalist in electronics and communication engineering.

The author has been working on multiple projects involving design and development of telecom management solutions for a variety of telecom clients in OEM and service provider space. He has multiple filed patents in the telecom domain. Jithesh has published several white papers in standardizing forums, such as TeleManagement Forum that specializes in defining standards on telecom management. He is an active member of many telecom management standardizing forums. He has also contributed a chapter entitled "Role of OSS/BSS in Success of IMS" in *IMS Handbook: Concepts, Technologies, and Services* published by CRC Press.

# I

# ELEMENT MANAGEMENT SYSTEM (EMS)

# Chapter 1

# What Is EMS?

This chapter is intended to provide the reader with a basic understanding of Element Management System. At the end of this chapter you will have a good understanding of what is a network element, the need for managing a network element, how an element management system fits into telecom architecture, and some of the major vendors involved in the development of EMS.

## 1.1 Introduction

Industries are highly dependent on their networking services for day-to-day activities from sending mail to conducting an audio/video conference. Keeping services running on the resource is synonymous with keeping the business running. Any service is offered by a process or set of processes running on some hardware. A set of hardware makes up the inventory required for keeping a service up and running. If the hardware goes down or when there is some malfunction with the hardware, this would in turn affect the service. This leads to the need to monitor the hardware and ensure its proper functioning in order to offer a service to a customer. This hardware being discussed is the element and an application to monitor the same constitutes the element management system. Since the element is part of a network, it is usually referred to as network element or NE.

> A network element is the hardware on which a process or set of processes is executed to offer a service or plurality of services to the customer.
>
> EMS is an application that manages a telecom network element.

Most element management systems are available in the market support management data collection from multiple network elements though they cannot be called a network management system. Hence an application is not necessarily a network management system just because it supports data collection from multiple NEs. The functionalities offered by the systems decide whether it is a network management system (NMS) or an element management system (EMS). Let us make this point clear with an example. A topology diagram in an EMS would have the nodes configured in the element while a topology diagram in an NMS will show all the configured network elements managed by the NMS. Again a fault management window on an EMS would show only the logs and alarms generated on the network element it is managing. The functionalities like fault correlation and decision handling based on events from various NEs are shown in an NMS.

The functionalities of an EMS are covered in detail in the chapters to follow. This introduction is intended to only give the reader a feel of what a network element and EMS is in telecom domain. In the sections to follow we will explore the architecture of EMS and take some sample EMS products to gain familiarity with EMS applications available in the market.

## 1.2 EMS in Telecom

As shown in Figure 1.1, an element manager (EM) collects data from the network elements (NE). An ideal scenario would involve one element manager to collect data from a single network element as in EM-2 and NE-n. It can be seen that this is not the case for an actual EMS product. Almost all the EMS products available in the market support management of a set of elements as shown for element managers other than EM-2 in Figure 1.1.

> Even when data is collected from multiple NEs, the management system is still an element manager when data collection and processing

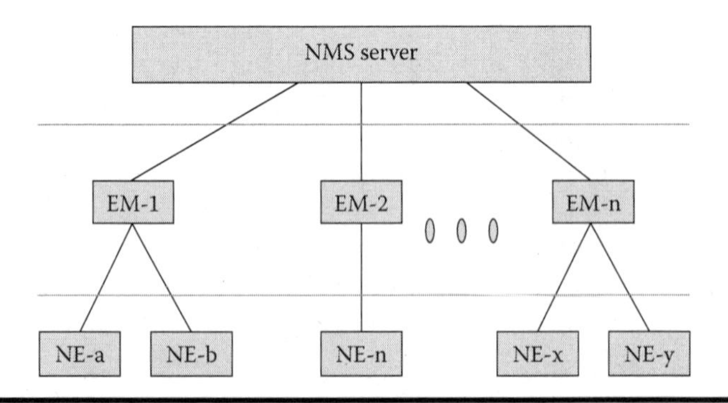


**Figure 1.1 Layered view of an element management system.**

is handled on a per NE basis and functionalities offered by the EM does not include any function that represents processed output from an aggregate of data collected from different NEs.

It is seen that while enhancing an EMS product or to make the product competitive in the market, many EMS solution developers offer some of the NMS related functionalities that involve processing of collected data from multiple NEs to display network representation of data. This point needs to be carefully noted by the reader to avoid getting confused when handling an actual EMS product that might include some NMS features as part of the product suite.

Figure 1.1 showing layered views closely resembles the TMN model with the network elements in the lowest layer, above which is the element management layer followed by network management layer. The TMN model is handled in much detail in Chapter 2.

The NMS server may not be the only feed that collects data from an EMS server. The EMS architecture is handled in more detail in the next section where the various feeds from an EMS sever are depicted pictorially and explained.

## 1.3 EMS Architecture

The element management system in Figure 1.2 is the EMS server that provides FCAPS functionality. The FCAPS is short for Fault, Configuration, Accounting, Performance, and Security. Basic FCAPS functionality at element level is offered in EMS layer. The FCAPS is also referred in conjunction with the NMS layer as an enhanced version of these functionalities come up in the NMS layer. While EMS looks at fault data from a single element perspective, an NMS processes an aggregate of fault data from various NEs and performs operations using this data.

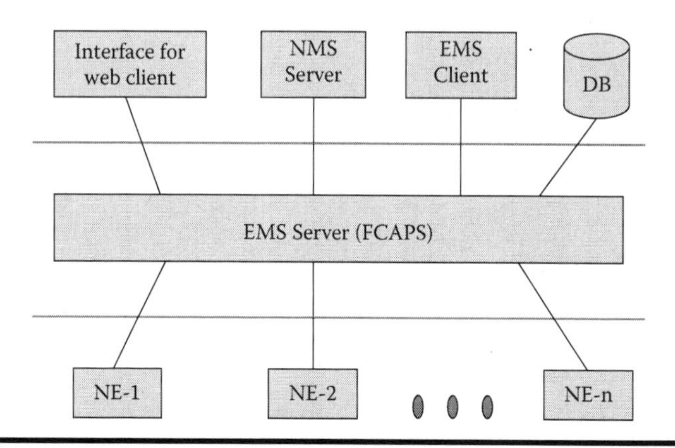


**Figure 1.2 Element management system architecture.**

Data collected by the EMS from the network elements are utilized by:

- NMS Server: The NMS uses collected as well as processed data from the EMS to consolidate and provide useful information and functionalities at a network management level.
- EMS Client: If data from EMS needs to be represented in GUI or as console output to a user then an EMS client is usually implemented for the same. The EMS client would provide help in representing the data in a user-friendly manner. One of the functions in 'F "C" APS' is configuring the network elements. The EMS client can be used to give a user-friendly interface for configuring the elements.
- Database (DB): Historical data is a key component in element management. It could be used to study event/fault scenarios, evaluate performance of the network elements, and so on. Relevant data collected by EMS needs to be stored in a database.
- Interface for web client: Rather than an EMS client, most users prefer a light-weight, web-based interface to be a replacement for the web client. This functionality is supported by most EMS solutions.
- The EMS output could also be feed to other vendor applications. These external applications might work on part or whole of the data from the EMS.

## 1.4 Need for EMS

The complexity of new services and the underlying network is increasing exponentially. At the same time there is a lot of competition among service providers offering new and compelling services. This leads to the need for minimizing the cost of network operations. This can only be achieved if the elements in the network are well managed. The increase in complexity can be handled cost effectively only by providing good user interface for managing the service and network thus hiding complexity. The other option of having more trained professionals is both expensive and risky.

Let us try to list the major issues in telecom industry that necessitates the need for a user-friendly and functionality-rich EMS:

- There are an increasing variety of NEs: New network elements are being developed to provide new services. In the network itself there are multiple telecom networks like GSM, GPRS, CDMA, IMS, and so on with independent work being carried out in the development of access and core network.
- Need to support elements from different vendors: Mergers and acquisitions are inevitable in the telecom industry. So a single EMS product might be required to collect data from a family of network elements.

- Increase in complexity of managing NEs: With advances in telecom industry there is a major trend to combine functionalities performed by multiple elements to a single element. For example, most vendors now have a single SGSN performing functionality of U-SGSN and G-SGSN and a media gateway is embedded with functionality to handle signaling protocols.
- Shortage of trained telecommunications technicians: Telecom industry is evolving so rapidly that telecom technicians are not able to keep pace with the same. This leads to a shortage of trained professionals in maintaining the network. With a functionality rich and user-friendly EMS solution, the effort to train technicians is reduced.
- Need to reduce cost of managing network and high-quality innovative services: Cost of managing the network and service is one major area where companies want to cut down cost as they do not directly generate revenue, unlike customer management and research activities that could help to get and retain business.

Meeting these challenges requires element management systems (not just EMS but good NMS, OSS, and BSS also) with the maximum efficiency and flexibility with respect to accomplishing tasks.

## 1.5 Characteristics Required in EMS

The characteristics required in an EMS solution are:

- User-friendly interface: The interface to work with the EMS (EMS client or web client) needs to be an intuitive, task-oriented GUI to allow operations functions to be performed in the shortest possible time with minimal training.
- Quick launch: This would enable a user working at the NML, SML, or BML to launch any EMS desired (when multiple EMS are involved).
- Troubleshoot NE: It should be possible for a technician to directly log into any NE from the EMS that manages it for ease in troubleshooting an issue. Single-sign-on is a popular capability that allows the same user id and password to be used when logging into the applications involved in different layers of TMN model that will be discussed in the next chapter.
- Low-cost operations platform: This would help in minimizing the total cost to own and operate the computing platform on which the EMS runs.
- Ease of enhancement: New features need to be added to an EMS product and the product might require easy integration and work with another product. Some of the techniques currently adopted in the design of such systems is to make it based on service oriented architecture (SOA), comply to COTS software standards, and so on.

- Context-sensitive help: This reduces the training required for technicians. Context-sensitive help can be associated with any functionality in EMS GUI. With event/fault, a context-sensitive help would involve a quick launch link for information on that fault. Another usual feature in applications is a "?" icon that can be dragged and dropped to any part of the application and a help is launched on how to work on the specific part where the icon is dropped.
- Access anywhere: This would require a person in any remote location to access the EMS. As already discussed most EMS now provide a web-based interface so that with secure connection technique the internet can be used to access company intranet and launch the web-based EMS.
- Effective database: All possible data collected from NE needs to be stored into the EMS database. This will improve service, save time, and save money. There should also be a functionality to generate standard reports and the ability to export data to other application tools that specialize in analysis and reporting.

Above all, the key to being a leader in EMS development is to make the product functionality rich and to ensure that there are key features that differentiate the product from its competitors.

## 1.6 Popular EMS Products

The data sheet below is a list of popular EMS products available in the market. Some or most of the product has capabilities that span outside the element management layer and might involve some level of network and service management. The information is compiled from data in the World Wide Web and does not present the authors opinion about a specific product. The products are listed in alphabetical order (see Table 1.1).

Some other EMS products from popular network equipment manufacturers are:

- Nortel's Integrated Element Management System
- Aveon Element Management Software
- Fujitsu's NETSMART Element Management System
- LineView Element Management System from Alcatel-Lucent
- NMC-RX EMS from Juniper Networks

## 1.7 Conclusion

This chapter thus provides a basic understanding of an element management system (EMS) and what a network element is from the perspective of EMS. The chapter also helps to understand the role of EMS in the telecom layered view and the EMS

**Table 1.1 Some Popular EMS Products**

| No | Product | Key Features |
|---|---|---|
| 1 | AdventNet Web NMS | • Fault management<br>• Configuration management<br>• Discovery<br>• Performance management<br>• Security<br>• Inventory management |
| 2 | AirWave: Wireless Network Management Software | • User and device monitoring<br>• Automated device discovery<br>• Group-based policy definition<br>• Centralized configuration and firmware management<br>• Automated compliance management<br>• Historical trend reports<br>• Automated diagnostics and alarms<br>• Automated RF management<br>• Multivendor management<br>• Multiarchitecture support |
| 3 | AudioCodes' Element Management System | • Client /Server architecture<br>• Java-based architecture<br>• API–OEM ready<br>• Audit logging<br>• Multi–language support ready<br>• Open solutions environment<br>• Operator individual views and profiles<br>• Supports gradual network deployment |
| 4 | BroadWorks Element Management System | • Auto-discovery<br>• Configuration management<br>• Centralized administrator management<br>• Fault management<br>• Performance management<br>• Capacity management<br>• Multirelease Support |

(Continued)

**Table 1.1 Some Popular EMS Products (Continued)**

| *No* | *Product* | *Key Features* |
|---|---|---|
| 5 | CABLEwatch EMS | • Visualization of the HFC network<br>• Network location tree view—hierarchy viewer<br>• Network location map view—topology viewer<br>• Alarm log view that displays the alarms in a table<br>• Integrated remote LMT viewer<br>• Alarm management |
| 6 | CatenaView EMS | • Full FCAPS and OAM&P functionality<br>• Scalable architecture<br>• OSS and NMS integration<br>• Remote service activation<br>• Templates for simplified provisioning<br>• Unified management of broadband<br>• Access systems |
| 7 | Cisco Element Management System | • Map viewer<br>• Deployment wizard<br>• Auto-discovery<br>• Event browser<br>• Object group manager<br>• Performance manager<br>• User access control |
| 8 | Convergent Element Management System from Vertel | • Advanced multilayer GUI's for all processes<br>• Terminal and browser-based client architecture<br>• Integrated with a wide range of databases<br>• Integration over a wide range of network devices<br>• A fully distributed architecture<br>• Mapping of protocol specific information to CIM<br>• Off-the-shelf standards-based management services<br>• Topology discovery and management<br>• Configuration and provisioning |

**Table 1.1 Some Popular EMS Products (Continued)**

| *No* | *Product* | *Key Features* |
|---|---|---|
| | | • Software administration<br>• Fault monitoring and management<br>• Performance monitoring and management<br>• Interfaces that can be changed, added, and deleted<br>• Off-the-shelf system services for distribution<br>• Security<br>• Database integration and database persistence |
| 9 | ENvision Element Management System | • Intuitive graphical interface<br>• Advanced network mapping<br>• Automatic device discovery<br>• Distributed architecture<br>• Color-coded alerting<br>• Flexible MIB tools<br>• Multiple login consoles |
| 10 | EdgeView Element Management System | • Data collection, monitoring, and management<br>• Fault management including trip data, multiple outstanding events, and summary alerts |
| 11 | expandsl Element Management System | • Java-based N-tier software architecture<br>• Automatic synchronization between EMS and NEs<br>• Web browser-based user interface<br>• Highly portable |
| 12 | MetaSwitch Element Management System | • Details of all installed hardware components<br>• Ability to configure the MetaSwitch call agent, signaling gateway, and media gateway components, as well as other<br>• MetaSwitch-managed devices (such as IADs)<br>• List of alarms currently raised on MetaSwitch network elements and managed devices, with correlation between primary and subsidiary alarms<br>• Interactive, context-sensitive help |

(*Continued*)

**Table 1.1 Some Popular EMS Products (Continued)**

| *No* | *Product* | *Key Features* |
|---|---|---|
| 13 | MBW Element & Network Management | • Network device discovery and mapping<br>• Device configuration and management<br>• Fault and alarm management/notification<br>• Element and network performance management<br>• Security management |
| 14 | Motorola's HFC Manager | • FCAPS support<br>• Easy-to-use graphical user interface (GUI)<br>• Device auto-discovery<br>• Remote access<br>• Customizable maps and containers<br>• Local language capability |
| 15 | NetAtlas from ZyXEL | • Auto-discovery and device viewer<br>• Templates provisioning for efficient configuration<br>• Traffic/performance monitoring and management<br>• Alarm and event management<br>• Batch Firmware upgrade<br>• Feature-rich Layer2 and Layer3 configuration<br>• Tight access control and log management<br>• Allow up to 10 concurrent users |
| 16 | NetBeacon Element Management System | • FCAPS model<br>• Service provisioning<br>• End-to-end service assurance<br>• MEF-defined services support<br>• Secure access |
| 17 | NetBoss Element Management System | • Java J2SE object-oriented technology<br>• Cross-platform compatibility<br>• Uniform and consistent look and feel across GUIs<br>• Client/server implementation<br>• Managed-object database compliant with DMTF CIM |

**Table 1.1 Some Popular EMS Products (Continued)**

| *No* | *Product* | *Key Features* |
|---|---|---|
| | | • Custom and standard external interfaces<br>• Textual presentation<br>• Explorer presentation<br>• Console control panel<br>• Online help |
| 18 | NetConductor Element Management System | • Equipment configuration<br>• Secure operating software downloading<br>• Alarm logging<br>• Configuration backup/restore<br>• Audit logging<br>• User access control<br>• Statistics collection<br>• Inventory reporting<br>• High availability<br>• Line diagnostic tests |
| 19 | NetConsul EMS | • Management of all device features<br>• Network-level alarm status by site<br>• Graphical alarm views<br>• Device templates for easy provisioning<br>• Drag-and-drop configuration<br>• Autonomous alarm monitoring via SNMP<br>• Security features including role-based permission hierarchy<br>• Context-sensitive online help |
| 20 | Netcool/Impact from Tivoli (IBM) | • Event enrichment with business and service context<br>• Real-time business activity monitoring<br>• Automated service impact analysis<br>• Advanced problem correlation, notification, and actions<br>• Advanced data visualization and metric calculation |

(*Continued*)

**Table 1.1 Some Popular EMS Products (Continued)**

| *No* | *Product* | *Key Features* |
|---|---|---|
| 21 | NetOp EMS | • Intuitive graphical user interface (GUI)<br>• Sophisticated user security<br>• Efficient inventory management<br>• Configuration management<br>• Superior fault management<br>• Auto node discovery<br>• Sophisticated alarm interface<br>• BGP attribute-based accounting |
| 22 | Net–Net EMS by Acme Packet | • Enables efficient deployment, configuration, and upgrade<br>• Monitors real-time SBC fault and performance status<br>• Delivers bulk performance data for capacity planning<br>• Control access for both individual users and user groups<br>• Provides interface to integrate with OSS applications |
| 23 | Nuera Element Management System | • Client/server architecture<br>• Java-based architecture<br>• API-OEM ready<br>• Audit logging<br>• Multilanguage support ready<br>• Open solutions environment<br>• Operator individual views and profiles<br>• Supports gradual network deployment |
| 24 | OccamView Element Management System | • Carrier class management<br>• Fault management<br>• Event management<br>• Configuration management<br>• Performance monitoring<br>• Inventory management<br>• Root cause analysis |

**Table 1.1 Some Popular EMS Products (Continued)**

| No | Product | Key Features |
|---|---|---|
| | | • Security<br>• Service activation<br>• Service level agreements |
| 25 | Prizm Element Management System | • Radio-level authorization and authentication<br>• Bandwidth management<br>• Tiered services<br>• Standards-based system integration |
| 26 | ProVision Element Management System | • Supports multiple devices<br>• Choice of Windows or Unix hardware platform<br>• Easy to install, learn, use, and administer<br>• Seamless integration with radio software and onboard craft tools<br>• Circuit visualization and diagnostics<br>• Bulk software downloads<br>• License management<br>• Network inventory reporting<br>• Ethernet performance monitoring |
| 27 | RADview-EMS | • Scalable management system<br>• Interface for integration with third-party NMS and systems<br>• Enhanced fault management capabilities<br>• Advanced security management functions<br>• Multiplatform, distributed element management system |
| 28 | ROSA EM—Element Management System | • Manages scientific Atlanta and third-party equipment<br>• Translates proprietary protocols to SNMP<br>• Highly reliable hardware and software solution<br>• Alarm notification with e-mail, pager, or SMS<br>• Easy to use, intuitive web browser interface |

*(Continued)*

**Table 1.1 Some Popular EMS Products (Continued)**

| No | Product | Key Features |
|---|---|---|
| | | • Provides easy integration with multiple client options<br>• Open standards-based interfaces<br>• Delivered with software already installed<br>• Software can be upgraded remotely over LAN/WAN<br>• Automatic remote backup and restore<br>• Dual temperature probes available as option |
| 30 | RetroVue Element Management System | • Status "Dashboard" UI<br>• Allows drilling down for detail<br>• Configuration management<br>• Web-based UI |
| 31 | SafariView Element Management System | • High availability<br>• Advanced functionality<br>• Fault management<br>• Configuration management<br>• Accounting management<br>• Performance management<br>• Security management<br>• Service/subscriber management |
| 32 | SkyPilot SkyControl Element Management System (EMS) | • Fault management<br>• Configuration management<br>• Performance management<br>• Security management |
| 33 | Sonus Insight Element Management System | • Ease-of-use for rapid service<br>• Provisioning<br>• User-friendly features<br>• Maintaining carrier-class reliability<br>• Advanced system security<br>• Seamless system integration<br>• Unlimited scalability. |

**Table 1.1 Some Popular EMS Products (Continued)**

| *No* | *Product* | *Key Features* |
|---|---|---|
| 34 | StarGazer Element Management System | • Configuration and provisioning<br>• Integrated fault management<br>• Security<br>• Performance and usage data |
| 35 | StarMAX EMS | • Fault management<br>• Configuration management<br>• Performance management<br>• Security management<br>• Bulk management operations support<br>• Service provisioning<br>• Telsima RMI bulk provisioning adapter |
| 36 | T-View Element Management System | • Field proven scalability<br>• Point-and-click graphical user interface<br>• Multilayer visibility<br>• Easy OSS integration<br>• Powerful bulk functions<br>• Secure operation<br>• Multiplatform support |
| 37 | TNMS from Siemens | • High capacity and scalability<br>• Ergonomically optimized graphical user interface<br>• Integrated management for transport equipment based on the technologies DWDM, SDH MSPP, SAN, and IP<br>• Seamless integration of element-layer and network-layer functionality<br>• Centralized monitoring and configuration of the whole network<br>• End-to-end configuration management<br>• Automatic, manual, and hybrid routing<br>• Management of protected connections and segments |

*(Continued)*

**Table 1.1 Some Popular EMS Products (Continued)**

| *No* | *Product* | *Key Features* |
|---|---|---|
| | | • Subscriber management<br>• Scheduled data in-/export via XML-files<br>• Scheduled export of all logfiles<br>• Data export in standard SDF formatted file<br>• Standby-solution for the whole system CORBA-based NML–EML interface (according MTNM Model) for integration in umbrella management |
| 38 | Tekelec Element Management System | • Reliable, centralized solution<br>• Reduced operating costs<br>• Scalable<br>• Automated alarm system<br>• Customized event viewing<br>• Advanced network mapping<br>• Smart agents<br>• Custom interfaces |
| 39 | Tellabs 1190 Narrowband Element Management System | • Simplified provisioning through a graphical user interface<br>• Auto-discovery of network inventory<br>• Complete fault detection, isolation, and correction<br>• Performance monitoring<br>• Configuration, alerts, collection, and storage<br>• Remote software download and remote backup and restore |
| 40 | Total Access Element Management System from ADTRAN | • Intuitive, user-friendly configuration and management GUI<br>• Automatically discovers ADTRAN devices<br>• Stores performance information<br>• Real-time and historical trend analysis and reporting<br>• Scalable architecture, multiuser system |
| 41 | Tropos Control Element Management System | • Configuration<br>• Network monitoring<br>• Network health |

**Table 1.1 Some Popular EMS Products (Continued)**

| *No* | *Product* | *Key Features* |
|---|---|---|
| 42 | Verso Clarent Element Management System | • Browser-based easy configuration<br>• Configuration and process management<br>• Real-time command center and call control monitoring<br>• Supports C5CM associations with third-party edge devices<br>• Multiple levels of user access permissions<br>• Subscriber portal providing subscribers<br>• Secure socket layer encryption<br>• Standardized interfaces: SNMP, XML, JDBC |
| 43 | Visual Guardian Element Management System | • Fault management<br>• Service provisioning<br>• Manufacturing<br>• Performance management<br>• Preventive maintenance |

architecture. The need for element management systems and the characteristics of an ideal EMS is covered. To give the reader familiarity with the EMS products and functionalities, a data sheet is also provided that shows some of the leading EMS products and their features. The information provided in the data sheet can be obtained from the company Web site that will give more information on the products.

## Additional Reading

1. Vinayshil Gautam. *Understanding Telecom Management.* New Delhi: Concept Publishing Company, 2004.
2. James Harry Green. *The Irwin Handbook of Telecommunications Management.* New York: McGraw-Hill, 2001.
3. Kundan Misra. *OSS for Telecom Networks: An Introduction to Network Management.* New York: Springer, 2004.

# Chapter 2

# TMN Model

This chapter is intended to provide the reader with an understanding of Telecommunications Management Network introduced by ITU-T (International Telecommunication Union-Telecommunication Standardization Bureau) and defined in recommendation M.3010. At the end of this chapter you will have a good understanding of TMN and its different management architectures comprising of functional, physical, information, and logical architecture.

## 2.1 Introduction

TMN stands for Telecommunications Management Network. The concept of TMN was defined by ITU-T in recommendation M.3010. The telecommunications management network is different from a telecommunication network. While the telecom network is the network to be managed, TMN forms the management system for the network to be managed. This is shown in Figure 2.1 and is explained in more detail in the next section.

The recommendations from ITU-T on TMN are listed in Table 2.1.

Recommendation M.3010 defines general concepts and introduces several management architectures to explain the concept of telecom network management.

The architectures are:

- Functional architecture: It defines management functions.
- Physical architecture: It defines how to implement management functions into physical equipment.
- Information architecture: Concepts adopted from OSI management.
- Logical architecture: A model that splits telecom management into layers based on responsibilities.

**Table 2.1 TMN Recommendations**

| *Recommendation Number* | *Title* |
|---|---|
| M.3000 | Overview of TMN recommendations |
| M.3010 | Principles for a TMN |
| M.3020 | TMN interface specification methodology |
| M.3100 | Generic network information model |
| M.3101 | Managed object conformance statements for the generic network information model |
| M.3180 | Catalog of TMN management information |
| M.3200 | TMN management services: overview |
| M.3207.1 | TMN management services: maintenance aspects of B-ISDN management |
| M.3211.1 | TMN management services: Fault and performance management of the ISDN access |
| M.3300 | TMN management capabilities presented at the F interface |
| M.3320 | Management requirements framework for the TMN X-interface |
| M.3400 | TMN management functions |
| M.3600/3602/3603/3604/3605/3610/3611/3620/3621/3640/3641/3650/3660 | TMN Recommendations for ISDN |

## 2.2 What Is TMN?

A telecom network is comprised of multiple network elements. When different vendors came up with their own custom main network elements and management solutions for the elements, there was a need to create or identify standard interfaces that would allow a network to be managed consistently across all network element suppliers. The TMN was a strategic goal from ITU-T toward defining these standards.

The relationship between TMN and a telecom network is shown in Figure 2.2. The figure gives the following information:

- A telecom network is comprised of switching systems (exchanges) and transmission systems.

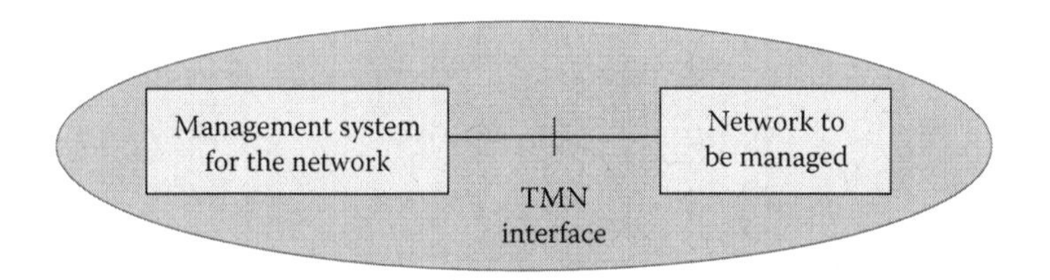


**Figure 2.1 TMN and telecom network.**

- TMN is comprised of multiple operations systems. For example a separate OS for provisioning, another for traffic management, and so forth.
- Operations systems perform most of the management functions.
- Data communication network is used to exchange management information between the operation systems. It thus forms the link between TS/SS in telecom network and OS in TMN.
- Work stations allow operators to interpret management information.
- A workstation can take input from a data communication network or from an operation system.

Hence we have the TMN interface that is comprised of the following elements:

- Roles and relationships of communicating TMN entities
- OAM&P requirements (operation, administration, maintenance, and provisioning)
- Application and resource information models
- Protocols for transport of information

Since the scope of TMN is clear from Figure 2.2, now let us identify the items that are defined by TMN inside its scope with TMN entities and the managed network.

TMN can be said to be:

- A set of architectures that:
  - Define management activities
  - Define tasks associated with the activity
  - Define data communication in the activities
  - Map relation between management entities and associated data, task, and activities
- A set of models that:
  - Separate technology, network, service, and business
  - Define processes associated with information and management tasks
- A set of protocols that:
  - Define communication between NE and OS
  - Define communication between OS and OS

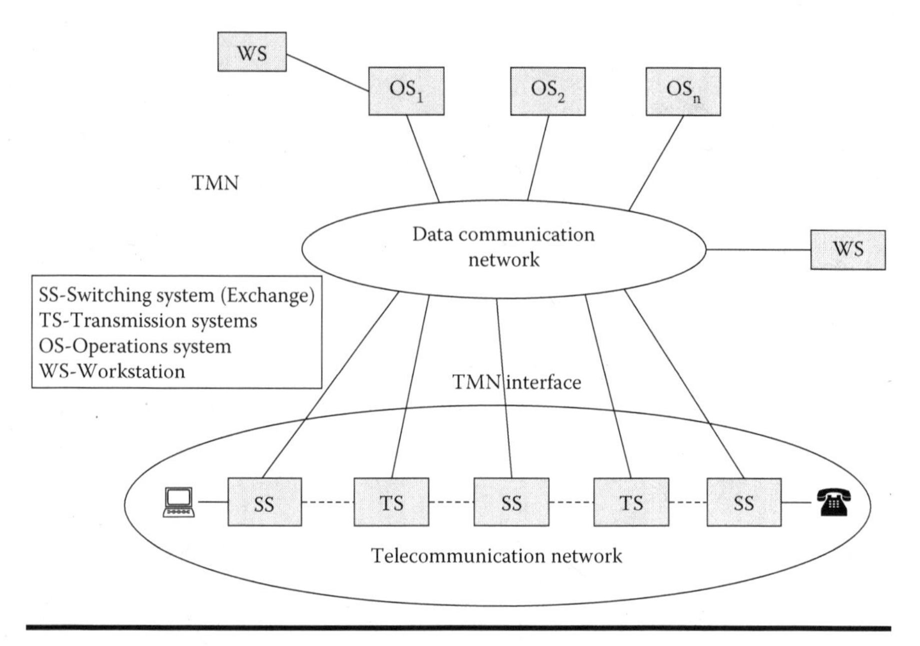


**Figure 2.2 Detailed relationship of TMN with telecommunication network.**

The TMN recommendation M.3010 does not make any reference to the Internet though it includes a number of concepts that are relevant to the Internet management community. There is also a strong relationship between TMN and OSI management. The discussion on the relation of TMN and OSI model will be covered in Chapter 5 after the concepts of the OSI model are discussed.

## 2.3 Functional Architecture

The functional architecture describes the distribution of functionality within the TMN to allow for the creation of function blocks from which a management network can be implemented. The functional architecture is composed of function blocks and reference points. Function blocks represent conceptual entities that make up TMN. Reference points are used to represent the exchange of information between function blocks.

The systematic decomposition of management functions in TMN functional architecture is useful for the following tasks:

- Analysis and description of existing operations systems
- Analysis and description of existing process flow
- Specification of new operations systems
- Specification of new process flows

- Analysis of operations plans
- Development of new operations systems

TMN functional architecture consists of the following function blocks:

- Operations system functions (OSF)
- Mediation functions (MF)
- Network element functions (NEF)
- Work station functions (WSF)
- Q-adapter functions (QAF)

The function blocks that are fully part of the TMN are:

1. Operations system functions (OSF): This relates to functions in the network manager that collects data from the network elements. When collecting data from multiple network elements the OSF represents the server and the network elements are the clients that provide data popularly called agents.
2. Mediation functions (MF): There is sometimes a need for the presence of a mediation component while communicating with OSF. Some of the possible scenarios include filtering of irrelevant information, conversion from one format to another, and so forth. Mediation functions block is the conceptual representation of the functions performed by this mediator.

The function blocks that are part of the TMN are:

1. Network element functions (NEF): The functions performed by the network elements constitute the NEF block. Network element functions consist of both primary functions and management functions. The primary functions are outside the scope of TMN making NEF only part of the TMN. The management functions of the network element are the functions performed by NE as an agent in sending management data to the OSF.
2. Work station functions (WSF): The WSF block allows operators to interpret management information. The WSF also includes support for interfacing to a human user, which is not considered to be part of the TMN. Thus it is only part of the TMN.
3. Q-adapter functions (QAF): The QAF block is used to connect to the TMN those entities that do not support standard TMN interaction.

The TMN functional architecture also includes:

- Reference points
- Data communication function
- Functional components

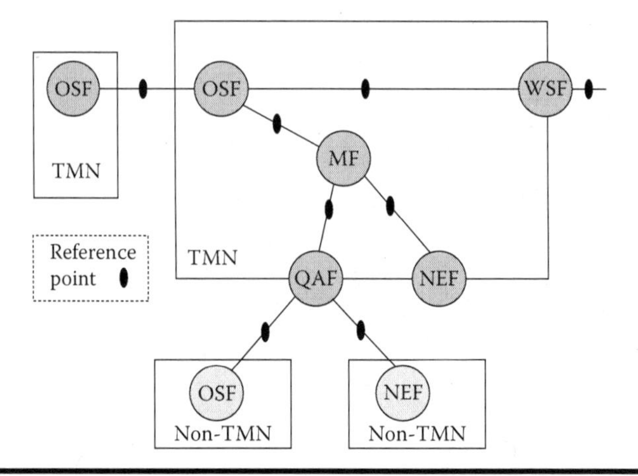


**Figure 2.3 TMN functional architecture.**

1. Reference points: The concept of the reference point is used to delineate function blocks.
2. Data communication function: DCF is used by the function blocks for exchanging information.
3. Functional components: TMN function blocks are composed of a number of functional components. The following functional components are defined in M.3010 and this standards document can be referred for details on the functional components:
   a. Management application function
   b. Management information base
   c. Information conversion function
   d. Human machine adaptation
   e. Presentation function
   f. Message communication function (MCF)

The interactions between the function blocks in TMN functional architecture are shown in Figure 2.3.

## 2.4 Physical Architecture

While the functional architecture defines the management functions, the physical architecture defines how the management functions can be mapped to physical equipment. When function blocks are mapped to building blocks (physical entities), the references are mapped to interfaces.

Hence the physical architecture consists of building blocks and interfaces. Building blocks represent physical node in the TMN, whereas interfaces define the information exchange between these nodes.

Physical architecture defines the following building blocks:

- Operations system (OS)
- Mediation device (MD)
- Q-adapter (QA)
- Network element (NE)
- Work station (WS)
- Data communication network (DCN)

1. Operations system or support system (OS): Perform operations system functions (OSF). The OS can also provide some of the mediation, q-adaptation, and workstation functions.
2. Mediation device (MD): Performs mediation between TMN interfaces and the OS. Mediation function, as already discussed in functional architecture, may be needed to ensure that data is presented in the exact way that the OS expects. Mediation functions can be implemented across hierarchies of cascaded MDs.
3. Q-adapter (QA): The QA is used to manage NEs that have non-TMN interfaces. The QA translates between TMN and non-TMN interfaces. For example, A TL1 Q-adapter translates between a TL1 ASCII message-based protocol and the common management information protocol (CMIP); similarly a simple network management protocol (SNMP) Q-adapter translates between SNMP and CMIP.
4. Network element (NE): An NE contains manageable information that is monitored and controlled by an OS. It performs network element functions (NEF) specified in functional architecture.
5. Workstation (WS): The WS performs workstation functions (WSF). WSs translate information to a format that can be displayed to the user.
6. Data communication network (DCN): The DCN is the communication network within a TMN. The DCN represents OSI layers 1 to 3. It performs data communication functions (DCF).

These building blocks generally reflect a one-to-one mapping with function block. However multiple function blocks can be mapped to a single physical block (building block). For example, most legacy network elements (NE) also have a GUI interface for technicians to configure and provision the NE. In this scenario a NE in a building block is having OSF and WSF functions in addition to NEF. There are different possible scenarios of multiple function blocks mapping to a single building block and a single function block distributed among multiple building blocks. The possible relations between function and building block are shown in Figure 2.4.

| Building blocks \ Function blocks | NEF | MF | QAF | OSF | WSF |
|---|---|---|---|---|---|
| NE | Mandatory | Optional | Optional | Optional | Optional |
| MD | Not applicable | Mandatory | Optional | Optional | Optional |
| QA | Not applicable | Not applicable | Mandatory | Not applicable | Not applicable |
| OS | Not applicable | Optional | Optional | Mandatory | Optional |
| WS | Not applicable | Not applicable | Not applicable | Not applicable | Mandatory |

**Figure 2.4 Relation between function and building blocks.**

## 2.5 Information Architecture

This architecture gives an object-oriented approach for TMN. It is based on the management information communications principles defined in the context of the OSI systems management standards.

The management view of an object consists of:

- Attributes: Characteristics of the object
- Operations: Activities performed upon the object
- Behavior: Response to operations
- Notifications: Data emitted by the object

Managed object is an abstraction of a physical or logical entity defined for the purpose of managing systems. It contains the attributes and behavior associated with the management of that entity. The design of management systems needs to support modularity, extensibility, and scalability. These characteristics are achieved in TMN using object-oriented design principles (see Figure 2.5). Operations are directed to manage objects and notifications are emitted from managed objects. Consistency constraints of the managed object are specified as a part of the behavior definition of the attribute. The value of an attribute can determine the behavior of the managed object.

## 2.6 Logical Architecture

TMN logical architecture splits telecom management functionality into a set of hierarchal layers. This logical architecture is one of the most important contributions of TMN which helps to focus on specific aspects of telecom management application development. Though new detailed models were built based on TMN as a reference, the TMN logical architecture still forms the basic model which reduces management complexity by splitting management functionality so that specializations can be defined on each of the layers (see Figure 2.6).

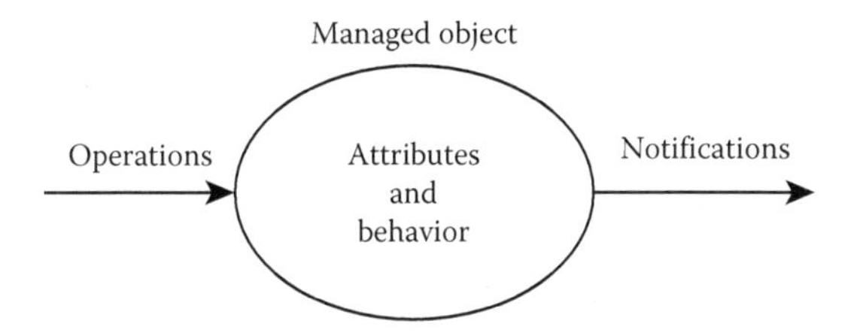


**Figure 2.5 Managed object.**

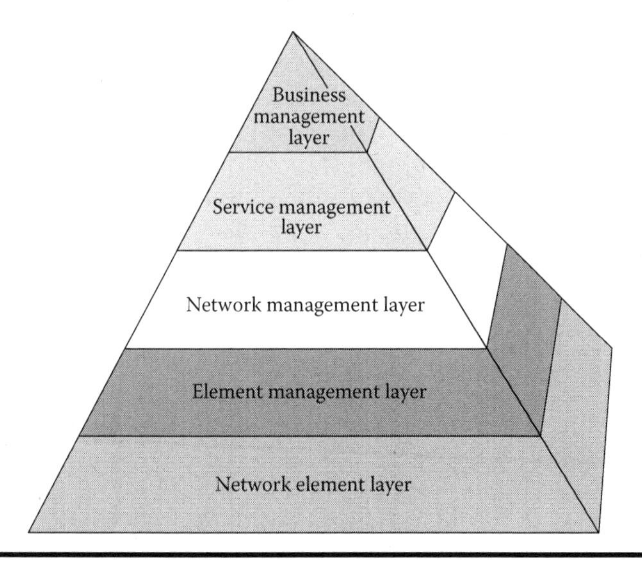


**Figure 2.6 TMN logical architecture (TMN pyramid).**

The layers in the TMN logical architecture are:

1. Network element layer (NEL): This is the lowest layer in the model and it corresponds to the network element that needs to be managed and from which management information is collected. This layer can also include adapters to adapt between TMN and non-TMN information. So both the Q-adapter and the NE in the physical TMN model are located in the NEL and NEF and QAF functions are performed in network element layer.
2. Element management layer (EML): This is the layer above the network element layer and deals with management of elements in the NEL. The development of an element management system (EMS) falls in this layer and the functionalities discussed in the first chapter performed by element managers are grouped in this logical layer of TMN. When mapping it with functional architecture the OSF of the network elements are associated with this layer.

Currently, the operation support system (OSS) mostly used in association with service management applications is supposed to incorporate all management functions as in OSF of the TMN functional architecture. When we segregate management functionality into business, service, network, and element management, the business management application got associated with BSS, service management application with OSS, network management application with NMS, and element management applications with EMS.

3. Network management layer (NML): The responsibility of this layer is to manage the functions related to the interaction between multiple pieces of equipment.
   Some of the functions performed in this layer are:
   - Creating a view of the network
   - Correlation of network events
   - Single-sign-on to elements in the network and the element managers
   - Traffic management based of attributes and behavior of network elements
   - Monitoring network utilization and performance

   A detailed discussed of network management is handled later in this book.

4. Service management layer (SML): This layer, as the name signifies, is concerned with managing services. Service is what is offered to the customer and hence this layer is the basic point of contact with customers. To realize any service there is an underlying network and so the SML uses information presented by NML to manage contracted service to existing and potential customers. The SML also forms the point of interaction with service providers and with other administrative domains.
   Some of the functions performed in this layer are:
   - Service provisioning
   - User account management
   - Quality of service (QoS) management
   - Inventory management
   - Monitoring service performance

A detailed discussion of service management functions is handled later in this book under OSS.

5. Business management layer (BML): This is the top most layer in the TMN logical architecture and is concerned with business operations.
   Some of the functions performed in this layer are:
   - High-level planning
   - Goal setting

- Market study
- Budgeting
- Business-level agreements and decisions

A detailed discussion of business management functions is handled later in this book under BSS.

Associated with each layer in the TMN model are five functional areas called FCAPS that stand for fault, configuration, accounting, performance and security. These five functional areas form the basis of all network management systems for both data and telecommunications. These five functional areas will be covered in detail in the next chapter.

## 2.7 Conclusion

The telecommunications industry is evolving rapidly. With emerging technologies, acquisitions, multivendor environment, and increased expectations from consumer, companies are presented with a challenging environment. There is a need for telecom management to support multiple vendor equipment and varied management protocols, as well as continue to expand services, maintain quality, and protect legacy systems. The telecommunications management network (TMN) provides a multivendor, interoperable, extensible, scalable, and object-oriented framework for meeting these challenges across heterogeneous operating systems and telecommunications networks. This has lead to a mass adoption of TMN standards in building telecom management applications.

## Additional Reading

1. CCITT Blue Book. *Recommendation M.30, Principles for a Telecommunications Management Network*, Volume IV: Fascicle IV.1, Geneva, 1989.
2. CCITT. *Recommendation M.3010, Principles for a Telecommunications Management Network*. Geneva, 1996.
3. Masahiko Matsushita. "Telecommunication Management Network." *NTT Review* 3, no. 4 (1991): 117–22.
4. Divakara K. Udupa. "*TMN: Telecommunications Management Network*." New York: McGraw-Hill, 1999.
5. Faulkner Information Services. *Telecommunications Management Network (TMN) Standard*. Pennsauken, NJ: Faulkner Information Services, 2001.

# Chapter 3

# ITU-T FCAPS

This chapter is intended to provide the reader with an understanding of FCAPS introduced by ITU-T (International Telecommunication Union-Telecommunication Standardization Bureau) and defined in recommendation M.3400. At the end of this chapter you will have a good understanding of FCAPS functionality and how it applies to different layers of the TMN.

## 3.1 Introduction

Associated with each layer in the TMN model are five functional areas called FCAPS that stand for fault, configuration, accounting, performance and security. These five functional areas form the basis of all network management systems for both data and telecommunications (see Figure 3.1).

The information in telecom management is classified into functional areas using FCAPS. It was introduced for telecom network management by ITU-T in recommendation M.3400. The ISO (International Standards Organization) made the FCAPS also suited for data networks with its OSI (open system interconnection) network management model that was based on FCAPS. Further in this chapter, each of the FCAPS functional areas are taken up and discussed in detail. All element management systems are expected to provide the FCAPS function or a subset of the same. At higher levels like business, service, and network layer, derivatives of basic FCAPS functionality can be implemented like a complex event processing engine that triggers mail on business risks when network elements carrying major traffic goes down or a service request is sent to the nearest technician or set of commands to the network elements to replace load or restart.

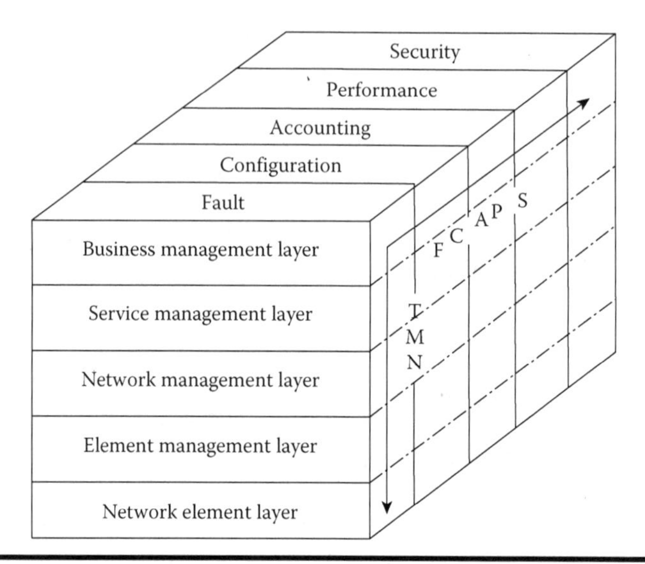


**Figure 3.1 FCAPS and TMN.**

## 3.2 Fault Management

Fault is an error or abnormal condition/behavior in a telecom network.

Fault management basically involves the detection, isolation, and correction of a fault condition.

The first part involves detection of an error condition by the element management system. This can be achieved in the following ways:

- The network element generates a notification when an alarm condition is identified with any of the node attributes associated with the element: Most network elements generate logs/alarms when an error condition occurs. Using a network management protocol these notifications can be collected from the network element.
- Executing diagnostics tests at regular intervals to check for issues: The element management can have its own set of rules to identify errors. For example the EMS can define severity for specific conditions of resource in the network element and do regular audits on the state of resources. When a state change happens, the new state of resource is compared with severity and a log is generated.
- Processing database information: The network element can have a database that stores static and dynamic information about the element. Querying this database will also provide information about fault condition. A detailed coverage of MIB (management information base) is later handled as a separate chapter in this book.

There are other methods that an EMS uses to detect fault also. Some of them include checking and filtering the system logs on the network element, adding code that defines thresholds for attributes of the network element and generating an event when threshold is crossed, and so on. Some EMS employ combinations of the above fault detection methods, like having the NE send event notifications as well as running routine checks if some data was lost or not sent by NE.

Once detected, the fault needs to be isolated. Isolation includes identifying the element in the network or resource (can be a process or physical component) in the element that generated the fault. If the fault information is generic, then isolation would also involve filtering and doing data processing to isolate the exact cause of the fault.

After the fault is detected and isolated, then it needs to be fixed. Correcting fault can be manual or automatic. In the manual method, once the fault is displayed on a management application GUI, a technician can look at the fault information and perform corrective action for the fault. In the automatic method, the management application is programed to respond with a corrective action when a fault is detected. The management application can send commands to the network elements (southbound), send mail to concerned experts (northbound), or restart processes on the management application server. Thus in automatic method interaction with northbound interface, southbound interface, or to self is possible.

Fault information is presented as a log or alarm in the GUI. A log can represent nonalarm situations like some useful information for the user. A log specifying that an audit was performed on a network element is an info log and does not raise an alarm. An alarm or log usually has several parameters (see Table 3.1).

A detailed set of alarm parameters for specific networks are available in 3GPP specification 32 series on OAM&P.

Having discussed the basic fault functionality, let us look into some applications that are developed using the fault data that is detected and displayed. Some of the applications that can be developed using fault data are:

- Fault/event processing
- Fault/event filtering
- Fault/event correlation

Event Processing: There are three types of event processing: simple, stream, and complex. In simple event processing when a particular event occurs, the management application is programed to initiate downstream action(s). In stream event processing, events are screened for notability and streamed to information subscribers. In complex event processing (CEP) events are monitored over a long period of time and actions triggered based on sophisticated event interpreters, event pattern definition and matching, and correlation techniques. Management applications built to handle events are said to follow event driven architecture.

**Table 3.1 Parameters in an Alarm/Log**

| *Parameter* | *Description* |
|---|---|
| Notification ID | The same alarm can occur multiple times. An easy way to track an alarm is using its notification ID. |
| Alarm ID | A particular error condition is associated to an ID. For example we can have CONN100 as the alarm ID to signify a connection loss and CONN200 to signify connection established. This way we can associate logs. In a fault database notification ID is unique while alarm ID can be repeated. |
| Generation time | This is the time when the error scenario occurred and the notification was generated on the NE. Some applications also have "Notification time," which signifies the time when the management application received an alarm notification from the NE. |
| Resource info | The information about the network element and the resource on the network element that caused the error condition is available in log body. |
| Alarm purpose | If the alarm is a new alarm, an alarm for overwriting status, or a clear alarm. For example, when a resource is having some trouble an alarm is raised, as the resource is restarted another alarm is send to overwrite the existing alarm and once the resource is back in service, a clear alarm can be raised. |
| Probable cause | This is another parameter for classifying data. Consider the example of CONN100 for connection loss. We might want to classify connection logs associated with a call server and connection logs associated with a media gateway in two different categories. This classification can be achieved with probable cause keeping same alarm ID. |
| Alarm type | In addition to classification based on cause, 3GPP recommends the use of alarm types for grouping alarms. Some of the alarm types recommended by 3GPP are: communications alarm, processing error alarm, environmental alarm, quality of service alarm, or equipment alarm. |
| Severity | The usual severity conditions are minor, major, critical, and unknown severity. |
| Alarm information | This part will contain complete problem descriptions, which includes the resource information and the attribute that gets affected. Sometimes the alarm information will even specify the corrective action to be taken when the log appears. |

Event Filtering: The corrective action for alarms are not the same. It varies with the probable cause. There could also be corrective action defined for a set of alarms as a single event and can generate multiple alarms on different network elements. For example, a connection loss between two NEs would generate multiple alarms not just related to connection but also on associated resources and its attributes that are affected by the connection loss. The relevant log(s)/alarm(s) needs to be filtered out and corrective action defined for the same.

Event Correlation: An integral part of effective fault handling is event correlation. This mostly happens in the network and service management layers. It involves comparing events generated on different NEs or for different reasons and taking corrective actions based on the correlated information. For example, when the connection between a call server and media gateway goes down, alarms are generated on the media gateway as well as the call server, but the NMS handling both these NEs will have to correlate the alarms and identify the underlying single event. The process of grouping similar events is known as aggregation and generating a single event is called event aggregation.

Event processing usually involves event filtering as well as event correlation. Event processing in a service layer can generate reports on a service that could help to analyze/improve quality of service or to diagnose/fix a service problem.

Some of the applications where basic fault data can be used as feed are:

- Quality assurance application: The lesser the faults the better the quality
- Inventory management application: Faulty NEs would need service or replacement
- Service and resource defect tracker
- Event driven intelligent resource configuration and maintenance
- Product and infrastructure budget planning

Most EMS/NMS applications also have fault clearing, fault synchronization, fault generation, and fault handling capabilities.

## 3.3 Configuration Management

Configuration management involves work on configuration data associated with the managed device/network element.

Some of the functionalities handled in configuration management are:

- Resource utilization: This would involve the representation of data on how effective the configuration is in getting the best out of the network. Some examples include count of the number of trunks that are not working properly, plot of under utilization of a trunk with less traffic routing, and so on.

- Network provisioning: Involves configuring network elements and subresources in the element to offer a service, or in other words the network is set up using interactive interfaces for proper working.
- Auto discovery: When the management application comes up it scans the network and discovers the resources in the network, which is called auto discovery. As part of auto discovery the state of the resource (in service, in trouble, offline, busy, etc.) and its attributes that are to be displayed and used in the management application are also collected.
- Backup and restore: It is critical to backup configuration information and restore the data when required. Some possible scenarios include, installing a new version of software on the NE, upgrading a software, taking the NE down for some maintenance or fault correction, and so forth.
- Inventory management: This involves keeping track of the resources (both active and inactive) in the enterprise, so that they can be allocated, provisioned, rolled out, kept as backup, and so forth. The output of auto discovery is a feed to inventory management to synchronize the status of the resources allocated for a network.
- Change management: The change in status of the network elements needs to be continuously monitored and updated. Break down of an element without proper mitigation strategy can result in loss of business and in a military domain it could result in loss of life. So the change in status would trigger actions corresponding to the new state.
- Preprovisioning: A set of parameters needs to be set during provisioning. Effective decision on the parameters to set can be obtained by trend analysis of the network element. This would aid in determining parameters for provisioning and is termed as preprovisioning.

Configuration management (CM) involves continuous tracking of critical attributes of the network elements and a successful initialization only marks the starting point for CM.

## 3.4 Accounting Management

> Accounting management involves identification of cost to the service provider and payment due for the customer. Accounts being calculated based on service subscribed or network usage.

In accounting, a mediation agent collects usage records from the network elements and forwards the call records to a rating engine (see Figure 3.2). The rating engine applies pricing rules to a given transaction, and route the rated transaction to the appropriate billing/settlement system. This is different from customer account

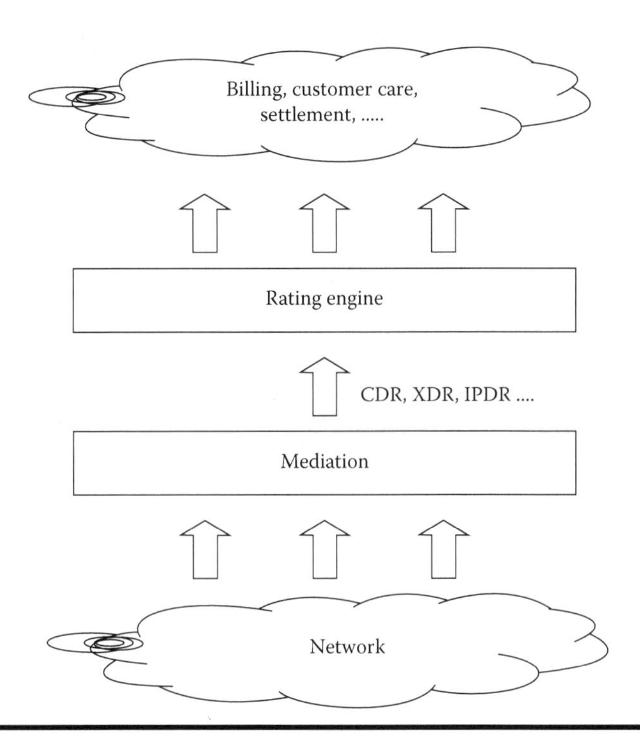


**Figure 3.2 Accounting data flow.**

management. (Any customer is represented in the billing system as an account. One account can have only one customer.)

Accounting management can be split into the following functions:

- Track service and underlying resource usage: The mediation agent collects data from NEs, which is used to determine charges to customer accounts. Account data processing must be carried out in near real-time for a large number of customers. On the data collected by the mediation agent, a tariff is applied based on the service level agreement the customer has with the service provider.
- Tariff: The tariff or rating of usage is also part of accounting. This involves applying the correct rating rules to usage data on a per customer basis and then applying discounts, rebates, or charges as specified in the service level agreement.
- Usage restrictions: Based on the subscription taken up by the customer there is a limit set on the usage of resources, for example, disk space, network bandwidth, call duration, services offered, and so forth.
- Converged billing: A single service is realized using multiple network elements. Traditionally accounting information is collected from all the NEs and a single bill is generated. With converged billing the customer can have multiple services and still get a consolidated billing for all the services.

- Audits: Accounting is a critical aspect of business and forms the revenue for physical network and service offered. Hence billing records and reports are handled carefully in management applications. Schedule audits and reports check the correctness of information and helps in archiving of data. An internal scheduler can perform report generation. Reports give consolidated data that form input to business and technical planning.
- Fraud reporting: Fraud management in telecom has evolved to be an independent area of specialization and there are applications to detect fraud that are delivered independently without being part of a management application. As part of the initial definitions of ITU-T, fraud reporting was part of accounting. Some of the telecom frauds are: Subscription Fraud, Roaming Subscription Fraud, Cloning, Call-Sell Fraud, Premium Rate Services (PRS) Fraud, PBX Hacking/Clip-on Fraud, Pre-paid Fraud, Internal Fraud, and so on.

The account management system should be such that it can interface to all types of accounting systems and post charges.

## 3.5 Performance Management

> Performance management involves evaluation and reporting the behavior and effectiveness of network elements by gathering statistical information, maintaining and examining historical logs, determining system performance, and altering the system modes of operation.

Network performance evaluation involves:

- Periodic tests to measure utilization, throughput, latency availability, and jitter
- Identifying key performance indicators and setting thresholds for evaluation
- Collection of performance data as register counts on dropped calls, outages registered, threshold parameters crossed to judge the network health
- Analysis of collected data to give plots and reports for technicians
- Archiving performance data to identify trends and best configurations
- Use of the analyzed output to develop patterns for improving and effectively utilizing the network
- Replacing or correcting inventory for improving performance
- Generating logs to notify technicians when thresholds are crossed on predefined parameters for performance evaluation

The functionalities in performance management that are built into most management applications can be classified into:

- Performance data collection: For each network element or for a service offered by network there are a set of key performance indicators. For a network, collection of data on these performance indicators help in determining and forecasting network health and with a service it helps to indicate the quality of service (QoS). Collection of this performance data is a key functionality in most management applications.
- Utilization and error rates: The traffic handled by a trunk to the maximum traffic it can handle is an indicator of trunk utilization. This in turn shows how the trunk is performing independently and compared to other trunks in the network. Even when a trunk is getting utilized, the overall network may not be effectively managed.

  A typical example is under utilization of one trunk and over utilization of another. Thresholds need to be set and error rates determined so that there is optimal utilization at network level of all possible resources. Different algorithms are used to implement this and make sure there is minimal manual intervention.
- Availability and consistent performance: The most basic check of performance is availability of a resource. If the resource is generating reports then availability is confirmed. Performance data is usual for forecasting, trend development, and planning only when the performance levels are consistent.
- Performance data analysis and report generation: All performance management applications have data collection, data analysis, and report generation. Multiple reports are generated as graphs and plots of the performance data.

  Data analysis would also involve creating correlations and finding correlated output on threshold data. In addition to graphical representation, there are also features in management applications that permit export of data for later analysis by third-party applications.
- Inventory and capacity planning: Performance data is used as input for inventory planning on deciding what network elements need to be replaced or monitored and for capacity planning on deciding the routing path of trunks, how much traffic to route, and so on.

Performance data is usually collected as:

1. Bulk download from the NE: In this method PM data are generated and stored in a predetermined folder on the NE. The management applications collect the data at regular intervals of time.
2. Send from the NE on generation: The network element contains code to send the data to the management application using a predefined protocol and the management applications keeps listening for data from NE.
3. Queried from NE database: The network element can contain a management information base (MIB) where performance data is stored and

dynamic attributes of performance updated at real-time. In this method, the management application collects data from the MIB.

## 3.6 Security Management

Security management deals with creating a protected environment for core resources and for managing the network.

In telecom, security is a key concern. Monitoring and tracking for potential attacks need to be performed as part of security management. The basic functionality in security is to create a secure environment. This can be done by using AAA to authenticate, authorize, and account for user actions. Protocols like RADIUS are used to implement the security.

The communication between network elements can be encrypted and a secure channel may be used while communication uses management protocols. XML-based management protocols like NETCONF inherently supports SSH (secure shell) in its communication architecture.

Authenticated access is required for logging into the network element or the management application. It is also quite common to use security servers that implement rules and policies for access like the Sun One Identity Server, which can control centralized access.

A typical implementation would involve the interface to add, delete, or edit users. Each of the users can be associated to a group. For example there will be groups like administrator, GUI users, technicians, and so forth, with a separate set of access permissions for each group. Permissions could be granted to specific functionalities in an application, like a user in the technicians group can only modify configuration management functionalities and view fault management functionality.

Some applications even define deeper levels of permission where a user in a specific group might be able to view and add comments to a log-in fault management functionality but will not be able to clear the log.

Logs can be used to track security. The network element generates security logs for different access and usage scenarios. For example a log is generated with the log-in user details and source of the user (IP address of the user machine), when a user tries to log into a network element. Logs can also keep track of the activities performed by a user on the network element. Alarms on failed log-in attempts and unauthorized access attempts are also captured as part of security management.

Functionalities like call trace and call interception is done using telecom management applications. When a law enforcement agency wants to trace calls from a user, the security management application configures the network element to generate call trace records for the user. The trace records are collected from the network element, analyzed, and displayed in predetermined format. In call interception, a call in progress between two users is intercepted for listening by a law enforcing agency.

> Single-sign-on (SSO) is a popular method for access control that enables a user to authenticate once and gain access to multiple applications and network elements.

SSO functionality is now available in most service and network management applications. There might be multiple element management applications that make up a network management application. For example, network elements may be from different vendors and each have a separate element manager that can be launched from the single NMS application for an entire network. With centralized control of security management a single id can be used by a user to log into all the applications in the NMS.

Security management can be a stand alone application run on a security gateway that provides the only entry point for communication with core networks (see Figure 3.3).

The rapid spread of IP-based access technologies as well as the move toward core network convergence with IMS has led to an explosion in multimedia content delivery across packet networks. This transition has led to a much wider and richer

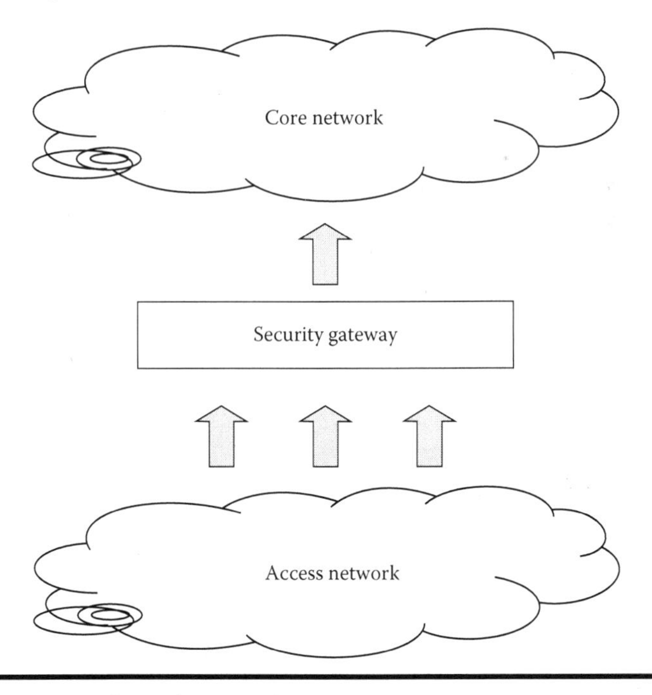


**Figure 3.3 Interaction using security gateway.**

service experience. The security vulnerabilities such as denial-of-service attacks, data flood attacks (SPIT), and so on that can destabilize the system and allow an attacker to gain control over it are applicable to video conferencing over IP, is true for wireless communication over IMS networks also. Currently, hackers are not limited to looking at private data, they can also see their victims. Videoconferencing systems can be transformed into video surveillance units, using the user equipment to snoop, record, or publicly broadcast presumably private video conferences. Network security and management of security has always been a key area in telecom. The latest concepts of convergence, value-added services, and next generation networks have increased the focus on security solutions.

## 3.7 Conclusion

The FCAPS forms the basic set of functionalities that are required in telecom management. All telecom management applications would use data obtained from these functionalities in one way or the other.

## Additional Reading

1. Vinayshil Gautam. *Understanding Telecom Management.* New Delhi: Concept Publishing Company, 2004.
2. James Harry Green. *The Irwin Handbook of Telecommunications Management.* New York: McGraw-Hill, 2001.

# Chapter 4

# EMS Functions

The motivation for writing this chapter is the four function model suggested by International Engineering Consortium in their EMS tutorial discussing how the EMS functions provide feed to the upper network and service management layers. At the end of this chapter you will have a good understanding of how the feed from EMS gets utilized by upper layers of telecom management.

## 4.1 Introduction

The basic FCAPS functionality offered by the element management system generates data that gives feed to the functionalities in network and service management layers. Some of the functionalities in upper layers that get input from EMS are:

- Network/service provisioning
- Network/service development and planning
- Inventory management
- Integrated Multivendor environment
- Service assurance
- Network operations support
- Network/service monitoring and control

The data and operations provided by an element management system should be based on standards so that service providers can add a high level network and service management functionality without any glue code or adapters. In most scenarios, though the business management, service management, or network management application used by the service provider will be from a single vendor, the underlying network used for realizing a service will consist of network elements from multiple vendors. Each operation equipment manufacturer provides a network element and an EMS solution to manage his element. Only when the EMS solutions from different vendors integrate easily will there be cost-effective deployment of process in layers above element management. This also offers significant cost and time reduction for integration of products and adding high level processes over the integrated EM.

## 4.2 Network/Service Provisioning

When an order is placed by a customer for a specific service, the service and underlying network needs to be provisioned to offer the service.

> Service provisioning involves all the operations involved in fulfilling a customer order for a specific service.

The first check is to verify if the existing inventory can offer the service. The inventory management system is used to identify and allocate appropriate resources to be used in the network. Next the network is provisioned, which would involve configuring the network elements in the network. The appropriate software and patches are loaded on the network elements and the parameters for proper operation are set. The backup and restore settings are then made. Once each network element is ready, the interconnections between the elements are made to make up the network. Service activation is the final stage when the network and elements in the network start operating to offer the service placed in the order. Activation signals are sent to a billing subsystem to generate billing records on the service used and the order is completed. (see Figure 4.1).

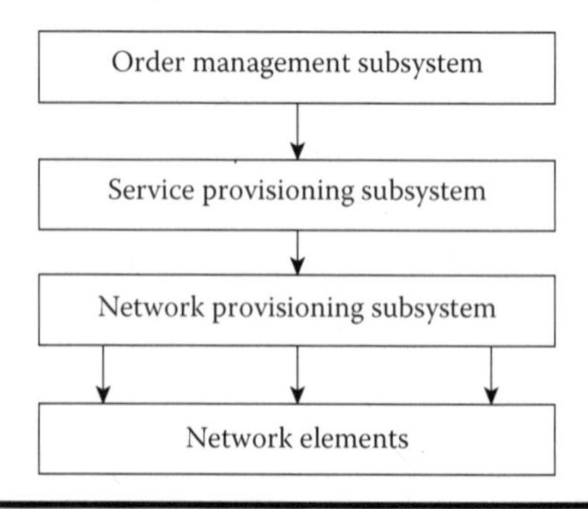


**Figure 4.1 Steps to complete a customer order for a service.**

Most of the operations involved in service and network provisioning need support from the element management system. Let us consider the EMS functionalities involved in provisioning:

- Using an auto discovery process, the EMS identifies its resources and updates its data base. This information is synchronized with the inventory management system so that the inventory management application can identify and allocate appropriate resources for a service.
- Installing software and patches for a network element is done by the element management application for the specific network element.
- Backup and restore is part of configuration management in FCAPS functionality of element management system.
- Configuration of a network element for stand-alone operation and after integration with other elements is performed by EMS.
- Generating billing records after provisioning and on service activation is an activity controlled by accounting in FCAPS.
- In addition to centralized security, ensuring that the service provisioning is in a secure environment for the individual NE, which is controlled by the NE using its secure access and security logging system.

In most scenarios the operator at the network operations center only works on a single screen that involves taking up the order and doing an automatic service provisioning. If the required inventory is not in place, the technicians in appropriate locations will be notified to set up the inventory. The EMS database will consist of parameters that promote auto configuring and sending messages to support its auto discovery. Most modern telecom management systems usually use just a single service provisioning application or a service provisioning functionality that is part of a bigger application and fills up data in a few graphical user interfaces. Based on the parameter messages that are send to the appropriate network and its network elements to auto discover the elements, configure the same and activate the service. Only when there are faults while provisioning will a technician be involved in using applications like NMS and EMS of the network elements to identify the fault and fix it. There are telecom management products available on the market that are solely dedicated to provisioning of specific services.

## 4.3 Network/Service Development and Planning

Service planning involves identifying a strategic view on the direction of service development. This can be based on external or internal research, knowledge of experts, interviews with customers, and so on. A major input is also market and technology forecasts that can trigger the expansion of existing services, investment on emerging services, and research for future services. The service plan would include details on service capabilities, standards, infrastructure, technical support, design elements, implementation, and budget parameters and targets. Service development is based on the analysis of service ideas that were decided in service planning, followed by management of service development. Service development also ensures evaluating service parameters like quality of existing services and using these inputs for defining specification of services under development.

Resource planning involves preparation of a strategic plan based on research on the technology trends in expanding existing resources, identifying emerging technologies, and resource capabilities to meet the needs of future services. The findings can be based on corporate knowledge, market analysis, and internal/external research. Resource development based on the planning would involve studying the performance parameters of existing network and its elements and making detailed design of the resource development based on the existing key performance indicators of resources in the network.

Some of the operations involved in service/network development and planning need support from the element management system. Let us consider the EMS functionalities involved in development and planning:

- Using fault management, the elements that require more maintenance and frequently show trouble conditions can be identified. For example, the call server that usually has a higher number of dropped calls compared to its peers, triggers frequent auto reboot, or needs a technician to reconfigure its parameters can be replaced with another call server that would give better results.
- Fault management also gives inputs on the common fault scenarios that can disrupt service and can be used to find out the software and hardware capabilities to improve or reduce the fault scenarios. For example, a particular algorithm might cause high congestion on some trunks and under utilization on other trunks. This information can be easily identified from logs generated showing percentage utilization of trunks during an audit of resources.
- Performance reports generated at regular intervals are a key input to forecasting those elements that would be best suited for investment on enhancing existing network capabilities. The performance reports also form input to identifying the quality of service. Users are requested to give feed back on their experience of the service.

  > The experience perceived by the user while using a service is termed as Quality of end user experience (QoE).

  The quality of service and quality of experience are an integral part of planning and development of service and resource.
- Configuration of an existing network element can be used as inputs for the detailed design of a network element for a purpose similar to the existing element.

- Billing records help in identifying the services that are of key interest to subscribers and are used to find usage patterns and forecast the future trends and services that can capture the market. Thus billing records are analyzed and the statistical data generated from the analysis are used for service and resource planning.

## 4.4 Inventory Management

Inventory management involves a centralized repository on network assets for tracking and managing the assets. While a network management system gives information about a fault that occurred in the network and the resource that generated it, the inventory management system is used to map this fault to resource location, connectivity, affected users, service failures, and so on. This is just a typical example of how data from the element and network management system can be complemented with functionality and data in inventory management to make a detailed map of information that can assist effective service and business management including planning.

An inventory management system typically includes:

- Asset database
- Inventory analysis functions
- Inventory reporting functions

Asset database is the repository of information about the inventory in the network. In addition to location and connectivity information that can change, the asset database even stores intricate static data like part number, and manufacturer of the equipment in the database. The analysis functions would include tools for manipulation, optimization, and allocation. When a fault occurs, based on the location of an inventory the nearest technician can be allocated to service the equipment and fix the fault. This helps for better work allocation and in meeting service level agreements offered to the customer. The reporting functions provide a detailed view or report of the inventory and also would include customized functions for generating a report on inventory based on location, type, number of services, and so on.

Having a network management system helps to monitor the network, but an inventory management system is required for planning, design, and life cycle management capabilities of elements in the network. The auto discovery process triggered at an element management system identifies the resources in the network. The updated resource information is then synchronized with the inventory management system. This information should be available in a central repository with easy user access that provides the ability to access up-to-date network information required for high-reliability environments.

In addition to taking feed from lower layers, the inventory management system also gives feed to upper layers. For example, it gives feed to a trouble ticketing system that allows device specific alarm information to be provided, as a part of the dispatch process, to a network engineer for trouble resolution. This in turn reduces the mean time to repair (MTTR).

Some of the operations involved in inventory management need support from the element management system. Let us consider the EMS functionalities involved in inventory management:

- Auto discovery to update dynamic status of the resources when it is put to use in the network.
- Audit, at regular intervals, of network resources and synchronization of data with the inventory management system.
- Update on attribute or type of a resource after a software or patch install on the network element needs to be notified to the inventory management system.

There are products solely developed for inventory management in telecom industry.

## 4.5 Integrated Multivendor Environment

At the Network Operations Center (NOC) only a single interface is provided to the operator to work on the underlying network. This is an integrated multivendor environment where the underlying network would contain network elements from different vendors and the network management system would have the capability to invoke any element management system for a specific network element. In most scenarios when an element manager for a NE is launched from the NMS, the element manager and the network manager application can be open simultaneously. With hundreds of network elements and its resources that make up the network, it is not practical to have separate applications handling each of the NEs and an operator going to the trouble of browsing through these applications to find the network health. With an integrated view of the entire network and centralized control, much of the overhead involved in handling a multivendor environment is reduced.

There is normalization to be performed on the data from a different element management system before it can be handled by a single NMS application. This normalization is performed by the adaptation layer, which performs the conversion to a common format (see Figure 4.2). The adaptation layer can be a separate application or it can be implemented as part of the EMS or in the NMS. That is, each vendor who sells an EMS also provides the adaptation application or a probe for the NMS to collect data from EMS, or the NMS developer adds an adapter for a specific EMS each time it buys a new network element and its element management solution.

The functions in a network management system involve actions that can be performed on all the network elements. To have a common set of functions applicable

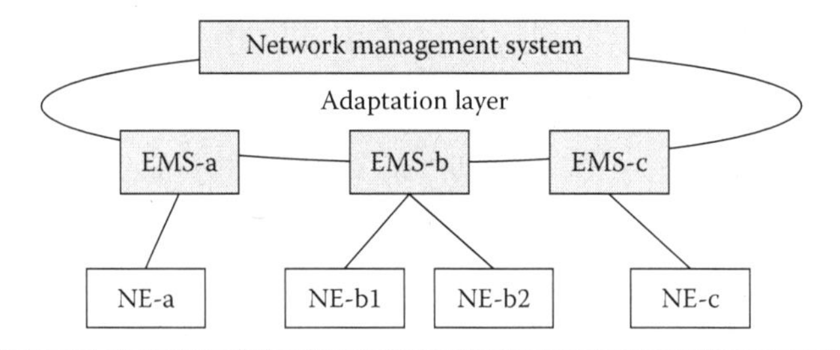


**Figure 4.2 Normalization at adaptation layer.**

to all the elements there must be consistency in the data handled at NMS. The adaptation layer performs the conversion required to a common format for handling the data at NMS.

Another method to solve this problem is to have well-defined standards of interaction. The set of queries that the NMS will send to an EMS or the underlying network element is to be defined. TeleManagement Forum (TMF) is working toward this initiative with its standards on multitechnology network management (MTNM).

## 4.6 Service Assurance

When a customer subscribes to a service, a service level agreement (SLA) is made with the service provider. Service assurance involves ensuring that the service provider is delivering the guaranteed level of service to the customer.

It is mainly comprised of:

- Service quality management
- Trouble or problem management

Quality of service (QoS) is a key component in service assurance covered in service quality management that ensures speed and reliability of service. The QoS requirements are used to derive the network element and its resource requirements for entities such as call processing, data communication, and storage. The QoS parameters relevant to requirements in service level agreement are identified, monitored, and controlled.

The QoS parameters may be oriented toward:

- performance (example: data rate)
- format (example: delay with different codec for media transmission)
- cost (example: different billing packages like first 100 calls free)
- user (example: quality of user experience QoE)

There is also a service agreement on fixing trouble in service and restoration of service within a specified time. It is here that trouble or problem management plays a key role.

Some of the operations involved in service assurance need support from the element management system. Let us consider the EMS functionalities involved in service assurance:

- When a trouble occurs in the delivered service, this would be associated with the occurrence of some event on a resource. At the service management level, an event is generated notifying the disruption of service. This is followed by an event trigger to automatically fix the trouble condition or a notification/assignment of work to a technician for looking into the issue and fixing it. A service level log/alarm will not be of much help to the technician. The log/alarm generated on the resource that caused disruption of service and displayed by the fault management system in the EMS would be the most precise fault notification with maximum information for the technician to fix the issue. For example, if there was a loss of service when a set of patches (bug fixes) were installed on a network element, then the exact patch that caused the network element to malfunction can be selected and removed. This helps fast restoration of service to meet SLA.
- As already discussed, performance management reports generated on the element management system is a key input on QoS. The periodic performance records collected on quality metrics and key parameters characterize the performance of the network resources over service intervals. Performance management also has functionality to set thresholds on specific parameters. The EMS generates events when thresholds are crossed, which helps to identify trouble scenarios before they arise and hence actions can be taken that would guarantee the commitments on service assurance specified in SLA.
- Customer support is also at times bundled with service subscription. There would be specific guidelines on the duration to complete a customer request. It could be setting up a service, fixing a trouble condition or answering a query. Good customer support is a key differentiating factor for better business and improved customer satisfaction and loyalty.
- The configuration and performance data can give information on the element utilization. Configuration parameters directly give the time the element was out of service, maximum load the element can handle, and so on and the performance data complements the configuration data in identifying utilization and performance matrix of the network element. This would help in resource planning for a network offering a service. Optimal configurations can be determined over time that would best suit the customer and meet the level of service assurance.
- Service assurance does not always mean a set of measures taken by the service provider to keep a customer happy. Assurance is to make sure that both

customer and service provider meets the terms and conditions specified in the service level agreement. In cases where the customer is deviating from the SLA, the service provider notifies the customer and takes corrective actions. A typical example where EMS functionality comes into play is for implementing security. The actions performed by the customer can be collected as security logs or using an event engine that listens for occurrence of a specific event and generates a report when an event occurs. When the customer performs an action that results in security violation, the customer service can be temporarily or permanently disconnected based on the SLA.
- To ensure a good quality of service that is essential for service assurance, there is a requirement to monitor network fault, performance, and utilization parameters to preempt any degradation in service quality. The role of fault, performance, and utilization parameters on service assurance has already been discussed earlier but a collective role of these functions is to ensure good QoS.

There are multiple service assurance solutions available in the telecom market that mainly provides overall network monitoring, handling of trouble, and isolation of issues. Service assurance is closely associated with business and is not completely a technical entity. Some of the business benefits in service assurance are:

- Increase in revenue as the network is better managed to meet expectations in SLA. With proper monitoring of performance, fault, and configuration data to resolve issues in a minimum amount of time the revenue is increased with optimal network utilization.
- Increased customer loyalty by managing service quality as perceived by the customer and effective customer support as guaranteed in the service assurance package.
- Reduction of capital expenditure (CAPEX) as new services will be introduced based on forecast and trends identified while end-to-end monitoring of the network for service assurance parameters.
- Customer centric approach is becoming popular with all aspects of telecom management and development. Service assurance is a customer centric approach to service management.
- Optimize performance of the service delivery platform. When the operational process is centered around providing good quality of service for customer service assurance, the performance reports can be used to optimize performance of the service delivery platform.
- Reduction of operational expenditure (OPEX) with improved operational processes to meet the parameters of service assurance.

There are products solely developed for service assurance in telecom industry.

## 4.7 Network Operations Support

Network operation support involves the maintenance activities involved with the network and its elements. Reduction in the cost of operations support is a major goal of every service provider. The cost for operations support is inversely proportional to the quality of network management. As the quality of network management increases the cost on operations support decreases and vise versa.

Some of the functions performed at the network operation center (NOC) providing network operation support are:

- Network monitoring and management
- Software and hardware maintenance
- Information security
- Troubleshooting of issues
- Call support: technical and customer
- Disaster recovery
- Migration, upgrade, and system engineering
- Deployment, replacement, repair, and restoration of elements

In addition to cost reduction, another issue that is of major concern is the lack of trained technicians to manage the network. The telecom network is evolving rapidly with new standards and enhanced hardware and software capabilities. The technicians need to keep pace with this change for managing the network effectively.

Minimum manual intervention is a requirement set out by standardizing bodies to be part of next generation network management. By making intelligent network and service management solutions, the manual intervention involved would be reduced, cutting down the cost of operations support.

Some of the operations involved in network operations support need functionalities in the element management system. Let us consider the EMS functionalities involved in operations support:

- Network management and monitoring would involve management and monitoring of the elements in the network. EMS is the key to manage and monitor elements in the network.
- Software maintenance on a network element would involve installing bug fixes, enhancement features, and so on. EMS provides functionality to remotely do software maintenance from the machine running the EMS at a NOC.
- Information security handled at NOC uses security logs captured by the EMS from the network element. These logs provide information on security at the element level, which can be critical when the element being managed is a security gateway.

- Troubleshooting of issues at element level is performed using an EMS. There would usually be a single NMS application that shows consolidated information of the network and from which element management system for a specific element can be launched.
- Most EMS application has the functionality that allows the user to log into the network element. This functionality can be used for performing operations like upgrade and migration of data from the network element remotely from some other machine and not physically going to the element location and logging into the NE.

## 4.8 Network/Service Monitoring and Control

Service monitoring and control is performed by an operation support system and network monitoring and control is performed by a network management system. The monitoring and control of a network or service has a subset of the functions performed by its management system. It mainly involves monitoring the health of the system and controlling the system with corrective actions in trouble scenarios. The network monitoring system discovers the state of the elements in the network at regular intervals of time. A trouble condition identified with a fault log can initiate a discovery to find the status of the element.

There are multiple network monitoring tools and systems that are available even as open source products. The basic monitoring tool follows server–agent architecture with a single monitoring server and multiple agents connecting and sending information to the server. The communication will be based on a predefined protocol that is supported in NMS. Some of the popular protocols are SNMP, CMIP, XMLP, SOAP, NETCONF, and so on. Some of the protocols will be handled in detail later in this book.

Network monitoring can also be seen as a specialized function in NMS. For example a security monitoring system would collect security logs, provide access permissions, create roles, and groups and performs the set of activities to ensure security in an NMS. Likewise, a fault monitoring system is a specialized fault handler that collects fault logs, helps to program actions for specific fault scenarios, allows the creation of user-defined logs and performs a set of actions that would be of importance from a fault perspective. As a result of the various functionalities, monitoring solutions or specialized functions are available as stand-alone tools.

Control functions associated with monitoring are usually simple controls like in security monitoring, the control function might involve preventing the user to log in by locking the account if there are three consecutive failed attempts with a wrong password or in a fault monitoring tool there controls to selectively clear certain faults. These levels of controls, though not featuring an elaborate set of functionalities in a full fledged NMS, does help to assist monitoring. For example a specific log might be generated at regular intervals during audit and may not be of much importance in fault monitoring. The fault control associated with monitoring then could add filters to selectively hide the log, not collect it from the element or clear the specific log.

Some of the operations involved in network/service monitoring and control need support from functionalities in the element management system. Let us consider the EMS functionalities involved in monitoring and control. First of all it can be seen that a monitoring and control system is a subset of functions in NMS and hence EMS being the layer below the NMS would also be the prime feed for a monitoring and control system.

The major difference here is that the EMS only sends specialized feed on a specific area like security, fault, and so forth, to the monitoring tool in most scenarios unless the application is more generic and it requests more information similar to NMS. Most of the functionalities of an element management system are of vital importance to the monitoring and control system. Based on the server–agent architecture discussed above, it can be seen that the design of a monitoring and control system is similar to an element management system that gets feed from multiple network elements.

In the functionality listing of most network management solutions, monitoring, and control is usually listed. So monitoring and control can be an independent tool, a specialized tool, or embedded into an NMS. Service monitoring is a layer above network monitoring, though the discussion of the association of EMS with network monitoring and control also applies to service monitoring and control.

## 4.9 Conclusion

Just above the network elements, the element management system layer (EMS) forms the first management layer collecting data from physical components. This chapter was intended to show the importance of the EMS in building up functionalities on the upper layers of the TMN model.

## Additional Reading

1. International Engineering Consortium (IEC). *Online Tutorial on Element Management System,* 2007. http://www.iec.org/online/tutorials/ems/

# Chapter 5

# OSI Network Management

OSI network management recommendations developed by ISO/IEC include a set of models for effective network management similar to TMN from ITU-T. Before the release of the initial M.30 recommendation from CCITT, there was very little cooperation between CCITT (now ITU-T) and ISO/IEC management groups. After the publication of M.30, the collaborations between these groups improved and many of the OSI ideas were incorporated into TMN. At the end of this chapter you will have a good understanding of OSI network management models and how it compares with TMN.

## 5.1 Introduction

The OSI network management architecture mainly consists of four models. They are:

- Organization model
- Information model
- Communication model
- Functional model

The OSI organizational model, also called OSI system management overview, was defined by ISO-10040. It defines the components of the network management system, the functions and relations exhibited by these components that make up the network management system.

The OSI information model defines how management information is handled in ISO-10165. It details the structure of management information with SMI (structure of management information) and storage of information with MIB (management information base).

The OSI communication model defines how management information is exchanged between entities in the network management system. The transport medium, message format, and content are also discussed.

The OSI functional model defines the functions performed by the network management system. The capabilities of the network management system are classified into a set of basic functions, with the capabilities being derived from these basic functions.

OSI network management concepts had a major influence in telecom management. Most of them form the basic principles of telecom management including the agent–manager framework that is used for data collection from elements. The object-oriented approach in OSI network management was appealing for managing the complex telecom network. The CMIP (common management information protocol), an application layer protocol, was developed based on OSI guidelines and is discussed later in this book. Though CMIP is not as popular as its counterpart SNMP (simple network management protocol) in implementation and deployment, the protocol forms the initial attempt on implementing a management protocol based on OSI, which can be used to replace SNMP used at that time for internet management.

The OSI management concepts that were adopted by TMN and the differences between TMN and OSI network management are also discussed in this chapter. The TMF (TeleManagement Forum), the successor of Network Management Forum (NMF) is now looking into development of network management standards based on both OSI network management and TMN.

## 5.2 OSI Organizational Model

The components that make up network management in the OSI organizational model are:

- Network objects
- Agent
- Manager
- Management database

The elements in a network are referred to as objects. It can be a router, gateway, or switch. The presence of an element in a network does not necessarily mean that there is a management system to support it. The elements or network objects that

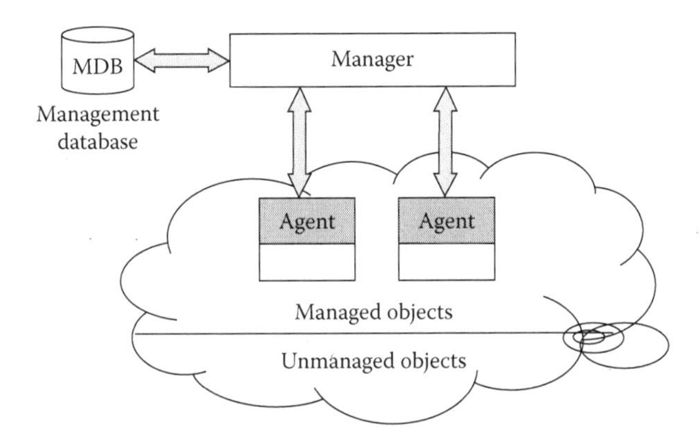


**Figure 5.1 Two-tier organizational model.**

have management functions associated to it, as defined by OSI, are referred to as managed objects and the remaining objects are called unmanaged objects.

Network objects can be of two types: managed and unmanaged objects.

The network management functions associated to managed objects makes up the management process running on managed objects. This management process is referred to as the agent. In the two-tier organizational model, the elements in the network that are managed does the regular network operation they are expected to perform as well as the management of the element by having the agent as part of the element (see Figure 5.1). The management functions as the agent are limited to functions like data collection, editing of data, and sending of event notification from the element.

The "manager" manages the managed objects in the network using data from the agents. It works aggregated information sent from multiple agents and performs the network level functions. The manager can:

- Send queries to the agent
- Receive response from the agent
- Send update and edit notifications to the agent
- Collect data from the agent (alarm notification, performance report, etc.)

The management database (MDB) is used to store management information. The data collected from the agents by the manager is processed to generate consolidated network level reports on fault, configuration, performance, and other management functions. These processed reports as well as raw reports from the agent can be stored in MDB. The MDB serves as an archive for historical information on the managed objects.

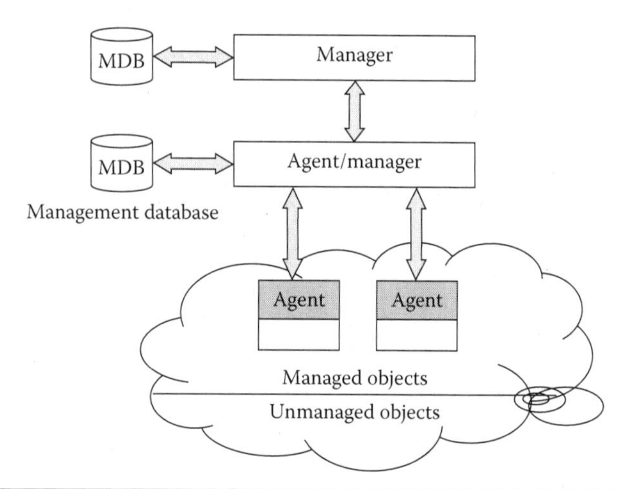


**Figure 5.2 Three-tier organizational model.**

There are three relational models to be discussed under OSI organizational models. They are:

- Two-tier model
- Three-tier model
- MoM model

The components, functions performed by the components and their relationship, have already been discussed for the two-tier model. The three-tier model is similar to the two-tier model except for the presence of an additional "agent/manager" with its own management database between the manager and the agent (see Figure 5.2). We will call this additional component the intermediate component to avoid confusion in the discussion below.

The manager in the intermediate component performs the role of collecting data from the managed objects. This manager can also process the data and store information in the management data base of the intermediate component. The manager in the intermediate component acts as a manager for the agents in the managed objects. The agent in the intermediate component performs the role of sending information to the top level manager above the intermediate component. It thus acts as an agent for the top level manager.

The concept would be clear if you view the intermediate component as an element manager and the top level manager as a network manager. The intermediate component like the element manager collects data from an element or a set of similar elements. It collects data from the agent on the element (this part acting as manager of the intermediate component) and provides feed to the network manager or top level manager (this part acting as the agent of the intermediate component). The feeder is always the agent and the component that is feed is the manager.

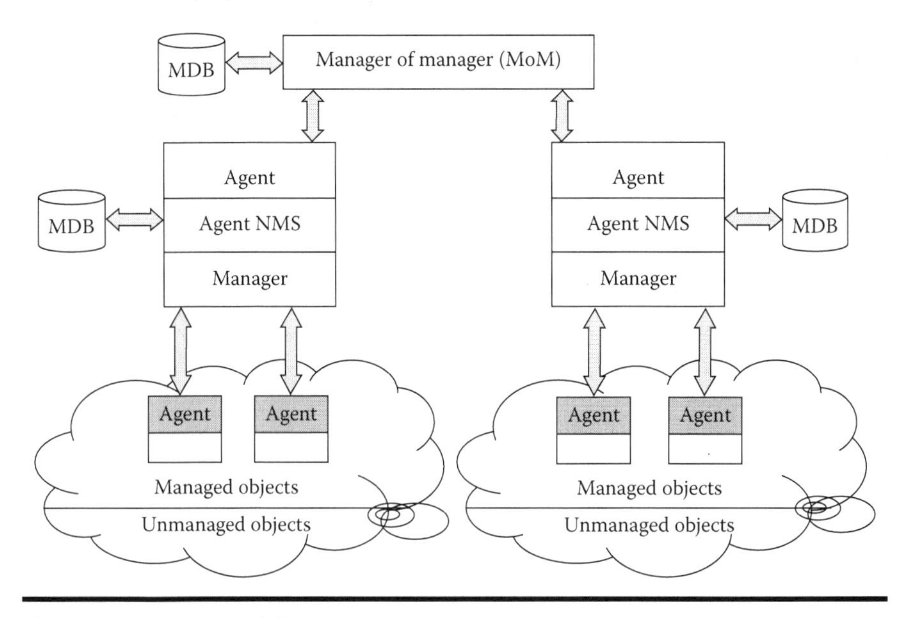


**Figure 5.3 MoM model.**

The three-tier model satisfies the needs of a network management system. Now consider the scenario where a network manager is required to manage multiple networks at different locations or networks from different vendors or two heterogeneous networks. This is where MoM fits in by providing a model for a converged network manager for multitechnology, heterogeneous network, mute vendor environments that is the current state of the telecom industry (see Figure 5.3). This model can handle network management of multiple domains as in an enterprise network.

The top most level in MoM model is the "manager of manager" (MoM), which is the manager that consolidates and manages a set of network domains by collecting data from the network managers that manage individual network domains. The MoM has a separate MDB for storing data. In between the MoM and the managed object is an intermediate component having a manager, agent NMS, and an agent. We will call it the intermediate component similar to the explanation for three-tier to avoid confusion. The manager in the intermediate component of MoM collects data from the managed object, the agent in the intermediate component feeds the MoM, and the "agent NMS" is the management process that deals with functions for managing network in a specific domain.

There is one more relationship defined in OSI organizational model that specifies:

An agent NMS and a manager NMS share a peer-to-peer relationship.

This is suitable for exchange of information between network service providers. Thus the organizational model defines the network management components, their functions, and the relationships possible with these components in various network management scenarios.

## 5.3 OSI Information Model

The OSI information model deals with structure and storage of information. It details how to form an information base using managed network objects and their relationships. Specified by ISO 10165, the OSI information model is an object-oriented data model.

It has two main definitions:

- structure of management information (SMI)
- management information base (MIB)

### *5.3.1 Structure of Management Information (SMI)*

The SMI defines the syntax, semantics, and addition information like access permission, status, and so on of a managed object. A managed object in the information model is similar to "class" in high level programming language (see Figure 5.4). The managed object definition like class provides a template to encapsulate attributes and management operations of the object. It also supports asynchronous interactions as events.

Consider the scenarios where the managed object class is representing a gateway. The rectangle is the MO, ellipsoidal shape represent attributes, rectangle with smooth corners represent behavior, and triangle represents event. There can be different kinds of gateways that share a common set of properties and methods of a gateway with specialized properties and methods based on the application that are put to use. A billing gateway would be different from a signaling gateway though both fall under the class of gateway. To support modeling for these scenarios the managed object like class supports inheritance. Hence a managed object class can be derived from some other class.

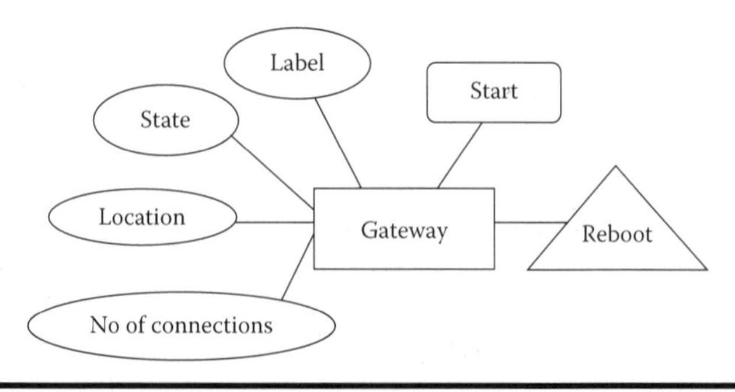


**Figure 5.4 Example of a managed object (MO).**

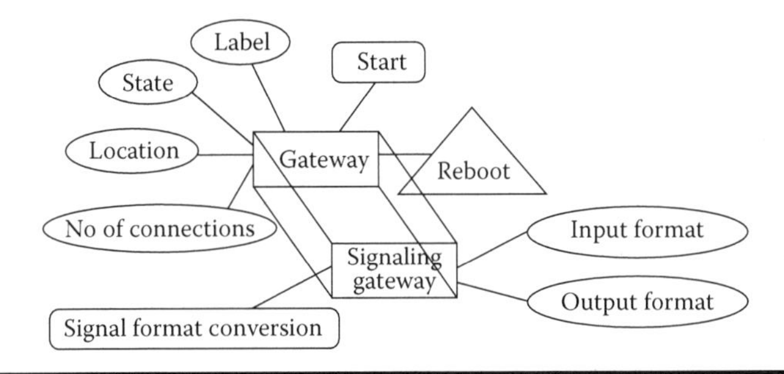


**Figure 5.5 Example of derived object.**

```
A sample Managed Object Template is given below:
<class-label> MANAGED OBJECT CLASS
[DERIVED FROM <class-label>;]
[CHARACTERIZED BY *;]
[ATTRIBUTES *;]
[BEHAVIOUR *;]
[NOTIFICATIONS *;]
REGISTERED AS <object-identifier>;
```

The managed object (MO) has ATTRIBUTES to define its static and dynamic properties, BEHAVIOR to represent the methods/actions performed by the MO and NOTIFICATIONS are for events. The class-label uniquely identifies the class and every managed object is to be registered as a unique object in the management information tree (MIT). The "registered as" part registers the MO definition on the ISO registration tree. The "derived from" section describes the base class whose definitions are inherited by the MO child class. A sample of the managed object template is given in Figure 5.5. The "characterized by" part includes the body of data attributes, operations, and event notifications encapsulated by the MO.

The SMI was introduced in ISO-10165 and it was elaborated in the guidelines for the definitions of managed objects (GDMO). The GDMO provides syntax for MO definitions. The GDMO introduces substantial extensions of ASN.l (abstract syntax notification) to handle the syntax of managed objects. The ASN will be covered in detail in the chapter on SNMP.

## 5.3.2 Management Information Base (MIB)

MIB is a virtual database that contains information about managed objects. The information is organized in a treelike structure by a grouping of related objects and relationship that exists between them (see Figure 5.6). A managed object instance in the tree includes attributes that serve as its relative distinguishing name (RDN). The RDN attributes are used to uniquely identify the instance among the derivatives or children of the managed object. By concatenating RDN with the MIT path from

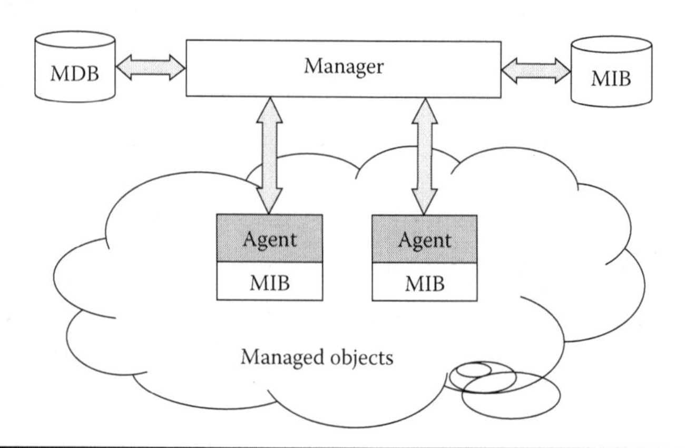


**Figure 5.6 Management information base (MIB).**

the root to a given node, a unique distinguishing name (DN) is obtained that identifies the node. The node can be any information about a managed object and the DN will be used by the manager to store and query any particular information.

Both agent and manager use the MIB to store and query management information. The agent MIB covers information only about the managed object it is associated with while the manager MIB covers information about all the network elements managed by the manager. While MDB is an actual physical database like Oracle or SQL, MIB is a virtual database.

The MIB implementation falls under of the following three types:

- MIB Version I: This first version of MIB was presented in RFC 1066 and later approved as a standard in RFC 1156.
- MIB Version II: This version of MIB was proposed to cover more network objects as part of RFC 1158. It was later approved by the IAB (Internet Architecture Board) as a standard in RFC 1213.
- Vendor specific MIB: These complement the standard MIB with additional functionality required for network management of vendor specific products.

This book has a chapter dedicated to the discussion of MIB considering the wide acceptance of this schema in information management.

## 5.4 OSI Communication Model

The OSI communication model defines the way management information is exchanged between entities in a NMS (agents/managers). The transport medium, message format, message content, and the transfer protocols are also detailed (see Figure 5.7).

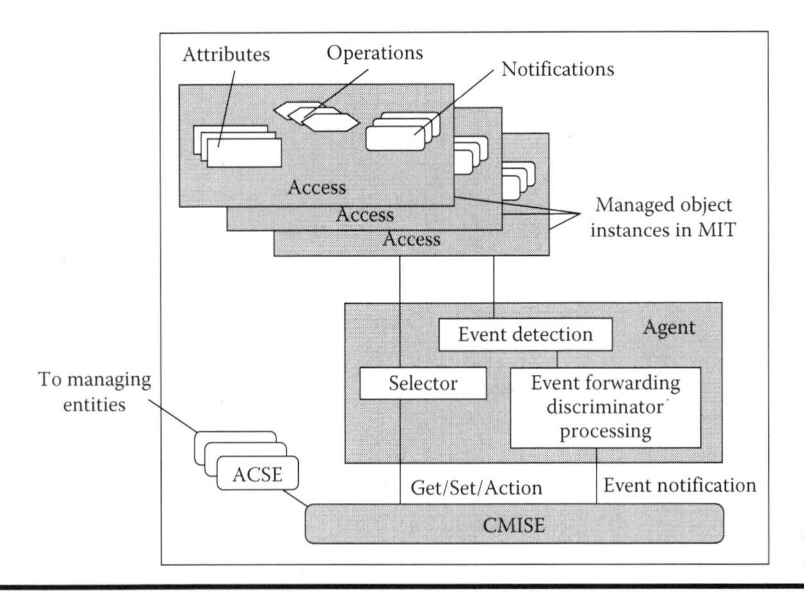


**Figure 5.7 OSI communication model.**

The entities in the communication model are:

- MIT (management information tree): It has the attributes, operations, and notifications defined for the managed objects.
- Agent: It communicates with the managed objects to collect data. The information in MIT is queried by the agent and updated using commands like Get, Set, and so on. It also has event detection and processing.
- CMISE (common management information service element): Has a set of operations providing communication. It works on CMIP protocol to achieve message exchange.
- ACSE (association control service element): Handles opening and closing of communication between the manager and the agent.

There are three kinds of messaging that happens between the manager and the agent as shown in the message communication model (see Figure 5.8). They are:

- Request/operations: These are message queries sent by the manager to the agent requesting information.
- Responses: These are messages sent by the agent to the manager as response to a query from the manager.
- Notifications/traps: When an event occurs that needs to be notified to the manager, the agent generates a trap with data on the asynchronous event.

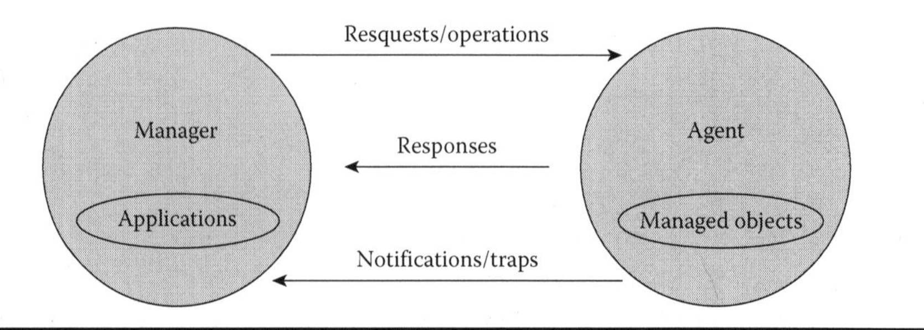


**Figure 5.8 Message communication model.**

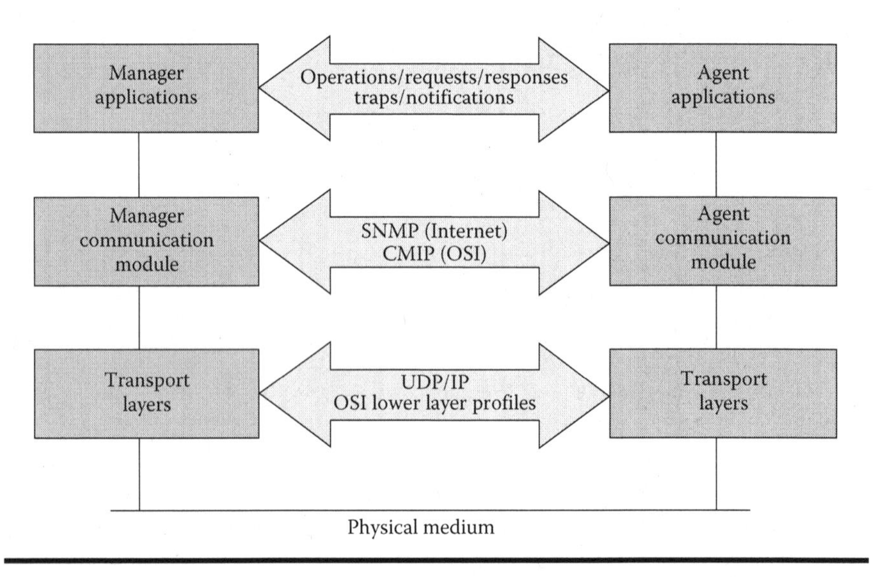


**Figure 5.9 Communication transfer protocols.**

The transfer protocols in the OSI communication model is detailed with three layers as shown in Figure 5.9 on communication transfer protocols. They are:

- Messaging between manager and agent application: The message communication model already discussed describes the supporting messages.
- Application layer protocol: The application layer protocol for interaction between agent and manager module is SNMP for the Internet and CMIP for the OSI network.
- Transport layer: The model specifies the use of UDP/IP and OSI lower layer profiles.

## 5.5 OSI Functional Model

The OSI functional model specifies that network management functionality can be grouped under five main categories (see Figure 5.10). They are FCAPS short for fault, configuration, accounting, performance, and security. The ITU-T introduced FCAPS as part of M.3400 for a telecommunication management network (TMN). The ISO used this concept for management of data networks by introducing it as part of OSI management model. These five functional areas form the basis of all network management systems for both data and telecommunications. Refer to chapter on FCAPS for a detailed explanation of the functional areas.

## 5.6 OAM

The OAM stands for operation, administration and maintenance. The OA&M and OAM&P are words popular in the telecom management world for network management functions. The five function areas are covered as part of OAM, hence it is quite common to refer to network management systems as OA&M systems. When the NMS involves a lot of provisioning functions, then the NMS solution is referred to as OAM&P where P stands for provisioning. It can be seen that the SNMP functional model adopts the concept of OAM (see Figure 5.11).

The ITU-T defined OAM in its document M.3020 and 3GPP uses the term OAM in its documents on network management standards for specific networks.

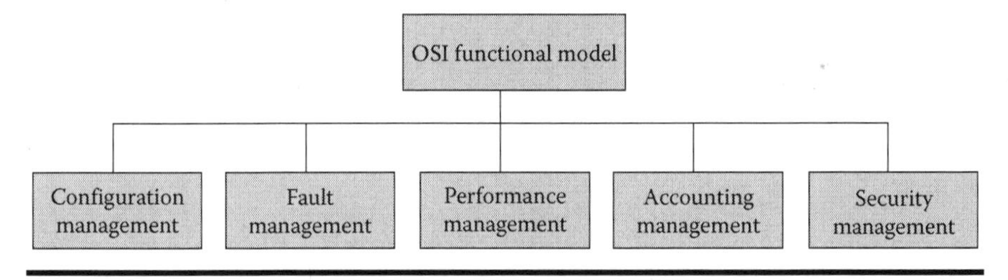


**Figure 5.10 OSI functional model.**

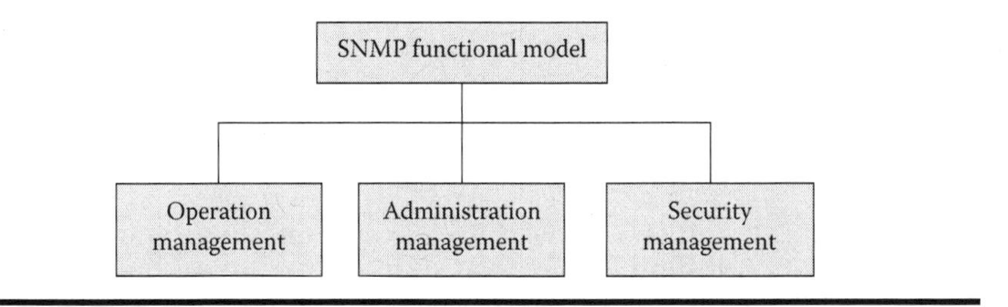


**Figure 5.11 SNMP functional model.**

Operations is the control center that coordinates actions between administration, maintenance, and provisioning. Administration deals with designing the underlying network, processing orders, assigning addresses, tracking usage, accounting, and other activities associated with network administration. Maintenance is to monitor, diagnose, and repair problem scenarios in the network. Provisioning handles installing equipment, setting configuration parameters, verifying that the status of service, updating the element with fixes/new software, and de-installation.

## 5.7 TMN and OSI Management

There was not much cooperation between OSI and TMN management groups initially when the first release of management standards were released by respective bodies. Later some of the concepts of OSI were incorporated in TMN.

The important changes to TMN based on OSI are:

- The manager-agent concept suggested in OSI was adopted by TMN. There is a reference to OSI document X.701 on OSI system management in TMN documentation.
- The object-oriented approach of OSI is another feature in the current TMN. The principles of TMN are now based on an object-oriented paradigm.
- Management domains introduced by OSI were adopted.

The difference in philosophy between TMN and OSI are:

- TMN has multiple architectures defined at different levels of abstraction while OSI has a single architecture for all levels of telecom management.
- The various management responsibilities are separated in the Logical architecture of TMN, while OSI model does not have such a detailed split of responsibilities which could help in application modeling.
- There is a separation between telecom and data network in TMN, where a telecom network is the network being managed and data network transfers management data.

## 5.8 Conclusion

The OSI and TMN do have some conflicts in concepts. But this does not make them two separate standards. Much of the concepts in one were adopted by the other and these models are the backbone for all developments that were made in telecom management. Now enhancements on standards from OSI management and TMN management are handled by TMF. This has resulted in a forum for unified standards creation in management space. This chapter gives an overview of OSI management model that was one of the most comprehensive documents initially available on standards in network management.

## Additional Reading

1. Mani Subramanian. *Network Management: Principles and Practice.* Upper Saddle River, NJ: Addison Wesley, 1999.
2. Vinayshil Gautam. *Understanding Telecom Management.* New Delhi: Concept Publishing Company, 2004.

# Chapter 6

# Management Protocols

The element management system interacts with network elements in its lower layer and network management system in its upper layer based on a TMN model. Proprietary protocols were initially used for the interaction between the EMS and the network elements. There was a requirement to have the same EMS used by multiple elements. Standard protocols were introduced to define the interactions of EMS with the network element. At the end of this chapter you will have an understanding of the protocols commonly used for interaction between EMS and NE.

It should be noted that management protocols are not limited to interaction between EMS and NE. It can be between an EMS and NMS, NMS and OSS, or even between two OSS applications.

## 6.1 Introduction

> Management protocol is a set of rules governing communication between components in the management system.

Telecom management paradigms are constantly evolving. Initial management systems were ASCII based, with most of the management logic placed in the network element. Since this model was not scalable, the management logic was then slowly moved to a management system. Proprietary management protocols were used for communication between the elements and the management system in initial legacy management systems. This trend changed when there was a need to support network elements from multiple vendors each having their own proprietary protocols for management. Standardized management protocols were introduced that defined a common vocabulary that could be used for communication.

There is no ideal standardized management protocol suited for all kinds of network elements and management applications. The management protocol suited for web-based network management may not be best suited for a client–server application and the management protocol for a packet-based network may not be best suited for a circuit switched telecom network. Hence various standardizing bodies worked on developing management protocols for various management applications and networks. These efforts and resources have resulted in a vast choice of standardized protocols, with vendors offering a wide range of tools and platforms based on these protocols. These tools and platforms greatly facilitate the development and integration of the management systems and network elements.

Application and network are not the only items that determine the choice of a management protocol. It can also be a trade off between complexity and simplicity. Some protocols are aimed at simplicity, which would ensure easy agent implementation and short turn around time in software development. These protocols like SNMP (simple network management protocol) are better structured and easy to understand. The simplicity in interaction often results in extra functionality to be inserted in mapping of the information model.

There are protocols like CMIP (common management information protocol) that are more functionality rich and have an object-oriented structure making it a complex protocol. The wide range of operations in the protocol results in complex but powerful agents. There can also be scenarios where different management protocols should be combined within the same network element yielding the best solution for a specific interaction paradigm. This chapter gives an overview of some popular standardized management protocols. Based on the information in this chapter, the reader can make an informed decision on what protocol to use for implementing the requirements elicited in a project.

## 6.2 Overview of SNMP (Simple Network Management Protocol)

The SNMP was developed by IETF in 1988 as a solution to manage elements on the Internet and other attached networks. SNMP was derived from its predecessor SGMP (simple gateway management protocol) and was intended to be replaced by object-oriented CMIP. Since CMIP was a complex protocol based on OSI it took longer to be implemented. Many vendors by this time had already adopted SNMP and were not willing to migrate to CMIP, which lead to a wide acceptance of SNMP. SNMP is still the most popular telecom management protocol.

The SNMP is an application layer management protocol that runs on UDP (user datagram protocol) in the transport layer. It follows the manager/agent model consisting of a manager having the management functions, an agent that is on the network element sending and receiving messages from the manager, and a database of management information. In SNMP, the manager and agent use a management information base (MIB) and a relatively small set of commands to exchange information. MIB is a collection of information that is organized hierarchically as a tree structure and uses a numeric tag or object identifier (OID) to distinguish each variable uniquely in the MIB and in SNMP messages.

The SNMP messages to exchange information are:

| *Message* | *Explanation* |
|---|---|
| Get [GetRequest] | This is a request from the manager to the agent, to get the value at a particular OID (object identifier). |
| GetNext [GetNextRequest] | This is a request from the manager to the agent, to get the value of the next instance in the MIB. GetNext is used to "walk" down the MIB tree. |
| Response [GetResponse] | The agent responds with information to the manager for a Get, GetNext, or Set command using the "response" command. |
| Set [SetRequest] | This is a request from the manager to the agent to set the value at a particular instance in the MIB. |
| Trap | This is an asynchronous message from the agent to the manager to notify the occurrence of an event. |

Advantages of SNMP

- Ease of implementation of agent and manger with few interaction commands
- Ease of adding program variables to monitor
- Expandability with minimum effort for future needs due to simple design
- It is widely implemented and supported by most vendors
- Simple to setup and use

Disadvantages of SNMP

- Initial version of SNMP had security gaps and SNMP version 2 fixed some security issues regarding privacy of data, authentication, and access control.
- Information cannot be represented in a detailed and organized manner meeting the requirements of next generation network and management due to the simple design.

## 6.3 Overview of TL1 (Transaction Language 1)

The TL1 was developed by Bellcore in 1984 as a man–machine language for network management. The basic intend was to create an open protocol for managing elements. It consists of a set of ASCII instructions that are human readable. Since TL1 does not require any decoders or debuggers, it has become the preferred CLI (command line interface) for equipment manufacturers. TL1 does not have any message encoding and it does not have an associated information model. Most of the documentation on TL1 standards was developed by Telcordia.

There is a wide range of standard TL1 messages out of which four types comprise the majority of TL1 communication. They are:

| *Message* | *Explanation* |
|---|---|
| Autonomous messages | These are asynchronous messages sent from the network element to notify the occurrence of an event. Similar to Trap in SNMP. |
| Input/command messages | These are messages sent from the management application to the network element. |
| Output/response messages | These are replies sent by the network element to the management application in response to an input/command message. |
| Acknowledgment messages | These are brief messages from the network element to the management application acknowledging the receipt of a TL1 command. |

Some more specialized standard messaging with TL1 includes provisioning messages and protection switching messages.

Advantages of TL1

- CLI: Operators and management applications can compose input message and directly interpret responses and events.
- Delay activation: This feature allows execution of a request at a predetermined time.
- ATAG (autonomous message tag): This tag helps to identify loss of messages in a sequence.
- Acknowledgment: As per TL1 standards, an acknowledgment needs to be sent by the network element if an output response message cannot be transmitted within two seconds of the receipt of an input command.

Disadvantages of TL1

- High bandwidth required for transmission of ASCII text.
- It is difficult to get syntax right for sending data due to message format.
- Strict message format makes it difficult to read returned result.

- Though a CLI can be used by operators to directly communicate with an NE that supports TL1, in most scenarios TL1 is usually hidden from user by management application.
- For using TL1 in daily management tasks scripting needs to be added.

## 6.4 Overview of CMIP (Common Management Information Protocol)

The CMIP is defined as part of the X.700 series of OSI management standards. It follows the manager/agent model similar to SNMP. CMIP was developed as a replacement protocol to fix the deficiencies in SNMP. By the time CMIP work was complete, SNMP had already gained huge popularity. CMIP was a strong well-defined management protocol but the additional features of CMIP made it complex to implement and maintain. Hence vendors were not willing to switch from SNMP to CMIP and so it did not get a wide acceptance. Unlike SNMP, CMIP uses a reliable connection-oriented transport mechanism. It also has built-in security features of access control, authorization, and security logs.

CMIP is based on an object-oriented (OO) framework and the CMIP objects support:

- Containment: Each object can be a repository of other objects, attributes, or combination of object and attribute.
- Inheritance: A class defining an object can be derived from another object.
- Allomorphism: Ability to interact with modules through a base set of interfaces.

Some of the message interactions in CMIP are:

| *Operation* | *Explanation* |
|---|---|
| GET (Request/ Response) | This is a request from the manager to the agent to get the value of a managed object instance followed by a response from agent to manager. |
| SET (Request/ Response) | This is a request from the manager to the agent to set the value of a managed object instance followed by a response from agent to manager. |
| DELETE (Request/ Response) | This is a request from the manager to the agent to delete an instance of a managed object followed by a response from agent to manager. |
| CREATE (Request/ Response) | This is a request from the manager to the agent to create an instance of a managed object followed by a response from agent to manager. |

| | |
|---|---|
| ACTION (Request/ Response) | This is a request from the manager to the agent to cause the managed object to perform an action followed by a response from agent to manager. |
| EVENT_REPORT | This is a message from the agent to the manager to notify of the occurrence of an event on the managed object. |
| CANCEL_GET (Request/Response) | Cancels an outstanding get request. |

CMIP for TCP/IP is called CMOT (CMIP Over TCP)

CMIP for IEEE 802 LAN's is called CMOL (CMIP Over LLC)

Advantages of CMIP

- Has built-in security
- CMIP variables can perform tasks in addition to relay of information
- Can have well-defined requests
- Can be programmed to perform actions and generate notifications

Disadvantages of CMIP

- Requires a large amount of system resources
- Difficult to program, deploy, maintain, and operate due to complexity

## 6.5 Overview of CLP (Command Line Protocol)

The CLP popularly known as server management command line protocol (SM CLP) was a protocol suggested by distributed management task force (DMTF). CLP is a protocol based on a command/response model for managing elements on the Internet, enterprise, and service provider space.

Some of the key features of CLP are:

- Ability to manage the server with a simple text console.
- A unified language that is the same across systems from multiple vendors.
- Scripting capability making it suitable for scripting environment.

In a command/response model, the CLP client initiates a management request and transmits it through a text message based transport protocol to a management access point that hosts a CLP Service. MAP processes the received commands and returns a response to the requesting client. The transport protocols for which mappings are defined by CLP include Telnet and SSHv2.

The three supported output data formats for CLP are XML (extensible markup language), CSV (comma separated value), and plain text. The CLP builds on the DMTF's common information model (CIM) schema, which has a well-defined information model.

CLP syntax is as follows:

<Verb> [<Options>] [<Target>] [<Properties>]

"Verb" is the command to be performed. Some of the verbs are:

| *Verb* | *Explanation* |
|---|---|
| "set" | To update data and modify state of managed object |
| "show" | To view data |
| "create" | To create data |
| "delete" | To delete data |
| "reset," "start," "stop" | To modify the state of a managed object |
| "cd," "version," "exit" | To manage the current session |
| "help" | To provide help on a specific command |

"Options" are used to modify the action performed by the verb.

"Target" is the address or path of the target for the issued command.

"Properties" are used to send values that are needed to process the command.

Advantages of CLP

- Lower training costs
- Ease of maintenance as the protocol is text based
- Lot of reuse for common operations using scripts

Disadvantages of CLP

- Not all CLP commands are mapped to CIM methods.
- It does not have enough functionality to support advanced network management.
- CLP relies on underlying transport protocol to provide authentication and encryption.

The CLP is mainly intended for operational control of the server hardware and rudimentary control of the operating system and it best suited for scripting environments to perform repeatable tasks.

## 6.6 Overview of XMLP (XML Protocol)

The XMLP is a set of standards developed by a working group of W3C (World Wide Web Consortium). XMLP uses XML (extensible markup language) to define a messaging framework that can be exchanged over a variety of underlying protocols.

The standards in XMLP are intended to define:

- Encapsulation of XML data in an envelope
- The content of the envelope when used for remote procedure call applications
- Mechanism for serializing data based on XML schema
- Using HTTP as transport protocol for XML data
- Format and processing rules of an XMLP message
- Basic interaction between agent and manager using XMLP for interaction

XMLP is not a conflicting standard for XML-based protocols like SOAP, NETCONF, and so on, but a set of useful standards on the attributes required in an XML-based protocol.

XMLP is more of a foundation toward development of a XML-based protocol. Its definition acts as a framework that allows subsequent extension of the design while leaving the foundation of the design intact.

XML-based protocols are much more popular today for both enterprise and telecom management. A major reason for this popularity is the features of XML that makes it well suited for data handling in network management. Some of these features are:

- Structure of management information can be defined in an easy, flexible, and powerful manner.
- Data transport uses widely deployed protocols.
- Third-party applications can intercept and parse the data if the structure is shared
- Interoperability and centralized validation of management data.
- Ease of integrating management data from disparate sources.

The XMLP message is defined in the form of an XML infoset, with element and attribute information in an abstract "document" called the envelope.

XMLP message envelope

| XMLP Namespace | XMLP Header | XMLP Body |
|---|---|---|

XMLP namespaces: This is used for a unique name that avoids name collision, helps grouping of elements, and for use as a version control scheme.

XMLP header: Used for adding semantics for authentication, transaction, context, and so on.

XMLP body: Data to be handled by the receiver.

XML-based protocols have their drawbacks too. Use of XML leads to an increase in traffic load for message transfer. XMLP-based protocols are slow and the time duration for an interaction between the agent and manger using an XML protocol is more compared to most non-XML protocols. Though initially XML-based protocol was intended for a web-based enterprise management and scenarios of network management on the web, now XML protocols have gained more acceptance. SOAP is the protocol suggested in MTOSI (multi-technology operations system interface) defined by TeleManagement Forum (TMF) and NETCONF is hailed as the next generation network management protocol.

## 6.7 Overview of SOAP (Simple Object Access Protocol)

SOAP is an XML-based lightweight protocol developed by XML protocol working group of W3C (World Wide Web Consortium). SOAP is intended to exchange structured information in a decentralized, distributed environment. It was mainly designed for Internet communication. Remote procedure calls (RPC) used for communication between objects like DCOM and CORBA were not suited for the Internet. Firewalls and proxy servers usually block RPC and hence it was not used for Internet due to compatibility and security problems. SOAP supports XML-based RPC and HTTP supported by all Internet browsers and servers made a good pair for Internet communication. SOAP RPC can pass firewalls and proxy servers as the SOAP envelope is encapsulated in the HTTP payload (see Figure 6.1).

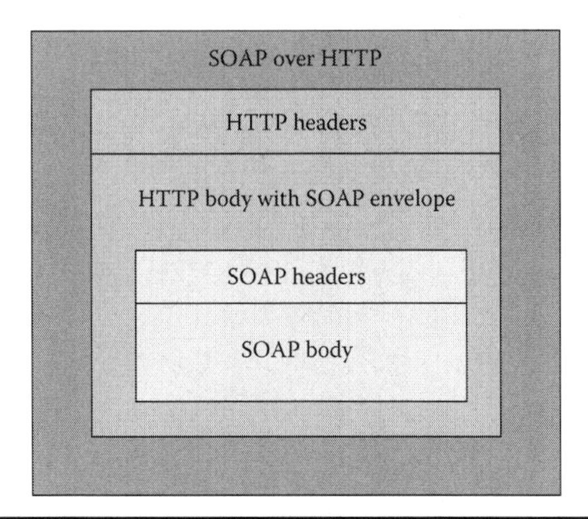


**Figure 6.1 SOAP envelope encapsulated in the HTTP payload.**

SOAP uses XML to define an extensible messaging framework that can work over a variety of underlying protocols, independent of the particular programming model used and other implementation specific semantics.

Advantages of SOAP:

1. Platform independent
2. Independent of language used for implementation
3. Inherits all the advantages of an XML-based protocol
4. Can easily get around firewalls

Disadvantages of SOAP:

- The performance issues of XML-based protocol are applicable to SOAP also
- It does not provide any activation mechanism
- There is no support for garbage collection of managed objects
- It does not implement any security as part of the framework

SOAP is inherently stateless and is based on request/response model. It can be implemented over transport protocols like TCP/IP and UDP. Hence it supports protocol binding with multiple transport protocols.

> Protocol binding is the specifications on how application layer protocol messages may be passed from one node to another using an underlying transport protocol.

## 6.8 Overview of NETCONF (NETwork CONFiguration)

NETCONF is an XML-based protocol developed by IETF. It is a protocol for configuration management with limited monitoring capabilities. Usual network configurations have their own specifications for session establishment and data exchange that resulted in the need for a standardized protocol. NETCONF supports operations like installing, querying, editing, and deleting of configuration data on network devices.

NETCONF is based on the agent-manager model and uses a remote procedure call (RPC) paradigm to facilitate communication. The agent(s) encodes an RPC in XML and sends it to a manager using a secure, connection-oriented session. The manager responds with a reply encoded in XML. The contents of both the request and the response are described using XML.

NETCONF can be conceptualized as consisting of four layers. The layers are:

1. Content layer: This is the topmost layer and corresponds to configuration data being handled using NETCONF.
2. Operation layer: Has a set of methods for communication between agent and manager.

| *Operation Tags* | *Explanation* |
|---|---|
| <get-config> <edit-config> <copy-config> <delete-config> | Used to manipulate configuration data |
| <get-state> | Retrieve the state of the managed object |
| <kill-session> | Terminates the NETCONF session |
| <commit> | Publishing a configuration data set to devices |
| <discard-changes> | Revert configuration to the last committed configuration data |
| <lock> <unlock> | Used to lock and unlock a configuration source |

3. RPC layer: Used for encoding the remote procedure call.

| *RPC Tags* | *Explanation* |
|---|---|
| <rpc> | Used by both agent and manager to send an rpc request |
| <rpc-reply> | Send by the agent in response to an <rpc> operation |
| <rpc-error> | Sent in <rpc-reply> went an error occurs during the processing of an <rpc> request |

4. Transport layer: NETCONF supports transport over SSH, SSL, BEEP, and console.

Advantages of NETCONF:

- Inherits all the advantages of an XML-based protocol
- Offers secure network management
- Leads to centralized configuration management
- Easy to implement and maintain

Disadvantages of NETCONF:

- The performance issues of XML-based protocol are applicable to NETCONF also.
- It does not have enough functionality to support advanced network management,

## 6.9 Overview of IIOP (Internet Inter-ORB Protocol)

The definitions for IIOP were developed by OMG (object management group). IIOP specification consists of a set of formatting rules and message formats for communication in a distributed environment. In the distributed environment the client is isolated from the server by a well-defined interface. IIOP is an underlying mechanism of CORBA (common object request broker architecture) where CORBA specifications detail creation and use of distributed objects. CORBA and IIOP are inseparable and together form a solid framework for distributed computing. The heart of this framework is the ORB (object request broker) that permits communication between applications in heterogeneous environments.

The IIOP is an implementation of abstract protocol GIOP (general inter-ORB protocol) over connection oriented (see Figure 6.2) TCP/IP for use on the internet. GIOP definition has three parts:

- Common data representation (CDR): This defines a set of data formatting rules based on data types supported by CORBA IDL (interface definition language).
- Interoperable object reference (IOR): This defines a common format for objects to publish their identities while transfer of data using IIOP. The IOR usually contains the protocol version, server address, and an object key to identify the remote object.
- Message formats: This defines the messages exchanged between the agents and the manager using this application layer for management.

Some of the messages are:

| *Message* | *Explanation* |
|---|---|
| Request | Used to invoke a remote method |
| Reply | This is the response to a "Request" message |
| LocateRequest | Used to check if a particular address is supported or request on the address of the remote object |

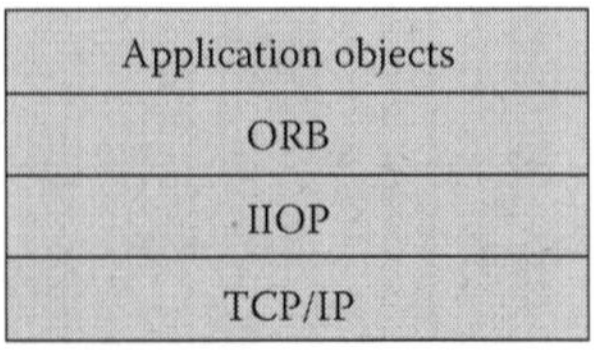

| Application objects |
|---|
| ORB |
| IIOP |
| TCP/IP |

**Figure 6.2 Layered view showing where IIOP fits in.**

| | |
|---|---|
| LocateReply | This is a response to the "LocateRequest" and contains the address of the object, in case the remote object has moved |
| CancelRequest | Used to cancel a request that was send previously |
| MessageError | Used to report an error in message |
| CloseConnection | Used to end a session between distributed objects |

Messages can also be split into fragments for communication. HTTP, which is widely used for internet communication, was not suited for remote method invocation (RMI) based on CORBA. Hence, the use of IIOP on CORBA can be used as an alternative to HTTP for Internet communication. IIOP is based on the client-server model.

Advantages of IIOP:

- Neutral architecture for communication
- Transparency in communication
- Scalability and code reuse
- Robust firewall navigation

Disadvantages of IIOP

- Complex standard
- The disadvantages of CORBA and RMI are combined in IIOP

## 6.10 Conclusion

This chapter only gives an overview on some of the popular management protocols. There are many other application layer protocols that are used for telecom management. For example, it is a common practice to have FTP (file transfer protocol) used to support performance management. This involves the element to be managed, writing performance records to files on the element. The manager collects the performance records generated by the element at regular intervals using FTP protocol and parses the records to collect performance data. Here the interaction between the element and the manager to get performance data is based on the FTP. Application layer protocols for connection like TELNET (TELecommunication NETwork) and SSH (Secure SHell) can be used to have a console for connecting to the element from the management system. Commands to be triggered on the element can be sent using these basic application layer protocols, which are simpler rather than using the pure management protocols like SNMP, CMIP, and so on.

Protocols SNMP, CMIP, SOAP, and NETCONF will be handled in detail later in this book. Different networks have different requirements and it can be seen that this resulted in the introduction of various protocols. While internet-based management like a web-based enterprise management would require a management protocol that works with HTTP, a telecom network would rather use SNMP with a well-defined MIB for its management.

> Evolution: The initial management requirements were limited to managing individual elements based on proprietary protocols. With standard protocols it was later possible to manage networks. Then a single management solution was expected to manage a set of networks in a domain. Now management solutions are expected to manage networks across multiple domains. An organization in the communication industry wants its management solution to handle enterprise management, telecom network management, service management, and management of its storage devices.

Most management solutions (Adventnet, IBM Netcool, etc.) support a variety of management protocols and can handle networks associated with different domains. Having support for multiple management solutions is just a stepping stone in developing a converged management solution. Having a common information model and set of standard interface definitions for interaction between management components are some of the other factors that affect the development of a converged management solution.

## Additional Reading

1. William Stallings. *The Practical Guide to Network Management Standards*. Reading, MA: Addison-Wesley, 1993.
2. Natalia Olifer and Victor Olifer. *Computer Networks: Principles, Technologies and Protocols for Network Design*. New York: John Wiley & Sons, 2006.

# Chapter 7

# Standardizing Bodies

Consider a crossroad in a busy street without traffic signals. The chaotic situation caused by this scenario is an example of the need for proper rules and regulations that everyone needs to follow in order to use the road. With multiple vendors coming with different network elements, element management systems, network management systems, and OSS solutions interoperability is a big issue and without defined standards communication is not possible. There are several organizations that bring out common standards in management systems to ensure interoperability. This chapter gives an overview on the different standardizing bodies in management.

## 7.1 Introduction

Most standardizing bodies used to work independently in defining standards for management system. For example 3GPP OAM&P standards were oriented toward specific networks, TMF and ITU-T were working on telecom management, DMTF was defining standards for enterprise management, and so on. With the demand for converged management systems that span multiple domains, the standardizing bodies are coming up with joint standards like the CIM/SID Suite, which is a joint venture of the Distributed Management Task Force (DMTF) and the TeleManagement Forum (TMF). This chapter gives an overview on the works of various standardizing bodies in management domain.

## 7.2 ITU (International Telecommunication Union)

The ITU is an international standards organization working in telecom domain. It started as International Telegraph Union in 1865. ITU-T is the standardization sector of ITU, which is a specialized agency of the United Nations (UN). The standards of ITU-T carry more significance in the international market compared to other standardizing bodies due to the link with the UN. Due to this reason well-accepted standards from other standardizing bodies are usually send to ITU-T to get their acceptance. An example of this is the e-TOM model of TMF that was adopted by a majority of service providers and later was accepted by ITU-T. It is based in Geneva, Switzerland and has members from both public and private sectors.

Standards from ITU-T are referred to as "Recommendations," the development of which is managed by study groups (SG) and focus groups (FG). While a study group is more organized with regular face-to-face meetings and adherence to strict guidelines and rules of ITU-T, the focus group is more flexible in organization and financing. To be more precise, a focus group can be created quickly to work on a particular area in developing standards, deliverables are determined by the group unlike guidelines from ITU-T and they are usually dissolved once the standard is released. There is a separate group called telecommunication standardization advisory group (TSAG) for reviewing priorities, programs, operations, financial matters, and strategies for the sector. The TSAG also establishes and provides guidelines to the study groups.

The recommendations of ITU-T have names like Y.2111, where "Y" corresponds to the series and "2111" is an identifying number for a recommendation or a family of recommendations. The document number is not changed during updates and different versions of the same document are identified with the date of release.

There is a separate study group in ITU-T to handle telecom management. This study group is responsible for coming up with standards on the management of telecommunication services, networks, and equipment. It should be noted that TMN, the basic framework for telecom management that is discussed in previous chapters was a standard from ITU-T.

Next generation network support and management is also an area ITU-T works on based on demands in current telecom management industry. There are special projects in telecom management that are also being prepared by focus groups in ITU-T. "Y.2401" is a recommendation on "principles for management of next generation network" with multiple sections discussing different aspects of NGNM. For example:

- Y.2401-6: Describes management areas and objectives of NGNM
- Y.2401-9: Next generation network management architecture overview
- Y.2401-10: eTOM business process framework
- Y.2401-11: Management functional view

The ITU-T participation requires paid membership. It also organizes a number of workshops and seminars on current trends and technologies to progress existing work areas and to explore new ones.

Web site: www.itu.int

## 7.3 TMF (TeleManagement Forum)

As the name signifies this is an organization dedicated to the development of standards in telecom management. It was initially started by eight member companies (Amdahl, AT&T, British Telecom, Hewlett-Packard, Northern Telecom, Telecom Canada, STC, and Unisys) in 1988 with the name Network Management Forum. After 10 years in 1998, there was more than 250 members participating in the forum and the name of the forum was changed to TeleManagement Forum, reflecting that the scope of work was on telecom management and not just network management.

The TMF has a set of technical programs that work on different aspects of telecom management. "Team Projects" is a part of the technical programs that develop standards. For example there is a project on eTOM (enhanced telecom operations model) for management process, SID (shared information and data) project for management of telecom data, NGNM (next generation network management) project for requirements in management of next generation network, MTNM (multi-technology network management) project on EML–NML interface, and so on. Membership in TMF is free and this has resulted in a huge participation from service providers, equipment manufacturers, and integrators in programs conducted by TMF to define standards.

The main concept of TMF is the NGOSS (New Generation Operations Systems and Software). The NGOSS is a framework that can be used for developing, procuring, and deploying operational and business support systems and software. NGOSS suggests a life cycle approach for telecom management and has four main components:

- Business process map: eTOM
- Information and data model: SID
- Management architecture: TNA (technology neutral architecture)
- Application map: TAM (Telecom application map)

In addition to technical programs based on NGOSS framework and its four main concepts, TMF also works on business solutions. The business solution activities include work on service level agreements, revenue assurance, integrated customer centric experience, and so forth. Prospero is a portal maintained by TMF to assist companies to easily adopt the standards of TMF. Prospero has package and guidelines based on actual implementation of TMF standards including case studies and downloadable solutions. According to TMF, knowledge base consists of both standards and corporate knowledge. A company that wants to adopt a standard needs to look into both the definitions of the standard as well as make use of corporate knowledge published by companies that have already adopted the standard. Hence TMF document base has a good mix of standards, white papers, case studies, solution suites, review/evaluation, and testimonials.

The TMF conducts events to promote its studies and standards and these events are also a venue for service providers and system integrators to demo their products that comply with TMF standards. The TeleManagement Forum also offers certification. It certifies solutions that comply with its standards, where the solution could even be a simple implementation of an interface based on the definitions of TMF. It also certifies individuals based on an evaluation of their knowledge on specific standards.

"Webinar" or online seminars are also conducted by TMF. People around the globe can register and listen to seminars on new concepts that need to be worked on in standardization and studies on implementation of existing standards. Information about these seminars, hosted on the Web are published on the TMF Web site and also mailed to TMF members. "Catalyst Projects" are used by TMF to conduct studies and develop standards on emerging issues in telecom management. While team projects perform studies on base concepts, the catalyst project shows how the base concepts can be used to solve common, critical industry challenges.

Web site: www.tmforum.org

## 7.4 DMTF (Distributed Management Task Force)

The DMTF is a standards organization started in 1992 with the name "Desktop Management Task Force." The name was later changed to "Distributed Management Task Force" with their standards mainly on management of enterprise and Internet systems. The interoperability standards of DMTF help to reduce operational costs in a multivendor environment. DMTF is composed of three committees. They are:

- Technical committee: This committee is used to steer the work group creation, operation, and closure for developing standards and initiatives of DMTF.
- Interoperability committee: The work of this committee is handled by focus groups that work on providing additional data on interoperability in multivendor implementations based on DMTF resources.
- Marketing committee: The role of this committee is to ensure that the activities of DTMF get enough visibility and resources have industrial acceptance.

Some of the technologies that DMTF works on are:

- Common information model (CIM): It defines a common language for representing management information as objects and relations. The architecture is object oriented and based on UML (unified modeling language). CIM has two parts, the schema that describes the models and specifications that detail CIM integration with other information models.

- Web-based enterprise management (WBEM): It defines standards for remote application management on a distributed computing environment. OpenPegasus, OpenWBEM, SBLIM, WBEM Services, and WBEMsource are some of the open source implementation of WBEM. It incorporates a set of technologies including CIM representation in XML, CIM query, URI (uniform resource identifier) mapping and discovery using a protocol called service location protocol (SLP).
- Desktop and mobile architecture for system (DASH) initiative: It defines a set of standards on Web services management for desktop and mobile client systems. Using DASH, desktop and mobile systems on different platforms from multiple vendors can be managed based on standards in a remote and secure manner.
- Systems management architecture for server hardware (SMASH) initiative: Its standards are mainly on a command line protocol (CLP) with support for interfaces, Web services, and security protocols used in server management (SM). The servers being managed using SMASH can be virtual and redundant servers, blades and rack, or high end servers used in telecom and mission critical applications.
- System management BIOS (SMBIOS): Defines a standard formant for representing a layout of management information about the first program that runs when the PC is started (BIOS) and about the computer system (desktop). It gives enough information for writing BIOS extensions and generic routines for conversion from SMBIOS format to other technologies.
- Common diagnostic model (CDM) initiative: It defines a platform independent architecture to diagnose (discover, control, and retrieve) the health of computer systems in a multivendor environment. It is an extension of CIM in diagnostics.

Two forums in the interoperability committee of DMTF are the CDM forum and system management forum.

- CDM forum: The main activity of the CDM forum is to support the development of test programs and software development kits that will help in development and integration of diagnostic modules based on CDM into CIM framework.
- System management forum: A major activity to ensure interoperability is checks to evaluate products compliance to standards. This forum handles the development of programs that checks a products conformance to DMTF standards on system management like DASH and SMASH.

For an individual to register with DMTF, he must be the representative of a company that is already a member of DMTF. This means that the company in which the individual is working should have membership in DMTF. Company membership requires payment while representatives can register for free. DMTF conducts events to showcase their work. It also collaborates with other standardizing forums in defining standards. Certification for individuals and products is also offered by DMTF.

Web site: www.dmtf.org

## 7.5 3GPP (Third Generation Partnership Project)

The 3GPP, as the name signifies, is a collaboration project established in 1998. Some of the organizations that are part of the partnership are ARIB, CCSA, TTA, and TTC. It was formed with the intent of jointly developing standards and reports for a third generation mobile system. The current specifications cover aspects of both GSM based and CDMA technology, where 3GPP works mainly on GSM based standards including GPRS and EDGE, and 3GPP2 works on CDMA2000. Standards on IMS (IP multimedia subsystem), the backbone to all IP network, are also defined by 3GPP.

3GPP has a project coordination group (PCG) and a technical specification group (TSG). While the project coordination group handles management activities in 3GPP, the technical specification group develops standards and creates work groups (WG) for focused work on developing the standard in a specific area. To have "individual membership" with 3GPP, the individual must have membership with one of the organizational partners of 3GPP.

The 3GPP has many standards defined on OAM&P (network management) including detailed implementation guidelines on FCAPS. For example, consider the 1999 release of "3GPP TS 32.111-2 version 3.4.0" on the alarm integration reference point. It specifies that an alarm needs to have two parts, the notification Header and alarmInformationBody.

The parameters in the notification Header are:

- managedObjectClass: Class specific to managed object (an element or component of the element being managed)
- managedObjectInstance: Instance of the managed object
- notificationId: Unique id that can differentiate this alarm notification from another notification on same problem scenario of the managed object
- eventTime: The time at which the event occurred
- eventType and extendedEventType: Details of the type of event
- systemDN: The domain name of the system

Some of the parameters in the alarmInformationBody are:

- probableCause: It provides information that supplements the eventType
- perceivedSeverity: The severity of the alarm

- proposedRepairActions: How to fix the problem specified in the alarm
- additionalInformation: Information like the time the alarm was acknowledged

The depth of defining the standard goes even further to the set of values that can be taken by some of these parameters, like perceivedSeverity, which can have values critical, major, minor, warning, indeterminate, and cleared.

Another example of 3GPP specifications on FCAPS for a specific technology is "3GPP TS 32.409 version 7.1.0 Release 7," which gives information on the performance parameters for specific network elements in an IP multimedia subsystem (IMS). 3GPP is one of the leading organizations on standardization in telecom domain. The use of the term "3GPP" in this section can refer to both "3GPP" and "3GPP2."

It can be seen that the OAM&P specifications of 3GPP can be very useful in implementing a network management system because of the depth of defining standards. Though some of the specifications of 3GPP are generic, in most of the scenarios its specifications are for a specific technology unlike standards like TMF that are mostly generic and intended for a set of network technologies. Location services management and charging managing are some other areas in network management for which there are standards defined by 3GPP.

Web site: www.3gpp.org and www.3gpp2.org

## 7.6 ETSI (European Telecommunications Standards Institute)

The ETSI is a standardizing body based in France and officially recognized by the European Commission and the European Free Trade Association. It is responsible for developing standards in information and communication technologies including fixed, mobile, radio, converged, broadcast, and Internet technologies. Though the organization is based in Europe, it has members around the globe and the standards developed by ETSI have worldwide acceptance.

Most standards of ETSI in network management will be tagged ETSI/TISPAN. TISPAN (Telecoms & Internet Converged Services & Protocols for Advanced Networks) is a body of ETSI that works mostly on developing standards for fixed networks and internet convergence. ETSI bodies "Telecommunications and Internet Protocol Harmonization Over Networks" (TIPHON) and "Services and Protocols for Advanced Networks" (SPAN) decided to merge and jointly work on a focus area, forming TISPAN in 2003. Being a body of ETSI, it is mainly concerned with European telecom market but also works on worldwide standards similar to its parent organization.

The standards of ETSI are developed by technical bodies in ETSI, which can establish working groups to focus on a specific technology area. There are three types of the technical body in ETSI. They are:

1. Technical committee: ETSI calls this technical committee a semipermanent entity. The works of the committee may not end up as standards but are used by the other ETSI technical bodies. Its primary focus is to work on a specific technology area. The technical committee is established from technology perspective with multiple standardizing activities happening at all times around a specific technology.
2. ETSI project: This technical body is similar to the technical committee. The difference being is that it is established based on market sector requirements and has a predetermined life span.
3. ETSI partnership project: As the name signifies this technical body works on partnership project with other standardizing bodies.

The documents developed by the technical bodies or the ETSI deliverables fall under one of the below categories:

- ETSI technical specification (TS): More on specifications rather than technical definitions
- ETSI technical report (TR): Regular report on a specific aspect of technology
- ETSI standard (ES): Definitions and specifications for global market
- ETSI guide (EG): Detailed documentation that serves as a guide
- European standard (or European norm, EN): Definitions and specifications for European market
- ETSI special report (SR): Special report on a specific aspect of technology
- Miscellaneous Item: When none of the other deliverables apply, then the document is called a miscellaneous item.

ETSI normally requires paid membership for participation as a full or associate member. The members have access to the working documents of the technical bodies, while the standards once released can be viewed or downloaded free of charge by anyone. Each technical body has a chairman who is responsible for overall management of the technical body including its working groups and programs.

ETSI conducts meeting, workshops, and exhibitions for study of technology and development of standards. Some of the events are also available as a Web cast. Tools are another outcome of the ETSI work. The third generation of mobile was developed by ETSI in collaboration with some other standardizing bodies under the umbrella of 3GPP. This makes GSM standards one of the major contribution areas of ETSI.

Web site: www.etsi.org

## 7.7 MEF (Metro Ethernet Forum)

The Metro Ethernet Forum (MEF) was formed in 2001 by Nan Chen and Ron Young. MEF works on developing technical specifications and implementation agreements on Carrier Ethernet. Its standards are intended to promote interoperability and deployment on areas related to Ethernet. Some of the standards released by MEF on network management include specifications on information models for EMS–NMS and service OAM framework and requirements.

Developing new standards is not the primary goal of MEF though they have worked on new standards. Its main goal is to create implementation agreements that extend the utility of existing standards developed by other standardizing bodies. MEF has a marketing committee that works on increasing worldwide awareness of Ethernet services and Ethernet-based transport networks and a technical committee for steering activities related to technical specification and standards.

The MEF mainly works on defining standards on Ethernet services for metro transport networks and Carrier-class, Ethernet-based metro transport technologies. It also looks into other transport technologies and non-Ethernet interfaces on which other standardizing organizations needs to work and does not have standards defined.

Some of the deliverables of MEF are:

- Whitepapers on business case studies, market, and technology
- Interoperability demonstrations
- Implementation agreements made by MEF members that use existing standards
- Test procedures that include testing methodologies and processes on interoperability
- Written requests to other standards bodies proposing additional technical work
- Technical specifications and standards

Telecom management is one of the four major areas of technical definition and development within MEF, with architecture, service, and test/measurement being the three others. MEF has generated substantial deliverables on telecom management of Ethernet networks including areas of element management, network management, and OAM. The work involves development of requirements, models, and definitions to ensure effective management and interoperability across Metro Ethernet platforms. This has lead to a decrease in the CAPEX and OPEX of Ethernet network management.

Certification of products and services is another offering of MEF. Some of the certifications provided to ensure compliancy to MEF specifications are:

- Service Certification for Equipment Manufacturers (MEF 9/MEF 18)
- Service Certification for Service Providers (MEF 9)

- Traffic Management Certification for Equipment Manufacturers (MEF 14)
- Traffic Management Certification for Service Providers (MEF 14)

Only corporate level membership is permitted with MEF. Only corporate can join MEF by paying a predetermined annual fee and no paid individual membership. An unlimited number of employees from a corporate member company can participate in technical or/and marketing activities of MEF. Completed/released specifications of MEF can be downloaded or viewed by anyone from the MEF Web site while working documents can be viewed only by members.

Web site: www.metroethernetforum.org

## 7.8 ATIS (Alliance for Telecommunications Industry Solutions)

The ATIS is a standardizing body based in the United States and accredited by the American National Standards Institute (ANSI). It works on developing technical and operations standards in communication and information technology. Most of the standards of ATIS are intended to promote interoperability between networks, network equipment, and customer premise equipment.

The functional framework for organizing and marketing work of ATIS consists of 10 areas classified under two categories. The categories are:

- Universal functions: There are five functional areas under this category. They are performance, reliability, and security; interoperability; OAM&P; ordering and Billing; and user interface.
- Functional platforms: There are five platform areas under this category. They are circuit switched and plant infrastructure, wireless, multimedia, optical, and packet based networks.

The ATIS has made many valuable contributions in network management and has a dedicated area in its functional framework for OAM&P. There are two main function groups in ATIS that handles the standardization activity called "forums" and "committees." These main groups can form subgroups like subcommittees based on work programs and create "task forces" to work on a focused area for a specific amount of time.

A proposal to work on an item is referred to as an "issue" in ATIS. An issue (issue identification form) needs to define the problem to be looked into, propose a resolution, and suggest a time line for the completion of work. An issue process is defined by ATIS on the different activities involved in the life cycle of an issue. International Telecommunications Union Telecommunications Standardization Sector (ITU-T) is a standardizing body with which ATIS has worked closely for defining standards that needs world wide acceptance.

The deliverables of ATIS are:

- ATIS standards: This deliverable can comprise a variety of documents including technical requirements, specifications, reports, and whitepapers
- ATIS implementable end-to-end standard: This deliverable handles end-to-end solutions and hence can comprise of ATIS Standard(s) in combination with deliverables from other forums and committees external to ATIS.

The ATIS standard documents are named in a specific format. If the number is ATIS-GGNNNNN.YY.XXXX, then GG corresponds to the two digit functional group number, NNNNN is the five digit number for standard document, YY is the supplement or series number and XXXX is the year or it represents a specific release. For example ATIS-0900105.01.2006 would mean a standard with number 00105 and series 01 for group optical (09) and approved in year 2006.

Only corporate and not individuals can become a member of ATIS. An unlimited number of employees from a corporate member company can participate in activities of ATIS. Membership is not free and nonmembers need to pay for downloading some of the standard documents. ATIS conducts different kinds of events to promote work and visibility of standards. Some of the events include conference to discuss cutting edge technologies, expo to showcase standards and solutions and webinars (web-based seminars) to promote work or interest on a topic. ATIS is also a member of the Global Standards Collaboration (GSC), which facilitates collaboration work between standards organizations that are part of GSC.

Web site: www.atis.org

## 7.9 OASIS (Organization for the Advancement of Structured Information Standards)

The OASIS was founded in 1993 and was initially called SGML as it was dedicated to developing guidelines for interoperability among products that support the standard generalized markup language (SGML). The name was later changed from SGML to OASIS to reflect a broader scope of technical work with extensible markup language (XML) and other related standards. The work of OASIS is directed toward development and adoption of open standards.

The standards of OASIS mostly add value in enterprise management rather than telecom management. This is because the consortium produces standards for Web services, security, and e-business that are suited for enterprise networks rather than telecom domain.

OASIS operation and structure is much more transparent compared to other standardizing bodies. Here members themselves set the technical agenda and work is approved by voting. There are different levels of membership in OASIS like foundational sponsors, sponsors, contributors, and liaisons. Employees of any of the member companies can register and start participating in the technical committees without any payment.

Technical committees can be formed by members of OASIS to work on a specific area and to develop standards. There are different levels of participation when joining an OASIS group. The participation can be as an observer, member, or voting member. Observer requires only minimum participation while a voting member needs to be always active. All committees are not oriented toward developing standards. For example the OASIS Standards Adoption Committees provide forums to unite specific industries (users/groups), governments, vendors, and other standards bodies to evaluate, identify gaps and overlaps, publish and promote existing standards and requirements. OASIS also organizes events including demos and conferences to promote work and studies on existing and future standards.

Web site: www.oasis-open.org

## 7.10 OMA (Open Mobile Alliance)

As the name signifies, OMA is a standardizing body formed by alliance of a set of organizations to extend the scope of work and avoid duplication of work among organizations that were combined to form the alliance. OMA is a consolidation of the following bodies: Open Mobile Architecture initiative, WAP Forum, SyncML initiative, Location Interoperability Forum (LIF), MMS Interoperability Group (MMS-IOP), Wireless Village, Mobile Gaming Interoperability Forum (MGIF), and Mobile Wireless Internet Forum (MWIF).

Open Mobile Alliance is intended for developing mobile application standards that enables interoperable services across countries, operators, and mobile terminals. OMA is not directly linked with network or element management, but it does have standards on mobile device management and some levels of service management. OMA works with other standardizing bodies like 3GPP and 3GPP2 to improve interoperability and decrease operational costs for all involved.

Web site: www.openmobilealliance.org

## 7.11 SNIA (Storage Networking Industry Association)

Storage network management is also an important area in converged space along with enterprise and telecom network management. SNIA develops and promotes standards on storage networking and information management. It was incorporated in 1997 and is headquartered in San Francisco, California with regional affiliates around the globe. In addition to standards, SNIA also works on conducting events for developing and promoting technologies and educational services. The activities of SNIA are performed by its Technical Work Groups (TWGs).

One of the key specifications of SNIA in storage management is storage management initiative specification (SMI-S). It offers methods that would facilitate interoperability and management of heterogeneous storage area network (SAN). The interoperable interface and messaging is achieved using XML on an object-oriented paradigm.

Web site: www.snia.org

## 7.12 Conclusion

This chapter provides an overview on some of the standardizing bodies in telecom management. Their structure and activities are touched upon to give basic familiarity about the organizations. There are many more organizations and alliance forums that work on developing standards in network management. DSL Forum, Optical Interworking Forum (OIF), and MultiService Switching Forum (MSF) are just a few names of the bodies that were not covered in this chapter.

This chapter is intended to give the reader information on what forum to research when looking for a specific standard, to participate in developing standards on a specific domain, and to join a forum as a corporate or individual member. The organization to look for is determined based on requirements. For example TMF would be a good place to look for telecom management standards as it mainly focuses on this area and the documents of the forum would have information on collaboration work with other forums. In the same way DMTF would be a good choice for enterprise management and SNIA for storage area network management.

## Additional Reading

1. TMF 513 TR133 NGN-M Strategy Release 1.0 Version 1.2. www.tmforum.org (accessed December 2009).

# Chapter 8

# MTNM Compliant EMS

The NMS has to interact with multiple EMS and present an integrated view of data. Now each EMS can be from a different vendor and based on a different technology. Moreover the different EMS may not talk in a common vocabulary or have a common format for data. Then how is it possible to have a single NMS for multiple EMS? This chapter gives you an understanding on the requirement to have standards for interaction between element management layer (EML) and network management layer (NML). It has a detailed discussion on the MTNM (Multi Technology Network Management) standards from TMF (TeleManagement Forum).

MTNM is a copyright of TMF (www.tmforum.org)

## 8.1 Introduction

Consider that service provider X wants to use your EMS and NE with their NMS solution. To get fault data from an EMS, the NMS solution from X sends query "getAlarmList," which your EMS does not understand and cannot respond to. The obvious solution is to write an adaptor (a process that acts as mediator between two modules and performs format conversion in message/data for the two modules to interoperate) that converts message/data from the format of NMS solution X to a format that your EMS can understand and vice versa. Now there is another service provider Y who also wants to use your EMS and sends a different query "getFaultList" for the same operation. For interoperability another adaptor needs to be written. For NMS solution X to interact with an EMS other than yours will also require an adaptor.

So in a multitechnology (underlying protocol), multivendor environment, where interaction between EML and NML is not standardized, for an NMS to collect data from multiple EMS or for an EMS to be used with multiple NMS, the possible method of operation would require writing multiple adaptors. Writing adaptors is not a cost-effective option in the evolving telecom industry where the major investment could end up in writing adaptors. Hence there was a need to have standards for interaction between EML and NML.

For the interaction to be standardized, the following basic requirements need to be satisfied:

- Messages for communication must be predefined (e.g., getActiveAlarmList, getConfiguration)
- The objects must be well defined (i.e., a common naming convention needs to exist for addressing the managed and management objects)
- The definitions must be independent of the management protocol used
- It should be applicable for multiple networks based on different technology
- Vendors must accept to the standard and make their products compliant

Use of these standards leads to ease of integration, in addition to reduced cost and time of integration. Multi-Technology Network Management (MTNM) model of TMF ensures interoperability between multivendor Element and Network Management Systems. It mainly consists of a set of UML models and CORBA IDLs to define the EML–NML interface. This chapter handles the basics of MTNM by discussing the four main artifacts that were release by TMF describing MTNM.

## 8.2 About MTNM

The first initiative in the development of an EML–NML interface was SONET/SDH information model (SSIM) of TMF. When TMF509, the first version of SSIM, got industrial acceptance another team ATM information model (ATMIM) was formed for working on ATM networks. Later SSIM and ATMIM teams merged and formed a MTNM team for jointly defining standards in EML–NML interface.

Some of the functional areas in telecom management covered by MTNM are:

- Configuration management: This includes network topology discovery and network provisioning. Topology discovery includes discovery of network elements, ports, topological links, relationship among termination points, and discovery of subnetworks. As part of network provisioning the support includes setting up port related transmission parameters, protection parameters, and maintenance operations.
- Fault management: Some of the fault management capabilities in MTNM are handling of alarm notifications, active alarm list, setting up of alarm reporting and severity profiles, acknowledging alarms, alarm filtering, and support for OMG (object management group) log service.

- Connection management: MTNM supports both protected and unprotected subnetwork connections, with functionality to set subnetwork connection (establishment, modification, and deletion) parameters, configuration of ATM inverse multiplexing traffic descriptors and SONET/SDH virtual concatenation.
- Call management: Involves creation of calls, deletion, or modification of calls. It also has adding, deleting, or routing connections linked with calls.
- Equipment management: This includes equipment provisioning, inventory discovery, and equipment mismatch notifications.
- Performance management: Capability to retrieve current and historical performance data, creation/enabling/disabling of threshold cross alert, transfer of historical performance data, and support for tandem connection monitoring (TCM).
- Security management: Includes identification, authentication, detection, and session establishment based on authentication.
- Miscellaneous: Other capabilities include GUI cut-through where the NMS GUI interacts with EMS GUI and gets data from the EMS without involving the interface modules and support for software backups on the elements (see Figure 8.1).

Latest release of MTNM also includes connectionless and control plane management capabilities and the functional areas associated with these capabilities.

The deliverables of MTNM are:

- TMF513: MTNM Business Agreement
- TMF608: MTNM Information Agreement
- TMF814: MTNM Solution Set
- TMF814A: MTNM Implementation Statement and Guidelines

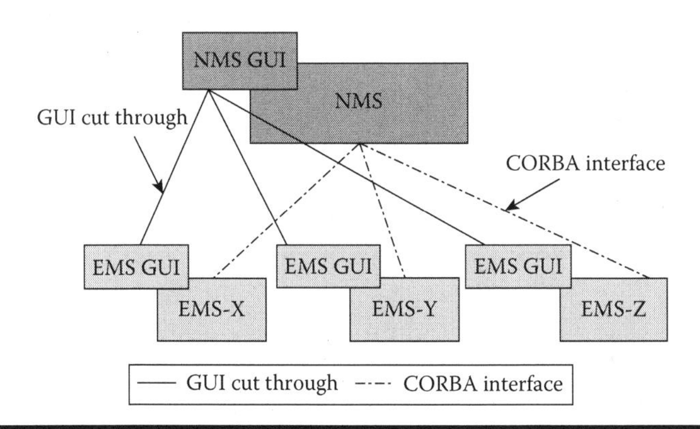


**Figure 8.1 GUI cut through in MTNM.**

## 8.3 MTNM Business Agreement (TMF 513)

TMF 513 defines the requirements for the interface between NML and EML in terms of the objects involved and operations that the interface is required to support for interaction between the objects. The MTNM Business Agreement has two parts. They are the Business Problem Statement and the Business Requirement Model. The problem statement defines the scope of the problem and identifies the problem needs or issues, while the requirement model handles an implementation independent description of requirements.

The aim of this document is to provide a single common solution to address interoperability between element management and network management layers that involves (see Figure 8.2):

- Multiple network technologies: Including SONET, SDH DWDM, and ATM
- Network elements that support multiple technologies
- Open standards
- Multivendor EMS and NMS

TMF513 is a set of use cases, and static and dynamic UML diagrams that can be used as:

- Foundation for information and requirement agreement of TMF and other bodies working in this area
- Definition of requirement agreement and input for information agreement for service providers
- Development of commercial off the shelf (COTS) products by vendor

According to TMF513, adopting MTNM will lead to:

- Lesser time for service activation
- Scalable NML–EML interface
- Reduced effort and time in integration

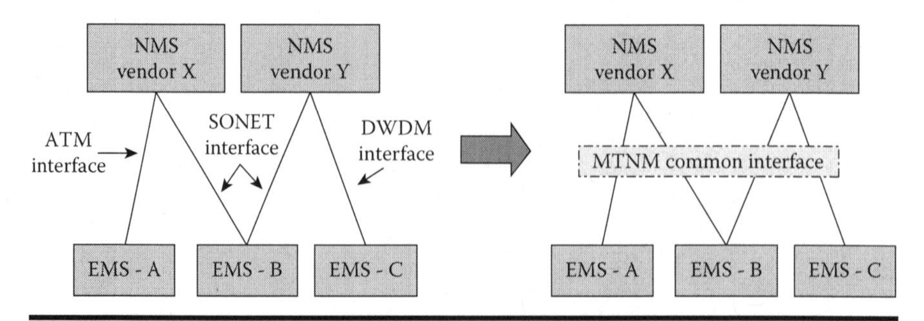


**Figure 8.2 Single solution based on MTNM.**

- More options for service providers in choosing products for interoperability
- Lesser cost to EMS vendors in making their product work with multiple NMS
- Rapid service delivery
- Easier to fix faults on a common interface

Most of the interface definitions (requirements are use cases) in TMF513 are centered on a basic set of objects that communicate using the EML–NML interface and share a set of common attributes. These objects are (note: there is cross reference between objects in the description below):

- Alarm severity assignment profile (ASAP): This object represents severities that can be assigned to alarm probable causes.
- Call: This object represents the association between endpoints in support of a service instance or an infrastructure trail.
- Connection: This object represents connectivity through an MLRA.
- Element management system (EMS): This object represents the management application associated with managing "elements or an abstraction of MLSN."
- Equipment: This object represents the physical component of an element that can be managed.
- Equipment holder: This object represents the resources that are capable of holding physical components.
- Equipment protection group (EPG): This object represents equipment protection.
- Flow domain (FD): This object represents a grouping of MFDs and TPs associated with MFDs.
- Flow domain fragment (FDFr): This object represents a transparent flow point logical connectivity.
- Group termination point (GTP): This object represent a sequence CTPs in the same element.
- Subnetwork point pool link (SNPP Link): This object represents a link between two SNPPs.
- Managed element (ME): This object represents the EMS view of a network element.
- Matrix flow domain (MFD): This object represents a collection of assigned connectionless port TP.
- MultiLayer routing area (MLRA): This object represents an MLRA that consists of MLSNs and provides a mechanism for the provisioning of connections across MLSNs.
- MultiLayer subnetwork (MLSN): This object represents the topology provided by the EMS.
- MultiLayer subnetwork point pool (MLSNPP): This object represents a list of SNPP.

- MultiLayer subnetwork point pool link (MLSNPPLink): This object represents a set of SNPP links in different networks.
- Performance monitoring point (PMP): This object represents an access point to provide performance data on monitoring and threshold parameters.
- Protection group (PG): This object represents protection schemes.
- Subnetwork connection (SNC): This object represents the relation between two end points.
- Termination point (TP): This object represents the logical termination point of either a TL, SNC, or FDFr.
- Termination point pool (TP Pool): This object represents a grouping of TPs or GTPs from the same subnetwork.
- Threshold crossing alert (TCA) parameter profile: This object represents a set of threshold crossing alert parameters associated with a set of TPs.
- Topological link (TL): This object represents a physical link between two termination points.
- Traffic conditioning (TC) profile: This object represents a collection of attributes to define the traffic conditioning parameters on a TP.
- Traffic descriptor (TD): This object represents a collection of attributes used to define bandwidth and quality of service characteristics on a CTP.
- Transmission descriptor (TMD): This object represents a collection of attributes to define multilayered transmission parameters and information parameters on a TP or MFD.

The common attributes shared by these basic objects are:

1. Name: A unique identifier for the object.
2. User label: A user friendly name that shall be provisionable by the NMS.
3. Native EMS name: The name of the object as presented on the EMS GUI.
4. Owner: An identifier for the owner of the object.
5. Additional information: Attributes that has not been explicitly modeled as part of the object.

The document defines requirements and use cases based on objects. Also provided in the TMF 513 business agreement is a set of supporting documentation (titled starting with SD) that contain details on various specific aspects of the NML-EML interface and are intended to help clarify some of the specific aspects of the interface that are discussed in the main document (titled staring with TMF513).

## 8.4 MTNM Information Agreement (TMF 608)

Information model is produced prior to the start of implementation of applications, system, and interfaces and it facilitates a requirement agreement between service providers and vendors. The TMN608 Interface Information Agreement details a protocol neutral model using UML that includes:

- Interface problem
- Interface information requirement
- Set of interface information use cases

Information exchanged over the interface needs to be structured based on the TMF 045 document on common information structure (CIS). CIS details how data attributes need to be structured in protocol neutral expressions for business and information agreements.

NamingAttributes_T structure is used:

- As a naming scheme between the NMS and EMS interface
- To define identifiers for managed entities
- To represent the containment relationship between objects

While the interface follows a NamingAttributes_T structure, the information model objects are defined as name-value pairs. For example when the object is "Example" and both name and value is a string. Then the representation is:

```
struct Example_T {
    string name;
    string value;
};
```

TMF 608 has multiple supporting documents (SD) for details on information exchange. Some of the most important supporting documents are discussed next.

Object naming: These naming rules will add consistency to the object names used by different vendors in designing the EML–NML interface. Some of the general rules and semantics for equipment naming discussed in "SD1-25_objectNaming" are:

1. A name–value pair is used to identify a component and has name=value syntax.
2. Component hierarchy will be separated by "/". That is an entry "Shelf-12/Slot-1/Card-0" and means that the Card-0 is placed in "Slot-1" that is in "Shelf-12." The left most thus represents the highest level of containment.
3. The equipment-holder class is identified before the circuit packs.
4. "Equipment" is used to identify the circuit-pack and there is a separate name associated with physical and logical components.

Document on attribute value change (AVC) notification or a state change (SC) notification: Every physical component has a state and a set of attributes associated to it. For example a link state would be busy, in service trouble, active, and so forth, and the state is different from the attributes like congestion level or number of connections. When the state of a component changes it is notified using SC notification and a change in any of the attributes will be notified from the EML to NML interface using an AVC notification. Details of AVC and SC are handled in the "SD1-2_AVC_SC_Notifications" document.

Document on performance parameters: The names to be used to identify some of the performance parameters are discussed in the "SD1-28_PerformanceParameters" document. The parameters to be monitored include average/maximum/minimum bit rate errors in an interval, count of received or transmitted cells/lost cells/misinserted cells, and count of cells discarded due to congestion/protocol errors/HEC violation.

Probable causes: An alarm notification usually contains the probable reason for the generation of the alarm. There are a set of fields required in an alarm notification, one of them being the probable cause of the alarm. "SD1-33_ProbableCauses" document includes some standard causes that can be used in the probable cause alarm field. It should be noted that TMF is not the only standardizing body involved in defining such standards. From FCAPS perspective the fields required in an alarm for a specific network are also defined by 3GPP. Some of these details will be covered in the next chapter.

The remaining supporting documents are for defining modes of operation, performance management file formats, traffic parameters, and so forth. Using UML, TMF 608 also gives the class diagrams to be used with the interface objects. The class diagram helps in converting the UML models to program code.

Some of the attributes in EMS Class are:

- "name": To represent the name of the EMS object
- "emsVersion": To have version control for installation and upgrade of new features
- "additionalInfo": For information that cannot be captured in the standard set of attributes defined by TMF for EMS and will be more vendor specific.

Some of the functions of EMS class are:

- getAllEMSSystemActiveAlarms( ): Get the active alarm list from the EMS
- getAllTopLevelSubnetworks( ): Get details on the Subnetworks from the EMS
- getAllTopLevelTopologicalLinks( ): Get Link information from EMS
- getAllTrafficDescriptors( ): Get the traffic descriptors from EMS

## 8.5 MTNM Solution Set (TMF 814)

While TMF 513 and 608 present a technology neutral representation of requirements and information agreement, the TMF814 gives a CORBA based (technology specific) IDL solution set for implementing the EML–NML interface. An EMS compliant to TMF 814 means that the EMS uses CORBA IDL for communication with the NMS. Each object in a CORBA interface is provided a unique address known as an interoperable object reference (IOR). While in fine-grained approaches each interface object has an IOR and a coarse-grained approach for client/server interface means that only a small number of interface objects have IORs. A lightweight object managed by a CORBA object (first-level object), is called a second-level object.

The object model in TMF814 follows a course grained approach. A coarse-grained approach with many-to-one mapping between object instances in the information model (TMF 608) and the objects (interfaces) in the interface model (TMF 814). But the operations vary from fine-grained as in getEquipment operation returns the data structure for a single equipment, to course grained as in getAllEquipments operation that returns the data structures for all the equipments identified by the management system. So certain interfaces can return multiple entities unlike a separate query for each network entity.

The objects in the network exposed using interface are based on the layered concepts and layer decomposition detailed in ITU G.805. While the managed objects having a naming convention of <Name>_T, the object manager or interfaces objects used to manage the managed objects including session objects are named <Name>_I.

The key manager interfaces are:

- EMSMgr_I: The NMS uses this interface to request information about the EMS. Active alarms, top level equipments, subnetworks, and topological links information can be obtained using this interface. In addition to collecting information, operations such as add/delete can also be performed using this interface.
- EquipmentInventoryMgr_I: The NMS uses this interface to request information about equipment and equipment holders. In addition to collecting information, operations such as provision/unprovision can also be performed using this interface.
- PerformanceManagementMgr_I: The NMS uses this interface to request current and historical performance data from EMS. File transfer is another mechanism to collect performance data.
- GuiCutThroughMgr_I: The NMS uses this interface to launch an EMS GUI from the NMS GUI and hence access data from the EMS.
- ManagedElementMgr_I: The NMS uses this interface to request information about managed elements. Information about terminations points like PTP or CTP and the cross connections can also be obtained using this interface.
- MaintenanceOperationsMgr_I: The NMS uses this interface to request information on the possible maintenance operations and to perform maintenance operations.
- MultiLayerSubnetworkMgr_I: The NMS uses this interface to request information about connections, managed elements, and topological links in the

subnetwork. In addition to collecting information, operations such as add/delete on a subnetwork can also be performed using this interface.

- ProtectionMgr_I: The NMS can use this interface to request information about equipment protection and its relationships.
- SoftwareAndDataMgr_I: The NMS uses this interface to request backup of configuration data.
- TransmissionDescriptorMgr_I: The NMS uses this interface to request information about transmission descriptors. In addition to collecting information, operations such as create/delete can also be performed using this interface.
- FlowDomainMgr_I: The NMS uses this interface to request information about flow domains. In addition to collecting information, operations such as add/delete can also be performed using this interface.
- MLSNPPMgr_I: The NMS uses this interface to request information about Multi Layer Sub Network and to set parameters.
- MLSNPPLinkMgr_I: The NMS uses this interface to request information about Multi Layer Sub Network Link information. It also supports assign/de-assign of signaling controller and set/modify actions related to signaling protocol and parameters.
- TrafficDescriptorMgr_I: The NMS uses this interface to request information about traffic descriptors (bandwidth and quality parameters). It should be noted that the TrafficDescriptor_T object is obsolete according to the latest definitions in TMF814.

A few examples of managed objects handled by the managers are:

- EMS manager (EMSMgr) is associated with EMS object, MLSN object, alarm object, topological link object, and so on.
- Equipment manager (EquipmentInventoryMgr) is associated with equipment object, equipment holder object, and so on.
- Managed element manager (ManagedElementMgr) is associated with managed element object, termination point object, alarm object, MLSN object, and so on.

The key session related interfaces are:

- EmsSessionFactory_I: This interface is the first instance used by the NMS to communicate with EMS. It is registered with the naming service and requires a user name and password to access the EMS.
- EmsSession_I: This interface is used to find interoperable object reference (IOR) for interfaces supported by the EMS including the notification service.
- NmsSession_I: This interface is supported by the NMS and used by EMS, unlike other interfaces, which are supported by EMS and used by NMS. This interface allows the EMS to notify the NMS about events like loss of alarms.

When a list of items is returned in response to a request from the NMS, an iterator interface can be used to retrieve items from a list. The iterator instance is destroyed after the client has finished retrieving data. Example: EquipmentOr HolderIterator_I.

The CORBA services that are intended to be used along with TMF814 are:

- Naming service: It involves mapping objects with a hierarchical naming structure. So that names can be used instead of objects in making queries for a specific interface. The server binds the object and name to a namespace. When the client makes a request with a name, the namespace is used to identify the associated object and then invoke a method on the object.
- Notification service: This service of CORBA allows a server to notify a list of registered clients about an event in a distributed environment. It also provides event filtering and QoS.
- Telecoms log service: This service of CORBA allows creation of events, review of past events, and recovery of events during a catastrophe.

## 8.6 MTNM Implementation Statement and Guidelines (TMF 814A)

The MTNM CORBA interface defined in TMF814 implemented in different ways. This leads to the need for a document that can guide in the realization of an MTNM interface that ensures interoperability among various vendor implementations. TMF 814A document serves this purpose by giving an implementation statement template and set of interoperability guidelines for the NML- EML interface.

According to the TMF 814A document, this can be used:

- For system vendors to specify in a standard format, the set of MTNM interface capabilities supported by their product
- For service providers to specific to the system vendors in a standard format, the set of MTNM interface capabilities required in the product
- To create an implementation agreement that has a standard format between multiple management system vendors.

TMF 814A has three main sections:

1. Functional Implementation Statements (FIS)
   This section describes templates to show that all functional capabilities of MTNM interface are supported. The functional capabilities include operations, notifications, and data structures related to objects in the information model.

The description of an FIS template is split into modules. Each module has:

a. Module name: This uniquely identifies each module.
b. Datatypes: The attributes associated with the module are listed in a table with attribute name in the first column in the table. The other columns are "Set By" describing whether EMS, NMS, or both can set the attribute, when and how the attribute was set, the format of the attribute that could be a fixed format or a list of values and a final column called "Clarification Needed" for specifying the length requirements of the attribute.

| Attribute Name | Set By | Set When and How | Format | Clarification Needed |
|---|---|---|---|---|

c. Interfaces: The operations associated with the module are listed in a table with the name of the operation in the first column of the table. The other columns include status to check if the operation is mandatory or optional, if any additional support is required in the operation, the exception or error codes that can occur in the operation, and the final column is to add any additional comments on the operation.

| Operation | Status | Support | Exception/Error Reason | Comments |
|---|---|---|---|---|

d. Notifications: Notifications can be of different types and each of them are discussed separately in TMF 814A. It starts with the alarm probable cause. The requirements of probable cause can be defined in a table with fields that even gives information on the severity of the alarm. Threshold crossing alert is another notification in which thresholds are set on the values of configuration and performance attributes. Crossing of the threshold value generates notifications of the requirements that can be defined in a specific way that both system vendor and service provider can understand. Some of the other notifications for which TMF 814A gives guideline to define specifications are attribute value change (AVC) for change in the value of an attribute, object creation in information tree, object deletion from the tree, and a generic discussion for any other form of notifications.

2. Nonfunctional implementation statements (NIS): As the title signifies this section in TMF 814A deals with nonfunctional templates of operations, notifications, or data structures. The discussion of NIS involves iterator implementation issues on the number of iterators that can be open at any point of time and timing issues that involve operation response time, time between two heart beat notifications, time between two heart beats, and so on.

3. Guidelines for using the MTNM interface: These guidelines give more clarity to MTNM specifications. Some of the guidelines include:
   - The need for NMS to understand how EMS packages managed elements to subnetworks.
   - How the MTNM service states can be mapped to other models like the ones from ITU-T.
   - The usage of resource names like userLabel, nativeEMSName, and so on.
   - Some examples of probable cause templates.
   - Set of implementation specific use cases.
   - Description of modes for state representation.

For a detailed understanding of the usage of MTNM, the TMF 814A document can be downloaded based on details given in Additional Reading.

## 8.7 Conclusion

This chapter may not be suitable for beginners and some of the terms and concepts expressed in this chapter might be difficult to understand for a novice in telecom management. However once the reader has completed the chapters on network management, then there will be a better understanding on the EML–NML interface. So the author would like to urge the reader to re-visit this chapter after covering the chapters on network management.

While the EML–NML involves some aspects of network and its management, the author decided to cover this chapter under Element Management System because, the major development to achieve compliancy with MTNM is to be done in the EMS. The EMS implements all the interfaces to respond for queries from the NMS. As already discussed in Section 8.5, there is only one interface supported by the NMS and used by EMS and all the other interfaces are supported by EMS and used by NMS.

## Additional Reading

Latest version of the documents specified below can be downloaded from www.tmforum.org


1. TMF 513, Multi-Technology Network Management Business Agreement.
2. TMF 608, Multi-Technology Network Management Information Agreement.
3. TMF 814, Multi-Technology Network Management CORBA Solution Set.
4. TMF 814A, Multi-Technology Network Management Implementation Statement (IS) Templates and Guidelines.

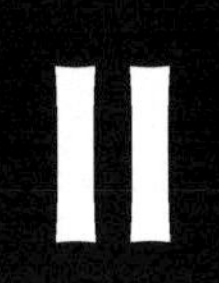

# NETWORK MANAGEMENT SYSTEM (NMS)

# Chapter 9

# Communication Networks

This chapter is intended to provide the reader with a basic understanding of communication networks. At the end of this chapter the reader will be able to work on the management of most communication networks by correlating components and functions with the networks discussed in this chapter. This helps in management data presentation, processing, and defining the management functionalities that will be useful when working on a specific network.

## 9.1 Introduction

A network is a collection of elements (equipments) that work together to offer a service or set of services. For example, the telephone service where you can call your friend's mobile from your phone is not a direct one to one interaction between two user equipments (elements). The call establishment, maintaining the conversation line, and termination of call in a telephone service is made possible using a mediator network like UMTS or CDMA that has a set of elements dedicated to perform a specific function and interoperate with other elements in the network to realize the call handling service.

The main operation of elements in a communication network is transmission of information (voice, data, media, etc.), where the information is transformed or evaluated to make communication between two equipments possible. Example of "Transformation" is a protocol gateway where the language for information exchange is changed from one protocol to another. "Evaluation" can happen at a call server that queries a register to find if the intended service can be fulfilled; or a billing gateway that evaluates the total bill a customer should pay based on service utilization.

Communication networks have evolved based on the need for more information transfer at a higher speed. There are different kinds of communication networks based on the area of coverage, type of modulation used, speed of data transfer, type of switching, bandwidth, and type of interface used for data transfer between elements. Discussing all the different kinds of communication networks is outside the scope of this chapter. This chapter is intended to provide an overview on some of the popular communication networks that are used in telecom. The architecture, elements, and a brief description of element functionality is handled.

The networks that are discussed briefly in this chapter are: ATM, UMTS, MPLS, GSM, GPRS, CDMA, IMS, and WiMAX. Based on the information presented in this chapter, the reader will be able to correlate components in any telecom network and hence have a basic understanding to manage the elements in the network as well as identify functionalities required in the management system.

Independent of the network being handled, a management protocol needs to be used for data collection and information exchange between the network and management solution. This will ensure that the management solution interacting with elements in the network is nonproprietary and scalable. It should be noted that a telecommunication network can be a simple computer network, the Internet or a PSTN (public switched telephone network) and not restricted to the networks discussed in this chapter.

## 9.2 ATM Network

The ATM (asynchronous transfer mode) is a packet switched network where data traffic is encoded into fixed size cells. It is connection oriented and requires a logical connection established between the two endpoints before actual data exchange. It can be used for both telecom and enterprise networks.

An ATM network consists of the following main components (see Figure 9.1):

- Set of ATM switches: The core layer of the network usually has only ATM switches.
- ATM links or interfaces: Logical or physical link are used to connect components.
- ATM routers/hosts: The routers are used to connect to connect a switch with external networks and connections. They are also referred to as the ATM end system.

There are two types of interfaces supported by an ATM switch:

- Network-node interfaces (NNI): This is the interface between two ATM switches.
- User-network interfaces (UNI): This is the interface between end systems and the switch.

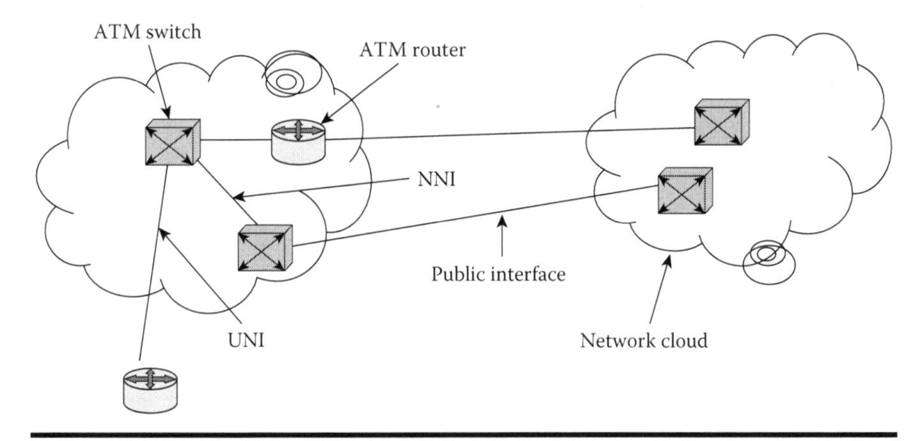


**Figure 9.1 ATM network simplified architecture.**

These interfaces can further be categorized as public or private interfaces where the private interfaces are usually proprietary (not standards based) and used inside private ATM networks, while the public interfaces confirm to standards. Communication between two ATM switches in different networks is handled by a public interface.

ATM cells carry connection identifiers (details of path to follow from one endpoint to another) in their header as they are transmitted through the virtual channel. These identifiers are local to the switching node and assigned during connection setup. Based on the ATM circuit, the connection identifier can be a VPI (virtual path identifier) or VCI (virtual channel identifier). The operations performed by the ATM switch include

- Looking up the identifier in a local translation table and finding the outgoing ports.
- Retransmission of the cell though the outgoing link with appropriate connection identifiers.

From a network management perspective, the basic FCAPS functionality is required in a management system for this network. Unlike a switched virtual connection (SVC) that is automatic configuration, the setting up of a permanent virtual connection (PVC) would require external intervention. The setting up of a PVC in the ATM network using the network management system is an example of specialized configuration management function specific to this network.

## 9.3 GSM Network

GSM stands for global system for mobile communication. It is a cellular (user coverage area defined in the shape of cells) communication network. It was developed

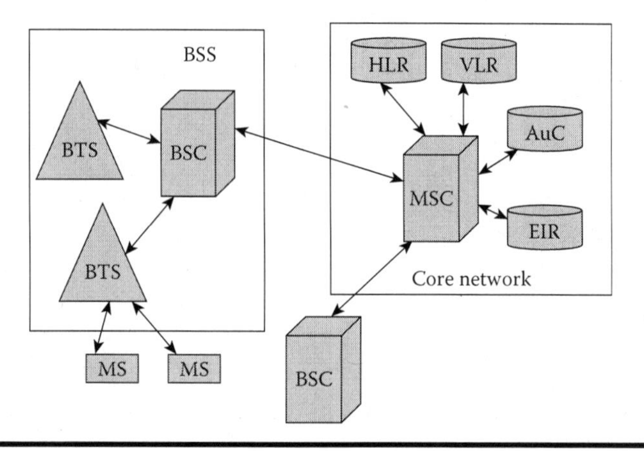


**Figure 9.2 GSM network simplified architecture.**

as an attempt to standardize the European mobile telephone network and is now one of the most popular standards for mobile phones in the world. GSM networks usually operate in the 900 MHz or 1800 MHz bands and in some countries they use the 850 MHz and 1900 MHz bands. The modulation schema used in GSM is Gaussian minimum shift keying (GMSK) and bandwidth is distributed among many users using a combination of time and frequency division multiple access (TDMA/FDMA). The GSM network consists of (see Figure 9.2):

- Mobile station (MS): This is the user/mobile equipment that helps the user to make calls and receive subscribed services.
- Base station system (BSS): It is linked to the radio interface functions and has two components.
  - Base transceiver station (BTS): The BTS is the signal handling node between an MS and the BSC. Its main function is transmitting/receiving of radio signals and encrypting/decrypting communications with the base station controller (BSC).
  - Base station controller (BSC): The BSC is the controller for multiple BTS and its main functionality is to handle handover (user moving from one cell to another) and allocate channels.
- Switching system/network switching subsystem/core network: It has switching functions and a main center for call processing. The components in the core network are:
  - Home location register (HLR): This is a database for storing subscriber information. The information includes subscriber profile, location information, services subscribed to, and the activity status. A subscriber is first registered with the HLR of the operator before he or she can start enjoying the services offered by the operator.

  - Mobile services switching center/mobile switching center/call server (MSC): This is the main switching node of the GSM network and provides functionalities like call switching to and from other telephone and data systems, call routing, subscriber registration and authentication, location updating, toll ticketing, common channel signaling, and handovers.
  - Visitor location register (VLR): It is a database for temporary storage of information about visiting subscribers. When a user moves to a location outside its MSC, the VLR in this new location queries the HLR and gets information about the subscriber to avoid querying the HLR each time the user wants to access a service. The user in the new location is called a visiting subscriber.
- Additional functional elements in GSM include:
  - Authentication center (AuC): Handles security using authentication and encryption.
  - Equipment identity register (EIR): Database to store the identity of mobile equipments to prevent calls from unauthorized equipments.
  - Gateway mobile switching center (GMSC): A gateway integrated with MSC to interconnect two networks. It can be used for routing calls from a PSTN (public switched telephone network) to a GSM user.

There can be many more components like message center (MXE), mobile service node (MSN), and GSM inter-working unit (GIWU) that can be seen in a GSM network. Some definitions specify operation and support system (OSS) as a part of GSM, though it is a more generic term that applies to operation and support for any kind of network.

From telecom management perspective, 3GPP lists a detailed set of specifications on operation, administration, and maintenance of GSM networks. The fault parameters and the performance attributes for network management of GSM network is given by 3GPP specifications. It should be noted that a telecom management solution for GSM network does not necessarily mean management of all components in a GSM network. There are element management solutions that specifically deal with MSCs, BSCs, or HLRs and network management solutions that only handle the core network in GSM.

## 9.4 GPRS Network

The general packet radio service (GPRS) network is more of an upgrade to the GSM network. The same components of the GSM network provides voice service and the GPRS network handles data. Due to this reason the GSM network providers do not have to start from scratch to deploy GPRS. GPRS triggered the transformation from circuit switched GSM network to packet switched network and hence is considered a technology between 2G and 3G, or commonly referred to as 2.5G.

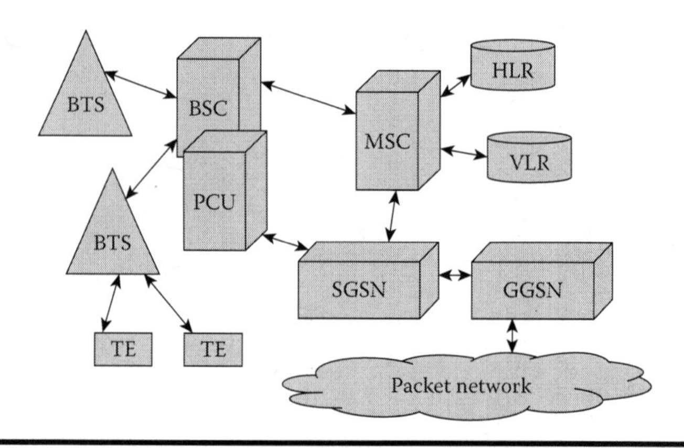


**Figure 9.3 GPRS network simplified architecture.**

Some of the benefits of GPRS over circuit switched networks include:

- Higher speed
- Instantaneous access to service (always on)
- New services related to data communication
- Ease of upgrade and deployment on existing GSM network
- Can support applications that does not require dedicated connection

Compared to GSM there are three new components that are required with GPRS (see Figure 9.3):

- Terminal Equipment (TE): The existing GSM user equipments will not be capable of handling the enhanced air interface and packet data in GPRS. Hence new terminal equipments are required that can handle packet data of GPRS and voice calls using GSM.
- Serving GPRS support node (SGSN): It handles mobility management functions including routing and hand over. SGSN converts mobile data into IP and is capable of IP address assignment.
- Gateway GPRS support node (GGSN): It acts as a gateway to connect with external networks like public Internet (IP network) and other GPRS networks from a different service provider. It can be used to implement security/firewall in screening subscribers and address mapping.

The remaining components in GPRS are similar to GSM with minor software and hardware upgrades. The BSC in GPRS needs installation of hardware components called packet control unit (PCU) to handle packet data traffic and some software upgrades while components like BTS, HLR, and VLR will only require software upgrades.

## 9.5 UMTS Network

UMTS (Universal Mobile Telecommunications System) is a 3G (third generation) network that delivers voice, data, and media services. The media services that include pictures, video, and graphics is a new feature of 3G compared to the 2.5G networks like GPRS. Some of the applications made possible with UMTS include: video and music download, mobile commerce, messaging, conferencing, and location-based services. The air interface for communication in UMTS uses wide-band code division multiple access (W-CDMA) and asynchronous transfer mode (ATM) is the data transmission method used within the UMTS core network.

W-CDMA used for UTRAN air interface is a modulation system where data is multiplied with quasi-random bits derived from W-CDMA spreading codes. These codes are used for canalization, synchronization, and scrambling. W-CDMA operates in both frequency division duplex (FDD) and time division duplex (TDD).

UMTS mainly has three major categories of network elements (see Figure 9.4):

1. GSM elements: Core network (MSC, VLR, HLR, AuC, and EIR) and BSS (BTS and BSC)
2. GPRS elements: SGSN and GGSN
3. UMTS specific elements: User equipment that can handle media and air interface and UMTS Terrestrial Radio Access Network (UTRAN) consisting of radio network controller (RNC) and "Node B."

Node B for the new air interface is the counterpart of BTS in GSM/GPRS. Based on the quality and strength of connection, Node B calculates the frame error rate and transmits information to the RNC for processing. RNC for W-CDMA air interface is the counterpart of BSC in GSM/GPRS. The main functions of the RNC include handover, security, broadcasting, and power control. The user equipment in UMTS network should be compactable to work for GSM/GPRS network.

From a network management perspective, 3GPP has come up with specification on interface reference points (IRP) for OAM&P in UMTS network. Some of the 3GPP specification documents on UMTS OAM&P include:

- TS 32.403: Telecommunication management, performance management (PM), Performance measurements: UMTS and combined UMTS/GSM
- TS 32.405: Telecommunication management, Performance management (PM), Performance measurements Universal terrestrial radio access network (UTRAN)
- TS 32.410: Telecommunication management, key performance indicators (KPI) for UMTS and GSM

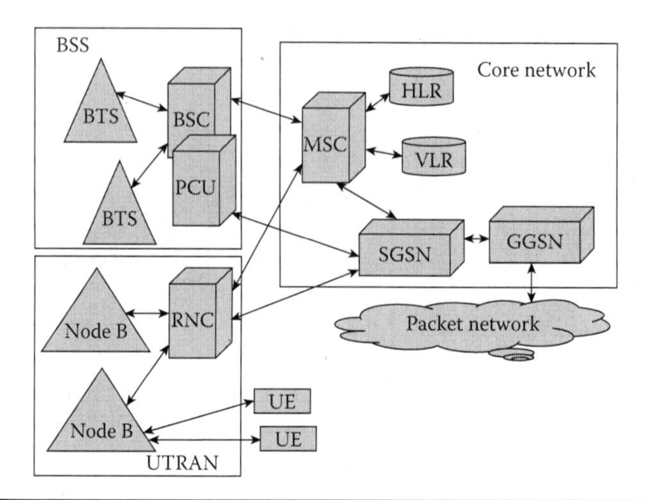


**Figure 9.4 UMTS network simplified architecture.**

## 9.6 MPLS Network

MPLS (multi protocol label switching) was developed with the intent of having a unified network for circuit and packet based clients. It can be used for transmission of a variety of traffic including ATM, SONET, and IP. It is called label switching because each packet in an MPLS has an MPLS header that is used for routing. The routing is based on label look up, which is very fast and efficient. Connectionless packet based network fails to provide the quality of service offered by circuit based networks that are connection oriented. MPLS connection oriented service is introduced in IP based networks. MPLS can be used to manage traffic flows of various granularities across different hardware and applications. It is independent of layer 2 and 3 protocols and is sometimes referred to as operating on "Layer 2.5."

There are two main kinds of routers in an MPLS network (see Figure 9.5):

- Label edge routers (LER): The entry and exist point for packets into an MPLS network is through edge routers. Incoming packets have a label inserted into them by these routers and the label is removed when they leave the MPLS network by the routers. Based on the functionality performed on the packets, these routers are classified in to Incoming (Edge) Router or Ingress router, and Outgoing (Edge) Router or Egress router.
- Label switch routers (LSR): Inside the MPLS network, packet transfer is performed using LSR based on the label. An additional functionality of the LSR is to exchange label information to identify the paths in the network.

Another terminology for routers is in MPLS based VPN network. Here the MPLS label is attached with a packet at the ingress router and removed at the egress

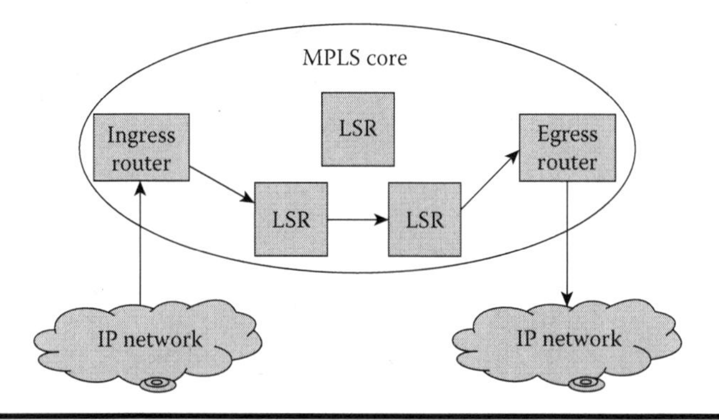


**Figure 9.5 MPLS network simplified architecture.**

router. The label swapping is performed on the intermediate routers. LER is also referred to as "Edge LSR." Forward equivalence class (FEC) is another terminology that is frequently used in discussions on MPLS. FEC represents a group of packets that are provided the same treatment in routing to the destination. The definitions of MPLS were given by IETF, which has an MPLS working group.

Management protocols like SNMP is widely used to gather traffic characteristics from routers in the MPLS network. This information can be used by management applications for provisioning and for traffic engineering of the MPLS network. Being an "any-to-any" network there are multiple challenges in management of MPLS. The biggest concern is to monitor performance and identify the service class when handling an application. The network management solution that offers traffic control should be able to identify existing paths and calculate paths that can provide the guaranteed quality of service. Some management solutions even reshuffle existing paths to identify optimal paths for a specific application. Multiprotocol support of a management solution will be useful when migrating to MPLS network.

## 9.7 IMS

IMS (IP multimedia subsystem) is considered to be the backbone of "all-IP network." IMS was originally developed for mobile applications by 3GPP and 3GPP2. With standards from TISPAN, fixed networks are also supported in IMS, leading to mobile and fixed convergence. Use of open standard IP protocols, defined by the IETF allows service providers to use IMS in introducing new services easily. With multiple standardizing organizations working on IMS, it will cross the frontiers of mobile, wireless, and fixed line technologies.

IMS is based on open standard IP protocols with SIP (session initiation protocol) used to establish, manage, and terminate connections. A multimedia session between two IMS users, between an IMS user and a user on the Internet, or between two users on the Internet are all established using the same protocol. Moreover, the

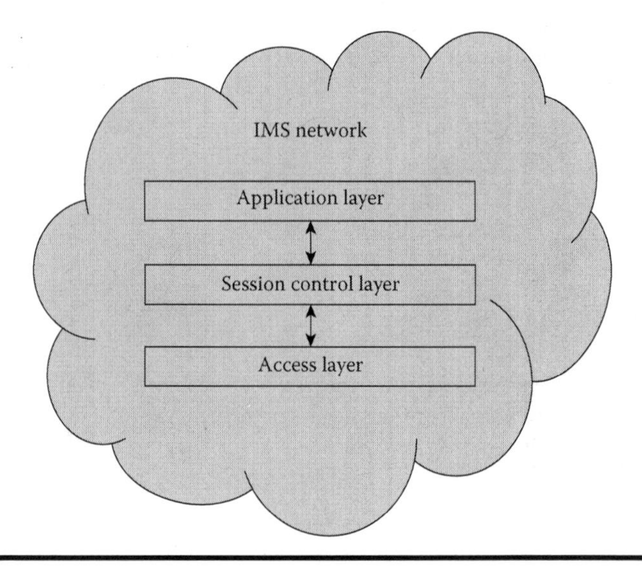


**Figure 9.6 Three layered IMS network.**

interfaces for service developers are also based on IP protocols. IMS merges the Internet with mobile, using cellular technologies to provide ubiquitous access and Internet technologies to provide appealing services. The rapid spread of IP-based access technologies and the move toward core network convergence with IMS has led to an explosion in multimedia content delivery across packet networks. This transition has led to a much wider and richer service experience.

The IP multimedia subsystem can be thought of as composed of three layers (see Figure 9.6):

- The service/application layer: The end services reside in the application layer. It includes a host of application servers that execute services and communicates with a session control layer using SIP. It can be a part of the service provider home network or can reside in a third-party network. With open standards defined on interactions with an application server, it is easier to build applications on application servers. The power of IMS lies in easily rolling in/out services on the fly in minimal time using application servers.
- The IMS core/session control layer: Session control and interactions between transport and application layer happen through the session layer. The main components of the core or session control layer are:
  - Call session control function (CSCF): It can establish, monitor, maintain, and release sessions. It also manages the user service interactions, policy of QoS and access to external domains. Based on the role performed, a CSCF can be serving (S-CSCF), proxy (P-CSCF), or interrogating (I-CSCF) call session control function.

- Home subscriber server (HSS): Database for storing subscriber and service information.
- Multimedia resource function plane (MRFP): It implements all the media-related functions like play media, mix media, and provide announcements.
- Multimedia resource function controller (MRFC): It handles communication with the S-CSCF and controls the resources in the MRFP.
- Breakout gateway control function (BGCF): It mainly handles interaction with PSTN.
- Media gateway controller function (MGCF): It is used to control a media gateway.

- The access/transport layer: This layer is used for different networks to connect to IMS using Internet protocol. It initiates and terminates SIP signaling and includes elements like gateways for conversion between formats. The connecting network can be fixed access like DSL or Ethernet, mobile access like GSM or CDMA, and wireless access like WiMAX.

Some other IMS components in IMS include signaling gateways (SGW), media gateway (MGW), and telephone number mapping (ENUM).

A network management solution for IMS usually handles components in the core network with limited support for elements in application and transport layers. The standards on OAM&P for IMS have been defined by 3GPP. The definitions include alarm formats and performance data collection attributes. Some of the specialized NMS functions of IMS include HSS subscriber provisioning and charging management based on 3GPP.

## 9.8 CDMA Network

During standardization activities for a 2G network, GSM was adopted in most parts of Europe and CDMA (code division multiple access) evolved during the same time, capturing markets in the United States and Asia. The first major effort in the development of CDMA network standard was from Telecommunications Industry Association (TIA/EIA) with an architecture named cdmaone that could be used for commercial deployment of CDMA networks. While IS-95A of Telecommunications Industry Association brought circuit switched services using CDMA, the revised version IS-95B gave subscribers packet switched data services. While networks based on IS-95B were considered 2.5G, the blue print for CDMA based 3G network came with CDMA2000 defined by ITU.

The components in a CDMA2000 network include (see Figure 9.7):

- Mobile station (MS): This is the client equipment or user equipment like a subscriber handset that provides interface for the user to access the services.
- Radio access network (RAN): It is the air interface components in CDMA for interacting with the core network. RAN is similar to BSS on GSM networks.

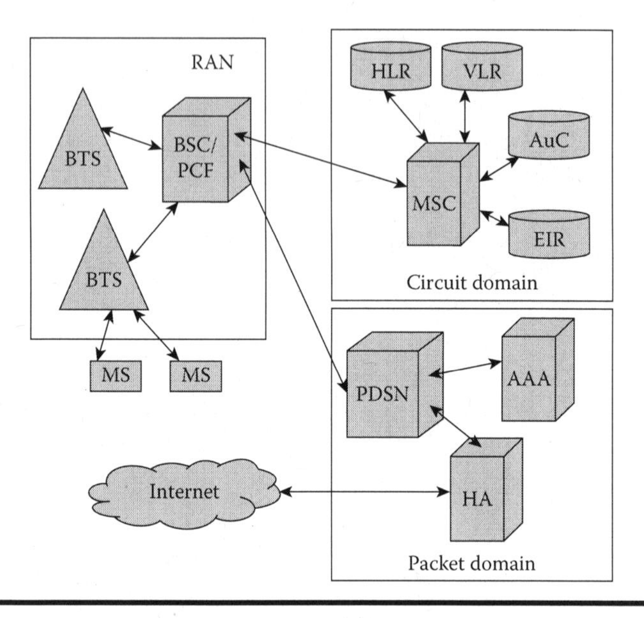


**Figure 9.7 CDMA network simplified architecture.**

It has the BSC and BTS found in GSM network. RAN also contains a packet control function (PCF) that is used to route IP packets to a PDSN (packet data serving node).

- Packet domain: Packet domain in a CDMA network consists of:
  - PDSN/foreign agent (FA): It acts as a gateway for the RAN by routing packets to an external packet network like the Internet. It can establish, maintain, and terminate a packet link.
  - AAA (authentication, authorization, and accounting): These are servers used to authenticate and authorize users for access to network and to store subscriber call details for accounting.
  - Home agent: Interface component to external packet network: Provides an IP for mobile messages and forwards it to the appropriate network. A PDSN can be configured to work as a HA.
- Circuit domain: The circuit domain is similar to the GSM network with components like the MSC, GMSC, HLR, AuC, and EIR. For details on these components refer to the section on GSM in this chapter.

The 3GPP2 is dedicated to the development of next generation standards for CDMA 2000. Some of the 3GPP2 documents relevant to network management of CDMA network include:

- S.R0017-0 v1.0 "3G Wireless Network Management System High Level Requirements"

- X.S0011-003-C v1.0 to v3.0 "cdma2000 Wireless IP Network Standard: Packet Data Mobility and Resource Management"
- S.S0028 series on "OAM&P for cdma2000"

In addition to standardizing forums like 3GPP2 that promote the evolution of CDMA, forums like the CDMA development group (CDG) work on adoption and ease in deployment of 3G CDMA wireless systems.

## 9.9 WiMAX

WiMAX (worldwide interoperability for microwave access) is a wireless broadband technology based on standard IEEE 802.16. It is a wireless alternative to cable modems, DSL, and T1/E1 links. The IEEE standard was named WiMAX by the WiMAX Forum to promote the IEEE standard for interoperability and deployment. It can support voice, video, and internet data.

The spectrum bands in which WiMAX usually operates include 2.3 GHz, 2.5 GHz, 3.5 GHz, and 5.8 GHz, with a speed of approximately 40 Mbps per wireless channel. Based on coverage, WiMAX is classified under MAN (metropolitan area network).

WiMAX can offer both non-line-of-sight and line-of-sight service. In non-ine-of-sight that operates at lower frequency range, an antenna on the personal computer communicates with WiMAX tower and in line-of-sight that operates at high frequencies, a dish antenna points directly at the WiMAX tower. It uses orthogonal frequency division multiple access (OFDM) as the modulation technique.

WiMAX architecture can be split into three parts (see Figure 9.8). They are:

- Mobile station (MS): This is the user equipment or user terminal that the end user uses to access the WiMAX network.
- Access service network (ASN): This is the access network of WiMAX comprising of base stations (BS) and one or more ASN GW (gateways). While the base station is responsible for providing the air interface with MS, the ASN gateways form the radio access network at the edge.
- Connectivity service network (CSN): This is the core network that offers the services and connectivity with other networks. It includes the AAA server for authentication, authorization, and accounting, the MIP-HA (mobile IP home agent), the services offered using supporting networks like IMS, operation support system/billing system that can be a part of the core or a stand-alone application and the gateways for protocol conversion and connectivity with other networks.

IEEE 802.16 has a network management task group that works on developing standards for management of WiMAX. The documents are part of project IEEE 802.16i. The standards are intended to cover service flow, accounting, and QoS management.

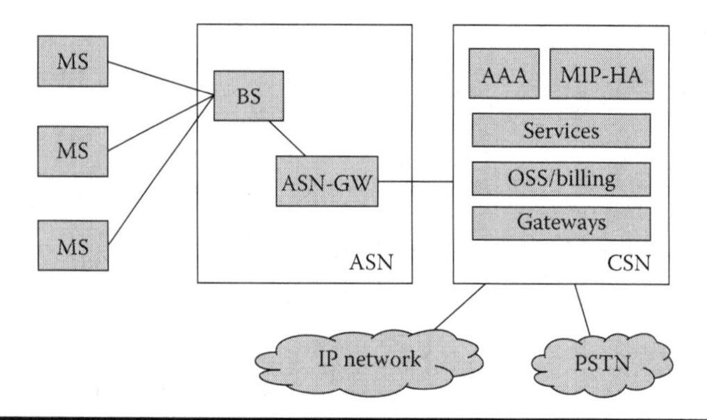


**Figure 9.8 WiMAX simplified architecture.**

## 9.10 Recent Initiatives like LTE and SAE

In this chapter we discussed some of the key 3GPP networks including the second generation (2G) GSM, the 2.5G network like GPRS, 3G network like UMTS, and 3G network extensions like IMS. The initiatives "Beyond 3G" taken up by 3GPP include LTE (long-term evolution) and SAE (system architecture evolution). The main difference between these two initiatives is that LTE is about enhancing the "access network" and SAE works on developing "core network" architecture to support the high-throughput/low latency LTE access system.

LTE is an effort to reduce the number of network elements, but offer higher data rates and better quality of service on a packet only network. It uses OFDM (orthogonal frequency division multiple access) for uplink and SC-FDMA (single carrier frequency division multiple access) for down link. LTE access network has two types of components (see Figure 9.9), the access gateways (AGW) and enhance node B (ENB). While the AGW does functionalities like mode management, data compression, and ciphering, the ENB works on all radio-related issues and mobility management.

SAE is expected to provide an All-IP network solution with all services on packet switched domain and no circuit switched domain. It will support mobility with multiple heterogeneous access systems both supported by 3GPP like LTE and non-3GPP access systems like WiMAX. SAE consists of (see Figure 9.10):

- MME (mobility management entity): It does functions like user equipment authentication and mobility management.
- UPE (user plane entity): It manages and stores user equipment context, including ciphering, packet routing, and forwarding.
- 3GPP anchor: Mobility anchor between 2G/3G and LTE
- SAE anchor: Mobility anchor between 3GPP and non 3GPP

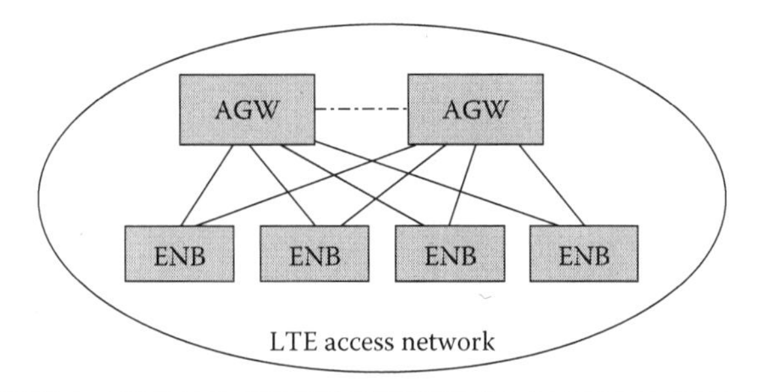


Figure 9.9 LTE simplified architecture.

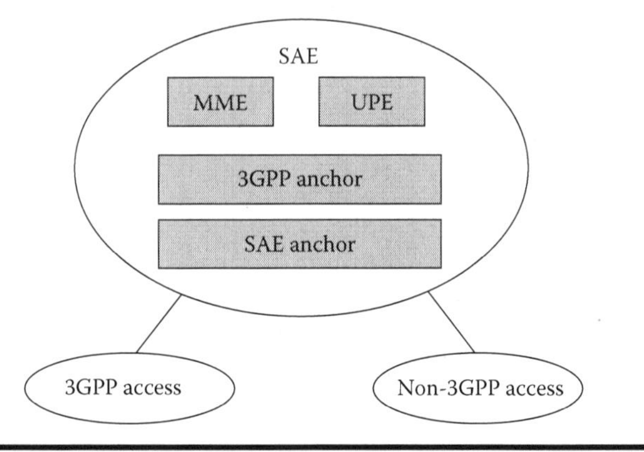


Figure 9.10 SAE simplified architecture.

While the initiatives for next generation networks development are specialized domains, by separating the access and core network the standards for network management of next generation networks are also handled by specialized forums like TMF where generic standards for managing a variety of networks are developed.

## 9.11 Conclusion

This chapter gives an overview on some of the most popular telecom networks. It can be seen that networks mainly have an access and a core component. Most network management solutions are developed for handling either access or core network. Technology is moving toward a converged network for enterprise and telecom with an All-IP infrastructure as the backbone. Network management solutions and standards are also moving in the direction of managing converged next generation networks.

## Additional Reading

1. Ronald Harding Davis. *ATM for Public Networks*. New York: McGraw-Hill, 1999.
2. Timo Halonen, Javier Romero, and Juan Melero. *GSM, GPRS and EDGE Performance: Evolution Towards 3G/UMTS*. 2nd ed. New York: John Wiley & Sons, 2003.
3. Miikka Poikselka, Aki Niemi, Hisham Khartabil, and Georg Mayer. *The IMS: IP Multimedia Concepts and Services*. 2nd ed. New York: John Wiley & Sons, 2006.
4. Luc De Ghein. *MPLS Fundamentals. 2006TMF 513, Multi-Technology Network Management Business Agreement*. Indianapolis, IN: Cisco Press, 2006.

# Chapter 10

# Seven-Layer Communication Model

This chapter is intended to provide the reader with an understanding of the ISO/OSI communication model. Any communication between two elements or devices is possible only when they are able to understand each other. The set of rules of interaction is called a protocol and the tasks in the interaction can be grouped under seven layers. At the end of this chapter the reader will have an understanding of the seven layers and the tasks and protocols associated with each layer.

## 10.1 Introduction

OSI (open system interconnect) seven layer communication model was developed by ISO as a framework for interoperability and the Internet working between devices or elements in the network (see Figure 10.1). The different functions involved in the communication between two elements are split as a set of tasks to be performed by each of the seven layers. Each layer is self-contained making the tasks specific to each layer easy to implement. The layers start with the physical layer at the bottom, followed by data link layer, network layer, transport layer, session layer, presentation layer, and application layer.

The lowest layer is referred to a layer 1 (L1) and moving up, the top most layer is layer 7 (L7). The physical layer is closest to the physical media for connecting two devices and hence is responsible for placing data in the physical medium. In most scenarios the physical layer is rarely a pure software component and is usually implemented as a combination of software and hardware. The application layer is closest

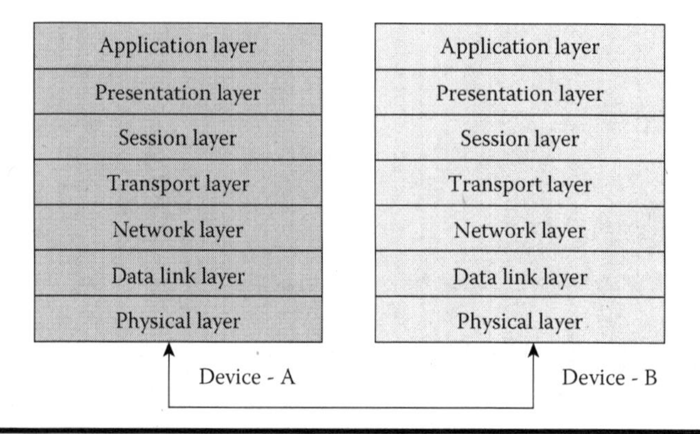


**Figure 10.1 OSI communication model.**

to the end user and is mostly implemented as software. Communication involves transfer of data between devices. There are two key areas of focus here: the data and the device. Layers 7–4 deal with end to end communications between the "data" source and destinations. Layers 3–1 deal with communications between network "devices." This chapter handles each of the layers in detail along with the protocols used to implement a specific layer. Network management protocols are usually application layer protocols but a basic understanding of the communication mechanism will help in understanding the different tasks involved in communication.

## 10.2 Physical Layer

This layer handles physical movement of data across the network interface. This layer is not just network cards and cables but defines a number of network functions. In the same way a network device like an Ethernet card is not just restricted to a physical layer but performs functions of both the physical and data link layer. Some of the physical can be of different types like open-wire circuits, twisted pair cables, coaxial cables, or fiber optic cables

> Physical layer is the lowest layer in the OSI model concerned with data encoding, signaling, transmission, and reception over network devices.

Some of the functions of the physical layer are:

- Encoding: This would involve transformation of data in the form of bits with the device into signals for sending over a physical medium.
- Hardware specifications: Both the physical and data link layer together detail the operation of network interface cards and cables.

- Data transport between devices: The physical layer is responsible for transmission and reception of data.
- Network topology design: The topology and physical network design is handled by the physical layer. Some of the hardware related issues are caused due to missed configurations in the physical layer.

The encoding can result in conversion of data bits to signals of various schemas. Some of the signaling schemas that the physical layer can encode to are:

- Pulse signaling: Where the zero level represents "logic zero" and any other level can represent "logic one." This signaling schema is also called "return to zero."
- Level signaling: A predefined level of voltage or current can represent logic one and zero. There is no restriction that zero should be used to represent the logic zero. This schema is also called "non return to zero"
- Manchester encoding: In this schema "logic one" is represented by a transition in a particular direction in the center of each bit, where an example of transition can be a rising edge. Transition in the opposite direction is used to represent logic 0.

Physical layer also determines the utilization of media during data transmission based on signaling. When using "base band signaling" all available frequencies (the entire bandwidth) is used for data transmission. Ethernet mostly uses base band signaling. When working with "broadband signaling" only one frequency (part of the entire bandwidth) is used, with multiple signals transmitted simultaneously over the same media. The most popular example of broadband signaling is TV signals, where multiple channels working on different frequencies are transmitted simultaneously.

Physical layer also determines the topology of the network. The network can have bus, star, ring, or mesh topology and the implementation can be performed with LAN or WAN specifications. The physical layer should ensure efficient sharing of the communication media among multiple devices with proper flow control and collision resolution. Some examples of physical layer protocols are: CSMA/CD (carrier sense multiple access/collision detection), DSL (digital subscriber line) and RS-232. The physical layer is special compared to the other layers of the OSI model in having both hardware and software components that makeup this layer. There can be sublayers associated with the physical layer. These sublayers further break the functionality performed by the physical layer.

## 10.3 Data Link Layer

This is the second layer (L2) in the ISO/OSI model. It is divided into two sublayers, the logical link control (LLC) and media access control (MAC) on IEEE 802 local area networks, and some non-IEEE 802 networks such as FDDI. LLC is the upper sublayer and handles functions for control and establishment of connection for logical links between devices in the network. MAC is the lower sublayer in the data link layer (DLL) and handles functions that control access to the network media like cables.

The functions of the data link layer are:

- Frame Conversion: The final data for transmission is converted to frames in the data link layer. Frame is a fixed length piece of data where the length is dependent on hardware.
- Addressing: This involves labeling data with destination information based on hardware or a MAC address. This ensures that the data reaches the correct destination.
- Error detection and handling: Error occurring in the lower layers of the stack are detected and handled using algorithms like cyclic redundancy check (CRC) at the DLL.
- Flow control: Data flow control and acknowledgment in the data link as performed by the LLC sublayer.
- Access control: This is performed by the MAC sublayer and is required to ensure proper utilization of shared media.

The data link layer offers the following types of service to the higher layer:

- Unacknowledged connectionless service: In this type of service there is no logical connection established or released between the source and destination. It is mainly used in scenarios where the error rate and loss of frames is low. A popular example of using this form of service is in streaming audio and video.
- Acknowledged connectionless service: In this type of service there is no logical connection between source and destination, but an acknowledgment message is send by the receiver for each input message received from source. If no acknowledgment is received at the source the corresponding frame is assumed to be lost and the frame is re-sent.
- Acknowledged connection-oriented service: In this type of service there is a connection established between source and destination for data transfer and an acknowledgment is issued by the destination to notify the server that a particular packet was received. The frames are numbered. In this type of service, error and frame loss is minimal.

The elementary data link protocols for simplex unidirectional communication that form the basis to dictate flow control and development of DLL protocol are:

- Unrestricted simplex protocol: The receiver is assumed to have infinite capacity. Sender keeps on sending data to receiver as fast as it can.

- A simplex stop-and-wait protocol: The receiver only has a finite buffer capacity. Care should be taken to ensure that the sender does not flood the receiver.
- A simplex protocol for a noisy channel: The channel is noisy and unreliable hence care should be taken to handle frame loss, errors, and retransmission while implementing this protocol.

The best-known example of the DLL is Ethernet. Other examples of data link protocols are HDLC (high level data link control) and ADCCP (advanced data communication control protocol) for point-to-point or packet-switched networks and Aloha for local area networks.

## 10.4 Network Layer

This is the third lowest layer (L3) in the ISO/OSI model. While the data link layer deals with devices in a local area, the network layer deals with actually getting data from one computer to another even if it is on a remote network. That is, the network layer is responsible for source to destination packet delivery, while the data link layer is responsible for node to node frame delivery. Network layer achieves end to end delivery using logical addressing.

Addressing performed in data link layer is for the local physical device, while the logical address given in network layer must be unique across an entire internetwork. For a network like the Internet, the IP (Internet protocol) address is the logical address and every machine on the Internet needs to identify itself with a unique IP address.

The functionalities of the network layer are:

- Addressing: Every device in the network needs to identify itself with a unique address called the logical address that is assigned in the network layer.
- Packetizing: The data to be communicated is converted to packets in the network layer and each packet has a network header (see Figure 10.2). The data to be packetized is provided by the higher layers in the OSI model. This activity is also referred to as datagram encapsulation. It can happen that the packet to be sent to the data link layer is bigger than what the DLL can handle, in which case the packets are fragmented at the source before it is sent to the DDL and at the destination the packets are reassembled.
- Routing: The software in the network layer is supposed to perform routing of packets. Incoming packets from several sources need to be analyzed and routed to proper destinations by the network layer.
- Error messaging: Some protocols for communication error handling resides in the network layer. They generate error messages about connection, route traffic, and health of devices in the network. A popular error handling protocol that is part of the IP suite that resides in the network layer is ICMP (Internet control message protocol).

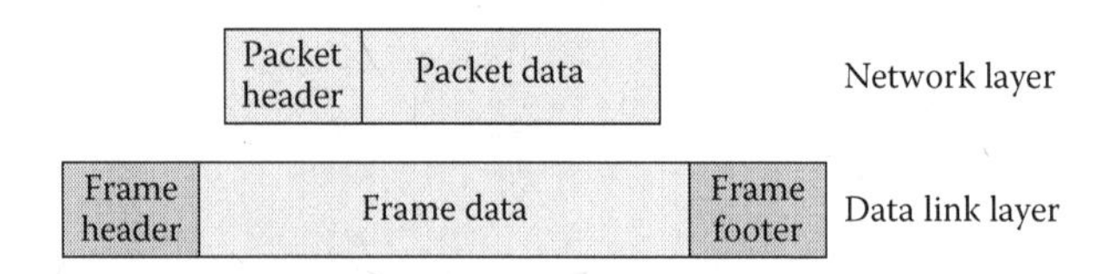


**Figure 10.2 Packet in the network layer.**

Though the network layer protocol is usually connectionless, it can also be connection oriented. That is, the network layer protocol supports both connectionless and connection-oriented protocols. When connectionless protocol like IP is used in the network layer, data consistency and reliability are handled using connection-oriented protocols like TCP in the transport layer.

IP (Internet protocol) is the most popular network layer protocol. There are several other supporting protocols of IP for security, traffic control, error handling, and so on that reside in the network layer. These protocols include IPsec, IP NAT, and ICMP. Another network layer protocol outside the realm of the Internet protocol is IPX (inter-network packet exchange) and DDP (datagram delivery protocol).

## 10.5 Transport Layer

This is the fourth lowest layer (L4) in the ISO/OSI model. While the physical layer is concerned with transmission of bits, the data link layer with transmission in local area, and the network layer with transition even with remote systems, the transport layer handles transmission of data between processes/applications. There will be multiple applications running on the same system and there would be scenarios when more than one application from a specific source machine wants to send data to multiple applications on a destination machine. Transport layer ensures that multiple applications on the same system can communicate to another set of applications on a different system. This would mean that a new parameter in addition to the destination and source IP address is required for application level communication.

So a transport protocol can have simultaneous connections to a computer and ensure that the receiving computer still knows what application should handle each data packet that is received. This task is accomplished using "ports." The parameter port is just an internal address that facilitates control of data flow. Ports that have specific purposes and used for generic applications across systems are known as well-known ports. For example data is routed through well-known TCP port 80 while using the Internet. While an IP address can be used to route data to a specific network element and the port number to direct data to a specific application, both of these tasks can be jointly done by an identifier called the socket. A socket is an identifier that binds a specific port to an IP address.

Transport layer supports both connection-oriented protocol and connectionless protocol. Connection-oriented protocol is the more common for reliable data communication as a connection needs to be established before data transfer. It allows for error checking functionality and stability. Connectionless protocol does not require a reliable connection established between the application during the entire duration of data transfer and it lacks extensive error checking. Connectionless protocol is considered to be much faster than connection-oriented protocol and is mainly used in intermittent communication.

The functions of the transport layer are:

- Connection management: In case of a connection-oriented protocol the transport layer ensures establishing the connection, maintaining the connection during data transfer and termination of connection after the transfer is complete.
- Congestion avoidance and flow control: There can be a scenario when the sender transmits data at a rate higher than the value that the receiver can handle. This scenario of flooding the receiver is called congestion. Regulating the flow of data or flow control can avoid congestion. Transport layer can implement flow control.
- Reliable data transfer: Acknowledgments can be used by the receiver to inform the sender about safe delivery or loss of data. Retransmission of lost data can also be implemented in the transport layer.
- Process level communication using ports and sockets.
- Data processing: Data segments from different processes can be combined to a single stream at the sender and again fragmented at the receiver. Also a large bulk of data can be split into segments at the sender and then combined again at the receiver.

The TCP (transport control protocol) and UDP (user datagram protocol) are the most popular transport layer protocols. TCP sends data as packets while UDP sends data as datagrams. TCP is connection oriented and UDP is connectionless. Connection-oriented services can be handled at the network layer itself, though it is commonly implemented at the transport layer.

## 10.6 Session Layer

This is the fifth lowest layer (L5) in the ISO/OSI model. All the functionality of identifying the destination and application to communicate is done in the layers lower to the session layer. So addressing and data processing is not done in the session layer. As the name suggests, session layer is mainly intended for session management, which includes establishing, maintaining, and terminating a session. Session is a logical linking of software application processes, to allow them to exchange data. A session is usually established between applications to transfer a file, send log-on information and for any type of data transfer.

Session can be of two types:

- Half duplex: In half duplex only one way communication is permitted at any point of time. So when a session is established between two network elements, each element has to communicate in turn as only one element is permitted to send data in half duplex mode.
- Full duplex: In full duplex session, both the network elements that have established a session for interaction can communicate or send data at the same time.

Some of the popular session layer protocols are session initiation protocol (SIP) and AppleTalk session protocol (ASP). Session layer software in most scenarios is more of a set of tools rather than a specific protocol. For instance RPC (remote procedure call) usually associated with the session layer for communication between remote systems gives more of a set of APIs (application program interfaces) for working on remote sessions and SSH (secure shell), another protocol associated with session layer also gives the implemented the ability to establish secure sessions. While the term session is usually used for communication over a short duration between elements, the term connection is used for communication that takes place for longer durations.

Let us consider a typical scenario of how a session layer works when handling a single application. The Web page displayed by a Web browser can contain embedded objects of media, text, and graphics. This graphic file, media file, and text content file are all separate files on the Web server. If the browser is a pure application object that cannot handle multiple objects together then a separate download must be started to access these objects. So the Web browser opens a separate session to the Web server to download each of these individual files of the embedded objects in the Web page. The session layer performs the function of keeping track of what packets and data belong to what file. This is required so that the Web page can be displayed correctly by the Web browser.

## 10.7 Presentation Layer

This is the sixth lowest layer (L6) in the ISO/OSI model. The role of the presentation layer is to ensure that communication between applications in two different systems occur even when the data representation format is not the same. As the name suggests, a common syntax or format that the two interacting systems can understand is presented to the systems at the presentation layer. The conversion of application specific system to presentation system happens both in the transmitting system and the receiving system. This common presentation syntax is called transfer syntax.

There can be abstract forms to define data types called abstract data syntax and presentation forms for transfer and presentation of data called transfer syntax.

In order to ensure that common syntax definitions are available for data recognition, ISO defined a general abstract syntax for the definition of data types associated with distributed applications. This abstract syntax is known as abstract syntax notation number one or ASN.1. The data types associated with ASN.1 are abstract data types. Along similar lines, transfer syntax was defined by OSI for use with ASN.1. This transfer syntax notation developed by OSI is known as basic encoding rules (BER).

The functions performed by the presentation layer are:

- Translation for representing data in a common format: A network will be comprised of machines running on different OS like Windows, UNIX, Macintosh, and so on. These machines differ in characteristics of data presentation. The presentation layer handles translation of data to a known format, thus hiding the differences between machines. This translation includes data transformation, formatting, and special purpose transformations.
- Request for session establishment, data transfer, and termination: Though session management is handled by the session layer, the request to perform session-related operations like the request for establishment of a session, the data transfer, and the final termination originates from the presentation layer.
- Negotiation of syntax: The presentation layer does the job of negation of syntax for data transfer. Multiple negotiations are performed before a common format/syntax is agreed upon between systems that want to communicate.
- Compression and decompression: This function is not usually performed in the presentation layer, but to improve the throughput of data, compression and decompression can be performed in this layer. Some functions, like compression to increase throughput and encryption to increase security, can be performed in almost any layer of the OSI model.
- Encryption of data: Implementing security for the data that travels down the protocol stack is a key functionality of the presentation layer. Encryption is performed on the data at the presentation layer and it can also be performed at the layers below the presentation layer. One of the most popular encryption schema associated with the presentation layer is the secure sockets layer (SSL) protocol.

The presentation layer is not always used in network communication. This is because most of its functionalities are optional. The encryption and compression can be formed at other layers in the OSI model. For data translation to come up with a common format is only required when handling heterogeneous systems. The remaining functions of the presentation layer can be performed by the application layer. Most of the protocol stack implementations may not have a presentation layer.

There are three syntactic versions that are important when two applications interact:

- Syntax of the application entity that requested the connection.
- Syntax of the application entity that receives the connection.
- Syntax used by the presentation entities in giving a common format.

The syntax of the application entity at the transmitting or requesting end is converted to syntax used by the presentation entity at the presentation layer of a requesting system. In the same way, the syntax used by the presentation entity is converted to the syntax used by an application entity at the receiving end at the presentation layer of the receiving system.

Another example of a presentation layer protocol is XDR (external data representation). XDR is used as the common data format mostly when interaction is involved with a system running UNIX OS. For a UNIX system to interact with a heterogeneous system running another OS, using remote procedure calls (RPC), the usual presentation layer format is XDR. XDR has its own set of data types and the application layer types will get converted to XDR format before transfer of data and the response data also needs to be converted to XDR. In a distributed system having heterogeneous systems, the presentation layer plays a key role to facilitate communication.

## 10.8 Application Layer

This is the top most layer of the OSI model. Being the seventh lowest layer in the ISO/OSI model it is also referred to as L7. The application layer manages communication between applications and handles issues like network transparency and resource allocation. In addition to providing network access as the key functionality, the application layer also offers services for user applications.

Let us consider how the application layer offers services for user applications with an example of World Wide Web (www). The Web browser is an application that runs on a machine to display Web pages and it makes use of the services offered by a protocol that operates at the application layer. Viewing the vast collection of hypertext-based files available on the Internet as well as providing quick and easy retrieval of these files using a browser is made possible using the application layer protocol called hyper text transfer protocol (HTTP).

Some of the other popular application layer protocols include simple mail transfer protocol (SMTP) that gives electronic mail functionality to be implemented in applications through which all kinds of data can be sent to other users of the Internet as an electronic postal system and file transfer protocol (FTP) is used to transfer files between two machines. The network management protocol SNMP is also an application layer protocol as it provides interfaces for network management applications to collect network data.

A program that implements the interfaces provided by the application layer protocol is required in the machine. That is, an FTP client should be running on the machine to transfer files from the FTP server and an SNMP agent should be running on the network element to source data to a network manager using the SNMP protocol.

The user application can operate without an application layer using protocol modules that initialize communication with other application, set up the presentation context, and transfer messages. In this mode of operation the application will interact directly with the presentation layer. In the same manner the application can implement modules that perform the application services offered by the application layer. These modules are called application specific elements.

Application layer protocols are classified into:

- Common application specific elements (CASE)
- Specific application specific elements (SASE)

As the name signifies, CASE elements are the integrated set of functions that are commonly used by an application while SASE elements are written on a need basis for a specific functionality. Some of the CASE related functions include initialization and termination of connection with other processes, transfer of message with other processes, or underlying protocol and recovering from a failure while a task is under execution.

It should be understood that not all applications use the application layer for communication and applications are not the only users of the application layer. The operating system (OS) itself can use the services directly from the application layer.

## 10.9 TCP/IP Model and OSI Model

The OSI model is the not the only network model available for implementing protocol stacks for communication between network elements. The four layer TCP/IP model was developed before the OSI model and is also one of the most popular models for implementing protocol stack (see Figure 10.3). The functions performed by the four layers in the TCP/IP model can be mapped to the functions performed by the seven layers of an OSI model.

The layers in the TCP/IP Model are:

1. Application layer: This is the top most layer of the TCP/IP model and performs functions done by application, presentation, and session layer in the OSI model. In addition to all the processes that involve user interaction, the

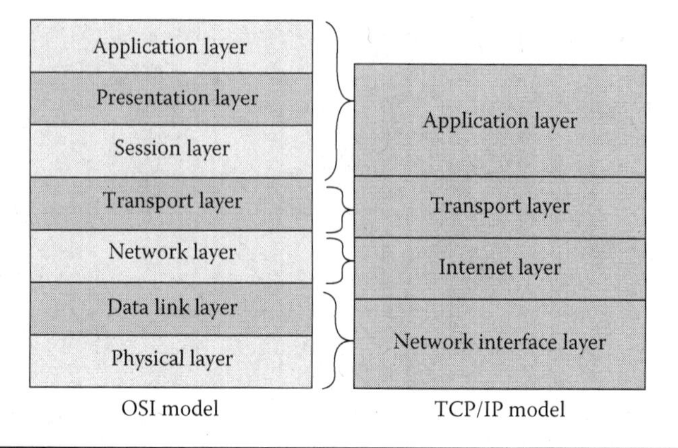


**Figure 10.3 Mapping layers in the OSI model to layers in the TCP/IP model.**

application also controls the session and presentation of data when required. The terms like socket and port that are associated with the session layer in an OSI model are used to describe the path over which applications communicate in TCP/IP model. Some of the protocols in this layer are FTP, HTTP, SMTP, and SNMP.

2. Transport layer: This layer of TCP/IP model performs functions similar to the transport layer of an OSI model. There are two important transport layer protocols, the transmission control protocol (TCP) and user datagram protocol (UDP). While TCP guarantees that information is received as it is a connection-oriented protocol, the UDP works on a connectionless setup suitable for intermittent networks and performs no end-to-end reliability checks. Flow control is also performed at this layer to ensure that a fast sender will not swamp the receiver with message packets.
3. Internet layer: This layer of the TCP/IP model performs functions similar to the network layer of an OSI model. The Internet layer is suited for packet switching networks and is mostly a connectionless layer that allows hosts to insert packets into any network and delivers these packets independently to the destination. The work of the Internet layer is limited to delivery of packets and the higher layers perform the function to rearrange the packets in order to deliver them to the proper application in the destination. The Internet protocol defines the packet format for the Internet layer. Routing of packets to avoid congestion is also handled in the Internet layer. This is a key layer of the TCP/IP model and all data has to pass this layer before being sent to lower layers for transmission.
4. Network interface layer: The physical layer and data link layer of the OSI model are grouped into a single layer called network interface layer in the TCP/IP model. This layer is used by the TCP/IP protocols running at the

upper layers to interface with the local network. The TCP/IP model does not give much detail on the implementation or the specific set of functions to be performed at the network interface layer except it being the point for connection of host to network and use of an associated protocol to transmit IP packets. Protocols like point-to-point protocol (PPP) and serial line Internet protocol (SLIP) can be considered protocols of the network interface layer. The actual protocol used in the network interface layer varies from host to host and network to network. The network interface layer is also called the network access layer. In some implementations of the network interface layer, it performs functions like frame synchronization, control of errors, and media access.

There are some significant differences between the OSI model and TCP/IP model.

- In the TCP/IP model, reliability is mainly handled at the transport layer. The transport layer takes on all functions for handling data transfer reliability, which includes capabilities like error detection and recovery, checksums, acknowledgments, and timeouts.
- The other difference is the presence of intelligent hosts when using a TCP/IP model. This means that an OSI model does not handle management of network operations while most TCP/IP implementations like Internet have capabilities to manage network operations, which make them intelligent hosts.
- OSI model was developed as a standard and serves more as a teaching aid. This is because it clearly demarcates the functionality to the performed when two systems interact over a distributed environment. On the other hand, the TCP/IP model was adopted as a standard due to more widespread application of the model in domains like the Internet. It should be noted that the TCP/IP model was developed before the OSI model.

## 10.10 Conclusion

The network model is required to have a good understanding of how two network elements communicate whether the elements are in the same network or in two separate networks. The OSI model is the most common model used in text book descriptions to give the reader an understanding of the functionalities in an interaction between elements spit into a set of layers while the TCP/IP is the most popular model for implementing protocol stacks of interaction. It can be seen that TCP/IP is a type of OSI model where a single layer performs multiple layer functions. From a functional model perspective, both the TCP/IP model and OSI model are the same.

## Additional Reading

1. Debbra Wetteroth. *OSI Reference Model for Telecommunications*. New York: McGraw-Hill, 2001.
2. Uyless D. Black. *OSI: A Model for Computer Communications Standards*. Facsimile ed. Upper Saddle River, NJ: Prentice Hall, 1990.
3. American National Standards Institute (ANSI). *ISO/IEC 10731:1994, Information Technology: Open Systems Interconnection: Basic Reference Model: Conventions for the Definition of OSI Services*. ISO/IEC JTC 1, 2007.
4. Charles Kozierok. *The TCP/IP Guide: A Comprehensive, Illustrated Internet Protocols Reference*. San Francisco, CA: No Starch Press, 2005.

# Chapter 11

# What Is NMS?

This chapter is intended to provide the reader with a basic understanding about the Network Management System (NMS). At the end of this chapter you will have a good understanding of the components in a legacy NMS, some of the open source products that implement full or partial functionality of these components, and where the network management system fits in the telecom architecture.

## 11.1 Introduction

The network management layer in the TMN model is between the element management layer and service management layer. It works on the FCAPS (fault, configuration, accounting, performance, and security) data obtained from the element management layer. The FCAPS functionality is spread across all layers of the TMN model. This can be explained with a simple example. Consider a fault data, which involves fault logs generated at the network element layer. These logs are collected from the network element using an element management system (EMS). The EMS will format the logs to a specific format and display the logs in a GUI (graphical user interface). Each log is generated in the network element to notify the occurrence of a specific event. Fault logs from different network elements are collected at the network management system in the network management layer. Logs corresponding to different events on different network elements are correlated at the NMS. An event handler implemented at the NMS can take corrective actions to rectify a trouble scenario. Since this is a network management layer functionality let us explain it in more detail with an example.

When the connection between a call processing switch/call agent (CA) and a media gateway (MGW) goes down due to installation failure of a software module on the MGW, a set of event logs are generated. The CA will create a log showing loss of connection with MGW and the MGW, which is a different network element, will generate its own log specifying loss of connection with the CA. There will also be logs in the MGW specifying failure of a software module on MGW and many other logs corresponding to the status of the links that connect the CA and MGW. While at EMS level only collection of data from a specific element like CA or MGW occurs, at the NMS level log data from both these elements are available. So the event logs from different network elements can be correlated and in this scenario the logs from the CA and MGW are correlated and the actual problem can be identified, which is not possible at the EMS level by just a connection down log at CA. Knowledge of the actual problem at the NMS level, leads to the capability of creating applications that can take corrective actions when a predefined problem occurs.

The fault logs generated at the network element is also used in layers above the network layer. The time duration for which a link is down can be used as input information for a service assurance application in the service layer, and this can be used to provide a discount in the billing module on the business management layer. There is a separate chapter that details the functions handled in the network management layer and this chapter is more about giving the reader a good understanding of a network management application and the usual modules required while implementing a network management system.

## 11.2 Network Management System

The network management system as the name signifies is used to manage a network. It provides a holistic view of the entire network. The different activities performed in a typical NMS when it is started and before any maintenance command can be issued are:

- Discovery of elements in the network: This can happen in different ways. The user of an NMS application can use a menu in the application to specify the elements that need to be managed as an IP address or a range of addresses like "X.X.X.100" to "X.X.X.201." The set of servers to be managed may also be taken from a domain server that has an IP address of all the elements in the network. Auto discovery is also a common feature in most NMS systems where the NMS automatically identifies all the elements in a particular network.
- Discovery of components that make up the element: Each network element will be made of hardware components that perform a specific functionality and needs to be discovered. For example a call agent will have a set of links and these will be plugged into link interface cards on the call agent. Fault

logs will be generated with information on the component associated to the fault condition.

- Download of fault data: This activity usually happens after the resource (element and its components) discovery, so that the fault data can be associated to the element and its specific component. The discovered resources are represented as a tree in most NMS applications and associating the logs to the specific resource helps in debugging at the resource level.
- Download of performance data: Unlike fault logs, which are usually send to the NMS as and when they occur, the performance data is usually dumped into files on the network element. These performance records are collected from the network element at regular intervals by the NMS, the data is then parsed and formatted for final display or processing input for other modules.
- Authentication of user: The user of the NMS needs to be authenticated to prevent unauthorized use of the network management application. Maintenance commands that can bring down key elements in the network can be performed from the NMS, which leads to high levels of security implemented at the NMS. There are a set of user groups and each user is assigned to the user group. Each group or user has a set of privileges with regard to functionalities in the application. For example, with fault functionality, some users can only view the logs, others can view and edit the logs, yet another group or user in a group can even delete logs. With single-sign-on (SSO) functionality implemented at NMS, a single authentication should permit the user working on NMS to jump to the console of any network element that is managed or any element management application that is associated with the NMS.
- Download additional records and configuration data: Any other record in the element that is required by some functionality at NMS is also downloaded from the network element. Predefined configuration data, like correlation between performance registers or rules for event handling are also downloaded.

Once the resources are discovered and the historical data downloaded, then the NMS application is ready for performing any maintenance level activities. The maintenance activities or network functions are performed with the downloaded data. These functions will include network provisioning, troubleshooting of network issues, maintenance of configured network elements, tracing the flow of data across multiple network elements, programming the network for optimal utilization of elements, and so on. During a connection loss between the NMS and an already managed element, the NMS will try for a connection establishment at regular intervals and synchronize the data when the connection is established. A rediscovery of the network resources and their states will also be performed at regular intervals to keep the most recent information at NMS.

## 11.3 Legacy NMS Architecture

The architecture of NMS can be broken down into a set of building blocks. Any management solution that can collect data from network elements and performs one or more management functionality is an NMS. The data collection may be limited to support for just a single protocol and the management function may be only fault handling or performance handling, but it is still a valid NMS when the basic building blocks or a subset of the building blocks is present.

The NMS architecture discussed in this section is called legacy NMS for a purpose. The old architecture used for the development of a network management system was not scalable, which led to work on new standards for developing NMS. AN NMS built using the guidelines of new standards is called next generation NMS and the old NMS systems that had become functionality rich with years of development is termed legacy NMS. Since this book will discuss next generation NMS, the term legacy NMS has been used from the beginning for the NMS architecture used currently in most solutions.

The building blocks in legacy NMS are (see Figure 11.1):

- Platform components: These components include object persistence, logging framework, libraries, and so on that are required by most of the management function implementation modules.
- Data collection components: Based on the NMS protocol supported on the network element, these components collect data from the network elements. In some management paradigms this component may not be part of the NMS and will be a separate solution called the EMS.
- Management function components: These components use the data collected from the network elements to provide FCAPS and other specialized management functionality.

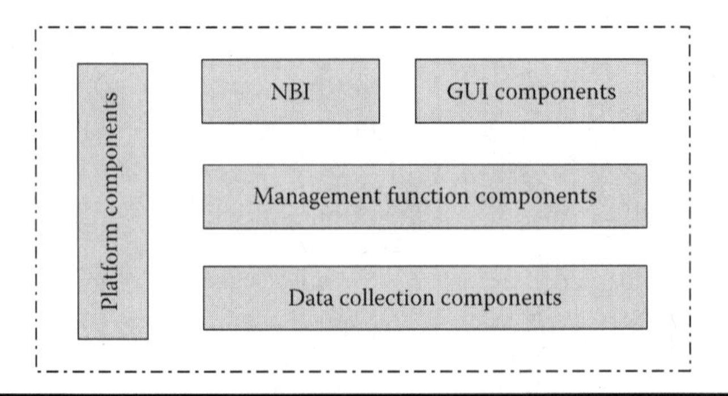


**Figure 11.1 Legacy NMS building blocks.**

- North bound interface (NBI) components: These components provide management data to the north bound interfaces like another NMS or operations support systems (OSS).
- GUI components: These provide user interaction screens that enable a user to perform operations on the various modules in the NMS.

These building blocks and the open source components that can be used in making these blocks are discussed in detail in the sections that follow. The data collection components can collect data directly from the network elements or from an EMS. In legacy NMS, solutions and products available from leading NMS vendors, the data collection was done usually from the network elements with interfaces that permit collection of data from an EMS or other NMS solution. Use of standard interface and shared data models as in next generation has led to much reuse of the NMS functional components.

A network management system will also have a database for storing the raw or formatted data. The presence of a database avoids re-query for historical data from an NMS to the network element. The historical data can be stored into a full-fledged database to which SQL-based commands are executed to send and receive data or it can be maintained as flat files or compressed flat files.

## 11.4 Data Collection Components

The data collection components collect data from the network elements using a specific protocol (see Figure 11.2). The protocol used can be a proprietary protocol specific to the company implementing the network element and the management application, but in most scenarios a standard management protocol is used. Use of a standard protocol promotes flexibility, so that the data collection agent implemented for the specific protocol can be reused to collect data from another network element that supports the same protocol. Also a standard protocol will allow agents from different vendors to be used to collect data from a network element.

An SNMP (simple network management protocol) agent is used for collecting data from a network element that supports SNMP. The agent will issue commands

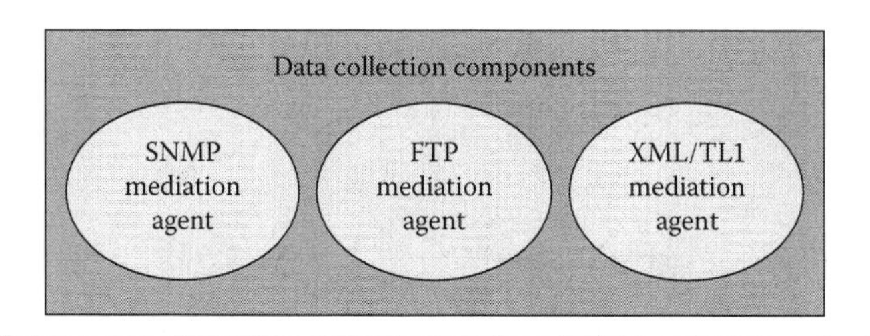


**Figure 11.2 Data collection blocks for specific protocols.**

like "Get" and "Get-Next" to browse the data in the network element. It should be kept in mind that not all network elements can respond to SNMP queries from the agent on the NMS. The network element should be maintaining a MIB (management information base) with configuration details and also generate logs/events as SNMP traps that can read by the agent.

There are many protocols that can be used for interaction between the management application and the network element. The SNMP, as the name signifies, is easier to implement and maintain and has been the most preferred network management protocol. With new requirements to support XML (extensible markup language) based data, which can be intercepted and parsed by any third-party application with sufficient authentication, XML agents based on XML protocols like XMLP (XML protocol) and SOAP (simple object access protocol) are becoming popular. SOAP is very effective for web-based network management. The TL1 is another protocol that is popular to provide command line based network management interface. For file transfers there is no better protocol than FTP (file transfer protocol). So the protocol agent available at the NMS data collection component is mainly dependent on the type of application for which the NMS will be put to use. Most of the popular NMS products have support for multiple protocols. Some of the open source SNMP agents are WestHawk SNMP agent, NetSNMPJ, and IBM SNMP Bean Suite. Similarly a popular open source FTP agent is edtFTPJ. Most of these open source products are built in Java programming language and can be downloaded from sourceforge.net

## 11.5 Platform Components

The platform components are the functional components and libraries that are used by the other components that make up the NMS. All items that are reused and form the basic blocks for building other components are aggregated under platform components (see Figure 11.3). Some of the common functionality that is required in any communication application like logging of events and encryption of data can be taken as example applications that fall under platform components.

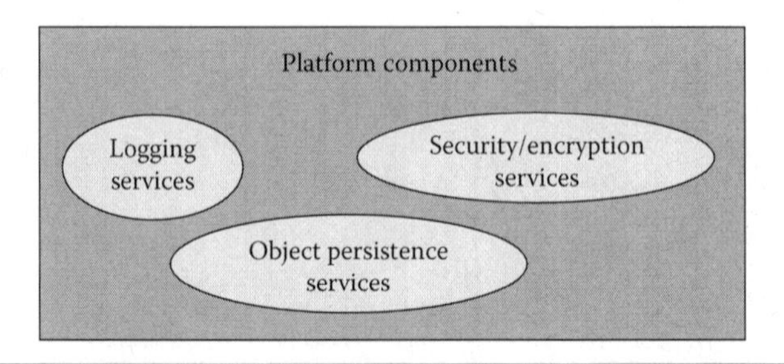


**Figure 11.3 Blocks in platform components.**

To have a better understanding of platform components, let us take a few sample services that are grouped as platform services that can be applied by other components:

1. Object persistence services: An NMS solution might be required to interact with multiple databases at the backend. An application that provides object persistence service is expected to generate SQL, do result in set handling, object conversion, and ensure that the application is portable with multiple SQL databases. All components that want to work with the database can use the services of object persistence, which gives transparent data persistence to all other components. "Hibernate," a persistence application using Java, is an open source product with LGPL license and can be downloaded from sourceforge.net.
2. Logging services: Logging is usually implemented as a debugging tool. While designing an application there would be requirements to provide debugging messages as "warning," "error," "information," and so on. Logging services provide applications in other NMS components with the interfaces that can be called to generate logs to a specific output. The output can be to a predefined file, socket, syslog, or console. The media to which the logging will be performed is based on parameters like the priority level. A warning can be written to a file while an error might be displayed in the user console. Log4J is a Java-based application that offers logging services. Log4J is an open source product with Apache License.
3. Communications security services: The NMS interacts with multiple EMS or network elements in the south bound interface (SBI) and with other NMS or OSS applications in the north bound interface (NBI). The interaction between the NMS and other applications is expected to be secure. Security of data is achieved by encrypting the data before transmission. There are different applications that provide interfaces for NMS to encrypt the data before communicating with NBI and SBI. The BouncyCastle package is a popular Java-based encryption package used in most network management solutions.

It should be understood that the NMS components discussed in this chapter are a logical grouping of functionality and there will not be specific process of package in an NMS application with these component names. The functionalities in the NMS are grouped under some main headings to explain the essential building blocks. Most open source packages for NMS are implemented in Java. The platform components can also include libraries and not just service packages like the examples discussed.

## 11.6 Management Function Components

The management function components are the heart of any network management solution. They implement the management functions like event handling and performance correlation, using the data collected from the network elements. These management functions can be split into program segments or logical blocks that can developed as individual components. There are multiple open source components that offer a set or subset of management functions in this component.

Some of the sample blocks in management function components are:

1. Discovery component: The discovery component will dynamically discover the network and its elements configuration using the interfaces offered by the data collection components. It would also use the interfaces of a data persistence component that is part of a platform component for database interaction. It reads, populates, and maintains the network object model using the discovery interfaces that would involve functions like reading from configuration files, reading from dynamically modifiable tables or configuration related event notifications.
2. Configuration component: It offers application program interfaces (API) to add, modify, and delete configuration information related to the network and its elements like nodes, links, and paths. The configuration component creates template objects that would be generic in nature and could be customized for any network element specific set of configuration variables.
3. Fault aggregation component: The role of this component is to collect fault data from the data collection components and convert them to events on objects in the topology model or network object model. This could be faults on the network elements or resources in the network elements. The aggregation logic would be given in the XML files as rulesets, which would have as inputs the different fields of the fault record. Each fault record is expected to have a specific set of fields giving data like the severity of the event that generated the record, the element in which the record was generated, the probable cause of the event, and so forth.
4. Fault correlation component: This component would be triggered on each event. Based on a set of predefined rules, the fault information is updated when each fault record or event is obtained. The update can be an add, modify, or delete operation in the database.
5. Fault delivery and query component: This component provides the application program interfaces for delivery of faults and query of faults from a database. This functionality is implemented using a listener module that registers for fault data. The registration as well as a fault related query can be filtered on any parameter of the fault record.
6. Performance aggregation and correlation component: This component collects performance records from the data collection components and provides object persistence by interacting with the object persistence modules in platform components. It can also perform correlation functions on the data based on predefined rules. These rules can be changed dynamically using APIs.

7. Performance query and threshold component: This component makes performance data queries to the database based on parameters in performance records. The threshold functionality permits setting, modifying, and deleting threshold alarm generation rules based on performance data. That is, a threshold alarm is generated when a performance parameter crosses a predefined rule value. Both time-based and multidata-based threshold rule setting is possible.

Open source applications like "OpenNMS" provide all these functions and can be used as a framework for making specialized NMS for a particular network. Open source applications that offer just one specific management functionality is also available, like "Mandrax" or "Drools" for rule-based fault correlation.

## 11.7 GUI Components

The GUI components are standard applications that convert the formatted data generated from management functional components to user-friendly graphical representations. For example performance records like link utilization data for a day monitored at regular intervals, would be easier to track and compare when plotted as in a graph. In similar lines the fault data will have a set of predefined fields, so when the logs have to be presented to the user, it is best to format the logs and show the log contents under a set of header in an MS Excel format based GUI. The flow of data from NE to user through GUI components is shown in Figure 11.4.

Some of the sample blocks in GUI components are:

1. Chassis viewers: The network elements usually in a COTS (commercial off-the-shelf) setup based on ATCA (advanced telecom computing architecture) specifications are usually placed in a chassis and connected using a physical medium to initiate interaction. The chassis will have a front end view and a back end view. On the NMS GUI, the user should be able to view where a particular element is placed in the chassis. So when a fault occurs on a particular network element it would be easier to isolate the correct element in the chassis and perform debugging or replacement operation using the chassis diagram created using the interfaces offered by chassis viewer application.
2. Map viewers: When the network elements are spread across various geographical locations as in a distrusted network environment, it would be

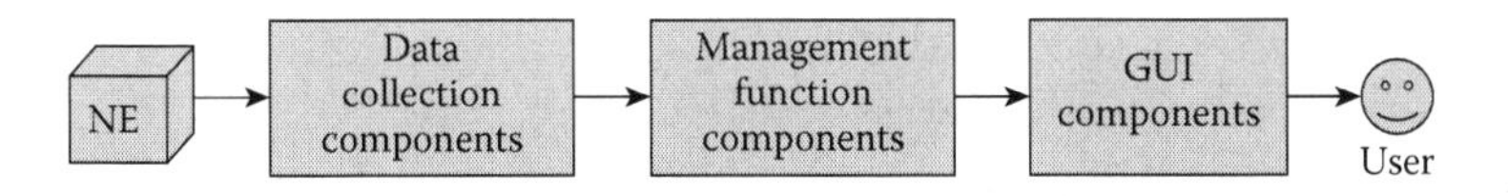


**Figure 11.4 Flow of data from NE to user.**

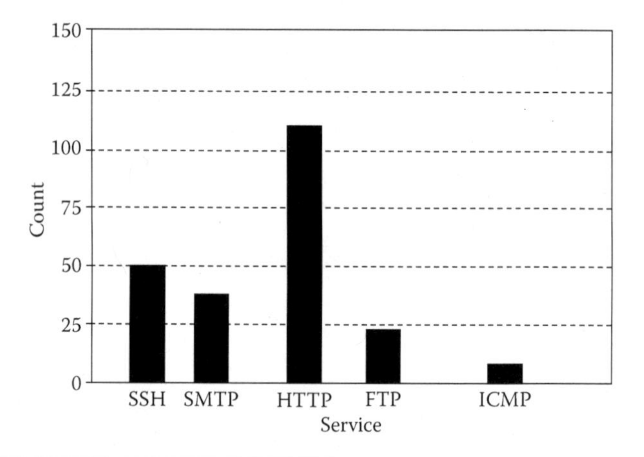


**Figure 11.5 Sample performance chart.**

easier to locate a faulty element when the elements are displayed over a geographical map of the local area, specific country or in some cases the world map. Tools like "MICA" and "Picolo" can be used for implementing the same.

3. Performance chart viewers: The formatted performance records can be represented graphically using applications like "JFreeChart," "iReport," and "DataVision." These applications provide interfaces to represent performance data to the user in a graphical manner. The figure of the sample performance chart (see Figure 11.5), shows the count of service outages for a specific period of time against the specific service.
4. Fault browsers: Similar to performance data, fault data also needs to be presented in a user friendly manner. The fault data generated at the network element is first parsed and formatted by the management function components that collect this data using data collection components. This formatted data has a set of fields and this data in GUI should be represented in such a way that the data can be filtered based on a specific field. A tabular representation is usually used with a fault data display in GUI.
5. User administration: The administration screen is a necessity to perform network related operations like adding a new node to monitor, deleting a node, clearing a fault record, adding comments to a log, defining a performance correlation, and so on. Authentication is required for the user to get access to the administration screen. Most NMS launch will result in a window or Web page requesting user name and password. Once the user is authenticated, based on the group and permissions exercised by the log in id, the user can access the administration screen and perform any of the operations that the user, based on privileges, is permitted to perform.

The building blocks in the GUI components, thus provide the interfaces that can be used during the NMS design of a user-friendly GUI. Most of the functional interface bundles can be downloaded along with code as opensource modules at sourceforge.net

## 11.8 NBI (North Bound Interface) Components

The NBI components provide the published interfaces for external north-bound applications to interact with the NMS. The external applications can be another NMS or an OSS/BSS solution. While the GUI of the NMS is used for working with network management related operations, some applications will work on formatted data provided by NMS. For example, the network outage report prepared by the NMS will be required by an OSS service assurance application. If the interface calls for interaction with the NMS then are published, more applications could be set up to collect and work with data from NMS.

The NBI components are composed of two major building blocks, the framework involving interface specifications and the server that provides the platform for publishing these interfaces.

Some examples of the framework involving an interface specification are:

- J2EE interface: This involves interface definitions based on Java for publishing the interactions permitted when working with the NMS. J2EE Interface is specifically for the Enterprise edition of Java and the opensource implementation framework for J2EE interface can be downloaded from the internet. "JBean" project provides the required interfaces for a Java-based implementation and the aim of this project is to utilize Java Beans and Ecomponent architecture, to easily assemble desperate software components into a single working program.
- CORBA interface: Different opensource implementations of common object request broker architecture (CORBA) specifications defined by the object management group (OMG) is available for developing interfaces that enable northbound interface applications written in multiple computer languages and running on multiple computers to work with a particular NMS solution. openCCM is an example of opensource CORBA component model implementation.
- .NET interface: .Net framework of Microsoft is a popular platform for application development that is not written for a particular OS. The .Net also provides interfaces for communication between desperate components or applications. Similar to Java, .Net also has a good bundle of interfaces that can make interfacing of NMS with NBI applications much easier.

Some examples of the server that provides the platform for publishing these interfaces are:

- Java application servers: Some opensource implementations include JBoss and JOnAS. Both these application servers are production ready and is compliant to Java Enterprise edition specifications.
- .Net Server: Window .Net Server is an example of the application server that fits in the platform category though it is currently not an opensource product.
- Objects request broker (ORB): This refers to a middleware technology for interaction and data exchange between objects. Opensource implementations of CORBA ORBs are "openORB" and "JacORB."

The north bound interface can also be proprietary modules or adapters that convert data from the NMS to the format required by the application that wants to interact with the NMS. For example IBM Netcool Network management product has a set of probes (generic, proprietary, and standards based), gateways and monitors that can virtually collect data from any network environment like OSS services or another EMS/NMS application. The gateway itself supports third-party databases and applications like Oracle, SQL Server, Informix, Sybase, Clarify, Siebel, Remedy, and Metasolv.

## 11.9 Design Considerations

In addition to the legacy NMS architecture discussed in this chapter, there are additional architectures that are put to use in the development of the NMS. As an example, event driven architecture (EDA) and service oriented architecture (SOA) and how it applies to development of NMS is briefly discussed here.

Event driven architecture in simple terms involves the development of component/application that can trigger or perform actions based on events. The EDA can be used to create event rule engines in NMS. The rule engines are programmed to respond with a specific action for an event or set of events. For example, when the event rule engine on the NMS receives a log from a network element stating failure of a service module during a software upgrade, then the event engine can trigger an action command that will cause the software upgrade operation to seize and fall back to the previous version or another action command will be issued to send a mail to the concerned network administrator to handle the problem.

The basic techniques used for implementing EDA include simple, stream, and complex event processing to provide event abstraction, aggregation, or triggering corrective actions. In simple event processing, an event is programmed to initiate downstream action(s). This is implemented as simple conditions involving an action triggered whenever an event occurs. In stream event processing, events are screened for notability and streamed to information subscribers. In complex event processing (CEP), events are monitored over a long period of time and actions triggered based on sophisticated event interpreters, event pattern definition and matching, and correlation techniques. An example of complex event processing in the event engine of NMS would involve the number of times a particular log that is issued when a trunk in a trunk group goes down, occurs over a time of one week, and generating a log or triggering a mail to network administrator at the end of one week showing those trunks in the trunk group going down more often than the others.

Next we have service oriented architecture. In SOA, data and functionality are exposed as reusable services described using a standards based format. A service refers to a modular, self-sufficient piece that performs a specific functionality expressed in abstract terms independent of the underlying implementation. Using SOA in NMS would involve developing the function blocks in management function components as independent services. Event rule engine and reporting engine are two example functional blocks in an SOA-based NMS framework that can act as independent services.

Consider the event engine in SOA. The event engine in this scenario will collect data from other functional blocks on fault, performance, security, or from data collection components, but it will always collect data in a specific format. This way the event engine works as an independent service that performs a set of actions based on rules, but can interwork with any module/service as it collects data and defines rules only in a specific format. Now consider the case of a reporting engine that exports data to certain standard formats like a pdf, doc, or html page. When a reporting engine is implemented to operate as a service in SOA, it can be used to create reports for multiple functional blocks. The same reporting engine will then generate a pdf document listing active logs from fault module or a doc on the graphical representation of the network.

## 11.10 Conclusion

Most books on network management discuss the NMS functions, standards, and protocols without giving a clear picture of the implementation perspective on how NMS architecture looks. This level of information is usually available in research publications created by developers who actually work on designing and coding the NMS. This chapter gives a solid base to understanding the components for making a network management system. It should be understood that there are enough opensource applications available for download, to build most of the components in the NMS. Some of these applications are also discussed in this chapter to avoid reinventing the wheel for working on developing an NMS for PoC (proof-of-concept) or other noncommercial purpose. Use of opensource products for commercial development should be done carefully after examining the license agreement of the product.

## Additional Reading

There are no major book references for the material discussed in this chapter. As an additional reference it is suggested visiting the web pages on the opensource components discussed in this chapter. This will give a better understanding of the specific component and how it fits in the complete NMS package.

1. openNMS: www.opennms.org
2. openORB: openorb.sourceforge.net
3. JacORB: www.jacorb.org
4. JBoss: www.jboss.org
5. JOnAS: jonas.objectweb.org
6. openCCM: openccm.objectweb.org
7. JFreeChart: www.jfree.org/jfreechart/
8. Log4J: logging.apache.org/log4j/1.2/index.html
9. BouncyCastle: www.bouncycastle.org
10. Hibernate: www.hibernate.org
11. edtFTPJ: www.enterprisedt.com/products/edtftpj
12. WestHawk: snmp.westhawk.co.uk
13. NetSNMPJ: netsnmpj.sourceforge.net

# *Chapter 12*

# NMS Functions

This chapter is intended to provide the reader with a basic understanding about the functions available in a network management system (NMS). At the end of this chapter you will have a good understanding of the categories under which network functions can be grouped in the eTOM (enhanced telecom operations map) model. This would include a discussion of the categories and how it maps to actual functions in the NMS.

## 12.1 Introduction

The TMF (TeleManagement Forum) expanded the network and service management layers in the TMN model to have more detailed mapping of the telecom operations and created the TOM model. Later the TOM model was revised by TMF to have three main sections: SIP (strategy, infrastructure, and product), OFAB (operations, fulfillment, assurance, and billing), and enterprise management. This revised model is called eTOM. The functions of the NMS mainly fall under the resource management layer of eTOM that spreads across SIP and OFAB. Though the NMS data can be used for planning phases in SIP, its functions can be directly mapped to real-time network configuring and maintenance in OFAB. So this chapter will mainly be about the network management functions from the categories of resource management in OFAB of the eTOM model.

The following categories are used to discuss resource management functions:

- Resource provisioning
- Resource data collection and processing
- Resource trouble management

- Resource performance management
- "Resource management and operation" support and readiness

A set of resources are required to develop a service. The term Resource as used by TMF could be any type of hardware, software, or workforce used to develop the service and includes within its scope a network or an element in the network. Hence resource management functions of TMF maps to the functions performed by NMS. It should be understood that the NMS functions in this chapter are generic and would be available in most network management solutions. There are also network management functions that are suitable specifically for a particular kind of network and high level capabilities that are implemented to satisfy specific requirements, both of which does not fall in the generic category.

An example of a management function specific to a particular type of network is HSS (home subscriber server) subscriber provisioning. This operation involves creating, editing, and deleting customer/subscriber related records on the HSS. Since the HSS is an element found only in an IP-based network, this provisioning functionality can be seen only in IP-based networks like IMS (IP multimedia subsystem). A high level functionality that is implemented only based on customer requirements is single-sign-on (SSO). Some NMS solutions where security levels do not require separate log-ins for working on an EMS or a NE from NMS, then SSO can be used to have a single user authentication for working with any applications or subapplications within the NMS, including log-in to other devices or applications that can be launched from the NMS.

## 12.2 Resource Provisioning

Network provisioning involves configuring the elements in a network to offer a service that is requested by a customer. The request for service is placed as an order by the customer and in most cases at a service provider space an automatic provisioning agent does the provisioning based on parameters specified in a predefined configuration file usually written in XML.

The NMS functions that are part of network provisioning category are:

- Allocate and install: The network elements that are allocated to satisfy a service by the inventory management system are discovered by the NMS and the required software installation is performed. The way installation is usually automated is to connect to a server that has the required software packages and initiate a command that will install the necessary packages on the target network element with or without a download of the packages to the target NE.
- Configure and activate: Once installation of packages are complete, the network element will be ready for configuring. When the modules in the installed

software are activated, they need to run using a set of predefined parameters to satisfy the service specifications. These parameters are read from configuration files. The process of setting these parameters before, during, and after activation is known as configuring the network elements. When the configuration involved at the preactivation stage is complete, the software modules can be started or activated.

- Test: A necessary function of provisioning is running some tests to ensure that the network and its elements are ready to offer a specific service. For example, if the element being provisioned is a call agent that provides call processing capability, then the test would involve running some test scripts that simulate an actual call and verifying the call records generated against an expected record set created for the simulated call. A test case can even be a simple check to ensure that no errors were generated during a software installation or verifying if the modules are running with the parameters specified as part of configuration.
- Track and manage: The network provisioning operations needs to be tracked and managed. Tracking involves ensuring that the various management functions involved in provisioning are executed properly. Tracking helps to identify errors or issues that are raised during provisioning so that the same can be posted to an administrator for taking corrective action. Managing involves interaction with other modules, for example if a particular element in a network is identified as faulty, the same needs to be communicated to the inventory management system for allocating a new replacement element and once network and service provisioning is complete, the order management system needs to be informed to close the request.
- Issue reports: The reports issued in network provisioning are quite different from the FCAPS reports in maintenance. The reports issued in network provisioning is intended to provide interested parties, information on the status of provisioning, efficiency achieved with current configuration parameters, total time for provisioning, elements that failed and the operation that caused failure, and so forth. This would mean that the provisioning reports are customized reports for a specific audience unlike a maintenance report.
- Recovery of element: Similar to allocation of resource, the unused resource or element needs to be deallocated for use in satisfying a new order or as a replacement for some faulty element in an existing order. The process of issuing a deallocation request, getting proper authorization, performing the deallocation and reallocating the same for some other order is performed as part of recovery of element.

In addition to understanding the functions performed by the network provisioning module, it is also important to have an understanding of the modules that interact with network provisioning. The main modules that interact with the network provisioning module are:

- Service configuration: This module is intended for configuration of service that can only be done once the underlying network to support the service is provisioned. So the service configuration module sends a "preservice request" and the provisioning module runs the preservice network level configuration and responds back with the results of the operation. Next the service configuration module issues a "network provisioning request" and the provisioning module does the network level provisioning and responds to service configuration module with the results and reports collected during provisioning of network. Next the service configuration module issues a "test request" for which the network provisioning module performs the predefined tests on the network and responds to the service configuration module with the results of the test showing the available capacity of the network to handle the service.
- Network planning and development: Provisioning plays a key role in network planning and development. During provisioning, key information is obtained on the functions that fail often and the elements that cannot handle a specific capacity. This capacity related information is used for network planning and developing existing configuration of the network. Based on precalculated results of a previous provisioning operation, the network planning and development module supplies the configuration requirements to the provisioning module. The provisioning module applies the supplied configuration.
- Network maintenance and restoration: The maintenance activity on a network might involve reconfiguring the network to fix errors or to restore the configuration of specific resources to a predefined value (like a fallback operation to previous configuration). Reconfiguration requests are issued to the provisioning module when multiple network element configurations need to be altered. Based on the configuration request from the maintenance-restoration module or the network monitoring module, the provisioning module executes the changes on the network resources and responds with a configuration complete message.
- Network data management: There are two representations of network data. One is the network's view dealing with actual configuration running on the network and the management view, which is the network data maintained or cached by the NMS. Having two different views of data lead to the need for synchronization. Discrepancy can occur in a large service provider space where the network view is expected to be a mirror of the master data in a management space that maintains information based on inputs from the inventory module. The discrepancy is corrected by reprovisioning the network that requires an interaction with the network provisioning module.

- Network inventory management: Provisioning of service for a new work order will require allocating of resources and updating the inventory management system. It is the network provisioning module that understands the capacity required for a specific service and sends a request to the inventory module for allocating a suitable resource or checking the feasibility of available resources for executing the work order. This process of interaction with the inventory module starts when a "preservice" request is given to the provisioning module by the service configuration module. With a change in inventory database, the inventory module can also trigger a request to the provisioning module to add a new element, change the configuration of an existing element, or replace an existing network element.
- Element management: A key functionality of the network provisioning module is to test the configured elements for proper working after configuration. The interaction with network elements happens through the element manager for the specific elements being provisioned. Using event and performance data collected by the element managers from the NE and from test execution results communicated to the provisioning module by the element managers, the provisioning module gets information on the success or failure of the provisioning operation to complete a work order.

## 12.3 Resource Data Collection and Processing

Resource data collection and processing consists of collection, processing, and distribution of management information and data records. The collection and distribution of data can be associated with resource, service instance, or an enterprise process. The data collection also includes auditing operations to synchronize static data and collect dynamic management data like the state of the network elements.

The NMS functions that are part of resource data collection and processing are:

- Collection of management information: This involves collection of management information from network elements, EMS solutions, and other service interfaces for processing and distribution to other modules. The collected data can be events, performance records, call records for accounting, configuration records for planning, and so on. Collection of management information can be from distributed systems or from processes in the same system. Different interface standards have been defined for data collection to and from the network management system.
- Processing of management information: The management information needs to be processed before distribution to other modules in the NMS or external applications. An example where formatting would be required after data collection before distributing it to another NMS module would be filtering of event logs based on the user role and aggregation of event logs from different

network elements to generate service level logs. Processing it in terms of formatting the management data would be required before distributing the data to another OSS or EMS application.

- Distribute management information: Distribution is concerned with functions that ensure that management data after processing is delivered to the intended recipient. This would involve sending the data to the concerned module, informing the sending module on the status of the delivery and freeing temporary storage in the sender once the delivery is complete.
- Audit Data: Auditing is the process of querying the network to identify the running or active configuration. The audit can reveal the current state of the network elements and other dynamic data like number of packets dropped, which forms a key input to service quality modules. As can be seen in the example of packets dropped, the audit is not restricted to just configuration. It can be used to get attributes on fault, security, accounting, and performance as well. Discovery is a process closely associated with auditing, though these operations are not the same. Discovery is about identifying a new resource or configuration, while audit is verifying the status of an already discovered resource or configuration. Though in a service provider set up any new resource or configuration is not expected due to flow through provisioning, there could still be scenarios where a discovery is required like inventory records might not be accurate, an operator might change things in the network and not always record those changes properly, discovering specific components the network might be more efficient than the current mechanism used, or inventory records might not be available.

Many NMS product specifications use the term "discovery" instead of "audit." In these scenarios, the process of initial discovery is explicitly expressed using the term "auto discovery."

In addition to understanding the functions performed by the resource data collection and processing (RDCP) module it is also important to have an understanding of the modules that interact with the RDCP module. The main modules that interact with the RDCP module are:

- Service quality management: The service management module dealing with quality, issues requests to the data collection module for performance reports. These reports are used for customer QoS (quality of service) management. This service module also requests customer usage records from the data collection module that can be used for rating and discounting. Based on the usage records, the customer is billed using rating or tariff policies of the specific customer and discounting is applied in the bill wherever SLAs (service level agreements) are not meet or as per the subscription policy of the customer.
- Network planning and development: Collection of management data is required for network planning and development. For example, consider

capacity planning of a resource in the network. Though the actual specifications for the resource might specify a capacity value, this would mostly be a response obtained in ideal conditions and not suitable while configuring the resource for a specific service. The throughput of the resource and generated performance records and events will provide information on the actual capacity of the resource for a specific service. While throughput is the actual output in a specific period of time, capacity is the maximum output possible.

- Network provisioning and inventory: The inventory module and provisioning module monitors the collected management data for changes in configuration. When a change in configuration is identified as part of an audit operation or during discovery, actions are performed to update the inventory and perform provisioning. Discrepancies in data between the inventory provisioning module and the actual value in the network can be fixed using one of the three methods:
  - Reconciliation: In this method the data in the network is always considered to be correct, so the other modules in the NMS are updated to reflect the values in the network. That is, configuration specified in the provisioning module and resource allocation in the inventory module is updated with the values identified from the network. This is the method usually used for synchronization.
  - Reprovisioning: In this method any discrepancy indicates that the network is not working as expected and needs to be reconfigured for optimal utilization. The underlying resources are reprovisioned to have the network values mirror the predefined values identified by the NMS. That is, the inventory module will reallocate the resources based on its original configuration and the provisioning agent will make the changes on the resources to remove any discrepancies in the network. This method of synchronization is usually used in large service provider setups.
  - Discrepancy reporting: Some management applications gives the operator the choice to select whether a reconciliation or reprovisioning is to be performed when a discrepancy is identified. The synchronization application only provides a report rather than taking immediate action. This gives the operator an opportunity to make choices on a case-by-case basis.
- Element management: The data collection module in the NMS usually does not interact with individual NEs, instead it interacts with the element managers for the NEs. Using predefined and standardized interface queries, the NMS module gets management data using the EMS. Both query for a specific data and bulk download are supported in most NMS.
- Other OSS solutions: Data from the NMS can be distributed to multiple applications. The processed data can be sent to another NMS, like a master NMS that provides a converged GUI for a set of networks managed using

different NMS. The distribution can be to OSS (operation support systems) applications like a service quality management module. The data from NMS could even be distrusted to higher layers like BSS (business support systems) like a billing module or SIP (strategy, infrastructure, and product) investment planning module.

## 12.4 Resource Trouble Management

Network trouble management involves handling the network related issues. This could be analyzing a trouble condition, fixing a bug or just sending an action report to the operator. Some of the activities in network trouble management like raising a trouble ticket and tracking the ticket to final closure cannot be performed at the NMS. These kinds of activities are handled in the OSS/BSS applications that in turn interact with network trouble management module.

There can be different ways in which a network trouble gets handled. The below list shows some sample scenarios on network trouble handling:

- Customer raises an issue that needs to be handled in the network: When the customer has an issue with a subscribed service; he/she contacts a customer care number to report the problem. The customer care representative tries to resolve the service problem based on available data. In case the service problem cannot be resolved using the operations of the executive, then a trouble ticket is raised to track and resolve the issue. The ticket is then forwarded to the appropriate team (service operators, network operators, application developers, etc.) for analysis and resolution.
- Service team identifies a trouble condition: When certain parameters like the quality of service deteriorates, the service team isolates the issue and asks the network operations team to resolve the issue or forward the request to the appropriate team in clearing the trouble condition. A ticket may be raised to track the issue.
- Network monitoring/maintenance professional raises a ticket: During the network monitoring operation, the network professional might encounter a trouble log that could require manual intervention from the site where the equipment is deployed to fix. The monitoring professional can raise a trouble ticket and assign the same to the concerned network operator for resolution of trouble. Alternatively, the network monitoring professional could issues commands to the network elements or take necessary action to resolve the issue when the trouble condition is easy to resolve and does not require participation from other teams.
- NMS takes necessary action for a trouble condition: It is possible to program the NMS to take a predefined action when a specific event log is received generated for a network trouble condition. The action taken by the NMS

could be to fix the problem by issuing necessary commands to the network elements or to generate a trouble ticket and forward it to the concerned network operator.
- Network element itself handling a network trouble: The network element can also be programmed to take actions for trouble conditions. For example, when an application on the network element is upgraded using upgrade software running on the NE and due to some reason the upgrade fails. In this scenario the upgrade software on the NE can be programmed to fall back on the application being upgraded to its earlier release.

The NMS functions that are discussed in this section as part of resource trouble management are:

- Report resource trouble: It is not necessarily that the customer will be the entity reporting a trouble condition. In most scenarios the customer might only have a service problem and not an actual resource trouble. An NMS that is used by the operator for monitoring resource trouble serves as the best report engine. The operator can use the event logs generated for a network trouble condition to take corrective action or put predefined rules to generate trouble reports whenever a specific trouble trigger a log that gets collected by NMS.
- Localize resource trouble: Network trouble reported by customer or identified by upper business/service layer needs to be localized. Localization involves identifying the specific element in the network that caused the trouble condition. Only after the trouble condition is localized to the lowest level possible, will it be possible to effectively resolve the issue. Localization is not just limited to identifying the network element that caused the trouble condition. For example, consider the scenario in which the customer reports degradation in quality of service and raises a ticket for the same. Localization starts with identifying the resources allocated from the inventory management system to offer the service. The resources offering the service can then be analyzed to see utilization and KPI (key performance indicator) values. This would lead to localization of trouble to the specific network element. Next the logs in the network element can be monitored or historical logs analyzed to identify the application in the network element causing the deterioration of service. Next the trouble ticket is assigned to an appropriate team or an operator to resolve the issue. The intent of a localized operation is to reduce the MTTR (mean time to repair) value by ensuring that a trouble condition is looked into by the correct team that can resolve the specific trouble as early as possible.
- Analyze resource trouble: Analysis of resource trouble is required first to localize a resource condition and then to resolve it. The NMS is equipped with multiple tools to help an operator in debugging a trouble condition. Most NMS products offer these tools as test suites to automate testing of the

product. For example, the test suite of AdventNet Web NMS comprises of simulation tool kits and utilities to perform functionality testing, scalability testing, performance testing, application monitoring, and client latency verification.

- Resolve resource trouble: Some trouble conditions will require change in code of an application running on the NE for resolution while most of the trouble conditions can be fixed by performing some operation on the NE. The NMS has interfaces to issue commands on the network element, perform a backup/restore, switch between active and passive elements for high availability (HA), and so on that could resolve or handle a network trouble condition.

Some of the modules that interact with resource trouble management module are:

- QoS management: Any SLA violations detected by a customer QoS module and service quality violations identified by a service QoS module that are associated with network trouble has to be reported, localized, analyzed, and resolved by a resource trouble management module.
- Service problem management: A service problem that needs a fix at network level needs to be notified to the network trouble management module. If on analysis, the network trouble module finds that the service problem will impact the customer, then a ticket needs to be raised by the network trouble management module to notify the customer and track the ticket to closure.
- Service/resource configuration: Fixing a network trouble might involve reconfiguring or altering current configuration of the elements in the network or it could also involve changing service configuration. To alter the resource or service configuration, an interaction is involved between the configuration module and trouble management module. The trouble management module issues the request to alter configuration to fix the trouble and the configuration module responds back on the status (success/failure) of the requested operation.
- Customer interface management: The customer interface management module can be a separate application, or performed as part of customer care/support processes. These modules directly interact with customers and translate customer requests and inquiries into manageable business events like creation of a work order, raising a trouble ticket, or generating an address change for sending bills. After raising a trouble ticket for a resource trouble, the issue needs to be communicated to the trouble management module. The trouble management module resolves the issue and provides information to a customer interface module for confirming the fix and closing the ticket.

## 12.5 Resource Performance Management

The performance management module works on data provided by the network data collection and distribution module. The performance data collected is used to create reports and is monitored to identify potential issues. If analysis of the performance records shows a violation of service agreement or degradation in quality due to network issues, then a notification is issued to the trouble management module. A performance management module needs to continuously monitor KPI (key performance indicators) and track a performance issue to closure, thus ensuring that the performance is maintained at a level required to support the services.

Performance data is usually represented in graphical plots on the NMS as the data is huge. Since performance analysis would require comparison with historical data, the graphical plots help to get results and make intelligent decisions faster. In most NMS solutions it is also possible to specifically monitor a particular performance parameter by setting thresholds. The process of "thresholding" involves identifying appropriate performance thresholds (values) for each important variable (KPI), so that exceeding these thresholds indicates a network problem worthy of attention. When the threshold is exceeded, an event (log) is automatically generated to notify the operator, helping the operator to keep track of multiple performance variables in the network at the same time. Rather than generating a log, the NMS can also take corrective actions when a threshold is exceeded.

Let us consider an example of the thresholding process where the variable to be monitored is the utilization for trunk-X. When the utilization of trunk-X is 50%, a log is generated with severity "minor," for 70% a log with severity "major" is generated, and for 80% utilization a log with severity "critical" is generated. Traffic routing algorithms are then applied to ensure that the utilization of trunk-X does not exceed 85% to achieve maximum efficiency of operation.

The NMS functions that are part of network performance management are:

- Monitor performance: After performance data collection, the key performance parameters can be monitored by an operator using the NMS. Graphical plots, aggregated records, and summary reports help in monitoring a large bulk of data. In most cases, data is collected off-line as a bulk download at regular intervals rather than polling on a regular basis. Streaming of performance data as events using protocols like Net-flow or IP flow information export (IPFIX) is also popular. Setting thresholds on specific performance variables as discussed earlier is also an effective method for performance monitoring.
- Analyze performance: Analysis starts with formatting the performance records collected from the network elements to get useful data specific to the performance applications in the NMS. All information in the original performance records generated from the NE may not be of use in the NMS. The NMS application will be using specific fields in these records in its reports

and processes. The main functionality of analysis is to identify trends in performance for planning, find degradation in quality to take corrective action, and generate reports of interest.

- Control network performance: The network performance control is also a part of performance management. Control mainly involves taking corrective action for issues identified during analysis and monitoring. The corrective action can be triggering action commands performed by a performance process on the NE or communication with other module like reconfiguration of the NE request to the resource configuration module.
- Report performance data: Customized performance reports as generated at the NMS. Customization can be at the user level or for user groups. For example, consider two user groups comprised of a service executive group and network operator group. The service executive would be interested in a graph showing the quality of service, while the network operator offering a resource for the same service would be interested in the throughput and utilization of resources. User specific customization is also required as different operators would have their own choice on the order and parameters in a performance report of the same data.

The modules that interact with network performance management module are:

- Network trouble management: If the performance issue identified by the network performance management module requires fixing in the network resources, then the same is performed by interacting with network trouble management module. Another scenario where the performance module interacts with the trouble module is when a ticket needs to be raised for tracking or for informing the customer about an event related to performance.
- Network data collection and distribution: Performance management module relies on the data collection and distribution module for its operation. The performance data first should be collected from the network elements by the data collection module before it can be analyzed by the performance management module. The reports generated by the performance modules using the collected data would be distributed to external OSS/BSS applications using the distribution module.
- QoS management: Performance management has a direct impact on the quality of service. While QoS is a term used in the service level, performance is mostly used when discussing the network layer. Degradation of performance from a resource would in most cases impact the service that the resource fulfills. The QoS itself depends on a set of network level KPIs that can be monitored by the performance management (PM) module. So the QoS module interacts and collects valuable data from the PM module. It can also request the PM module to collect records on specific performance parameters from the NE or create customized reports that the QoS management module can work on.

- Network configuration management: One of the functions of the performance management module is to control performance parameters. This might require configuration changes in the network resources including reconfiguring of the elements. Performance variables can also be collected by configuration module as attribute values during an audit or discovery operation. Some NMS solutions display resource attributes when they are discovered and represented in GUI. These attributes are updated with current status during an audit operation or from asynchronous event notifications as logs. Some of these attributes can indicate performance and hence these performance variables do not need a bulk report download to get the current status.
- Network planning and development: The performance degradation reports and violations are logged for future reference in network planning and development. Recording the results of performance analysis when working with an issue is another input to the planning module, in identifying the potential issues that might come up and ways to avoid the same in the future. Changes made to configuration in order to stabilize the performance of resource and keep it within predefined limits can be used to create future control and configuration plans that will result in a more efficient and predictable operation. The generated performance reports serve as valuable inputs to the network operator in optimizing the running configuration of the network and to plan for future network provisioning for supporting the same service on a different network or a new service on the same network.

## 12.6 Resource Management and Operation Support and Readiness

While the real-time management of the network is the main activity of the NMS, there are also functions of network management that works on available data to plan for a smooth operation and enhancement of the network. These functions make up the NMS support and readiness. A few examples of the planning functions in support and readiness of network management and operation are listed below:

1. Forecasting of resource requirements and outages: This would involve generating warning alarms on trouble conditions that might occur based on active configuration. Hence suitable corrections can be performed to avoid the problem scenario. Another scenario would be informing customers of a planned maintenance activity on the network or informing the operator using logs or message windows on what could be a possible way to better handle a particular resource requirement.
2. Version controlling: The NMS solution and the network it manages will have multiple software applications running in them. Maintaining the version

information of these applications is essential to modify or upgrade the applications. Version controlling is an integral part of the support activity associated with the network and its management solution.

3. Utilization and capacity planning: Underutilized elements are reallocated or reconfigured by interacting with the inventory management, provisioning, and configuration management applications. It is also possible that the planning applications collects information that shows nonsuitability of a particular element and need for an upgraded version of software or new hardware to offer service as per the SLA. In such scenarios the planning application will send notifications to the intended audience on the capacity actions to be performed.

## 12.7 Conclusion

There are multiple modules in an NMS application. Some vendors supply individual modules for managing specific network management functionality while other vendors provide functions as a fully functional NMS bundle. From a holistic telecom perspective, the network is just a resource or set of resources that operate to offer a telecom service. Hence resource management in the eTOM process model includes network management. This chapter uses the resource management functionalities given in eTOM to discuss functional modules that make up a complete NMS package. Only a few of the NMS functions are discussed with the intension of giving the reader a clear picture of what NMS is and how it is different from an element management solution that can manage a set of elements. Process-to-process interactions on how the different NMS modules interact is also discussed to show how the NMS modules can be implemented as independent components in line with SOA (service oriented architecture) and integration of open source components that will be discussed later in this book under the development of next generation network management solution.

## Additional Reading

1. Lakshmi G. Raman. *Fundamentals of Telecommunications Network Management.* New York: Wiley-IEEE Press, 1999.
2. Kundan Misra. *OSS for Telecom Networks: An Introduction to Network Management.* London: Springer, 2004.
3. Alexander Clemm. *Network Management Fundamentals.* Indianapolis, IN: Cisco Press, 2006.

# Chapter 13

# NMS Products

This chapter provides the reader with a basic understanding on some of the most popular network management products. This chapter is intended to give the reader practical exposure on actual network management products rather than the theoretical aspects of network management. This chapter presents the author's understanding of a few network management products and the content is not reviewed or endorsed by the product owners. Hence the author or publisher takes no position on the accuracy or completeness of data presented in this chapter.

## 13.1 Introduction

There are many NMS products on the telecom market. These products can be generally classified under five main types. They are:

1. NMS framework for customization: These are products that provide generic fault, performance, and other functions under the FCAPS model. It provides a generic framework on which specialized applications can be created as customization for a specific requirement of the client. An example product that falls in this category would be OpenNMS. Being a generic framework, products in this category usually are developed by integrating open source products. Hence the product sells for free and clients are charged for services offered on the product like bug fixing, maintenance, and customization of the product. In addition to creating new applications on the framework, customization could also be simple look and feel changes (GUI), re-branding, license administration, or integrating multiple management solutions with the framework to offer a common GUI.

2. NMS package with add-ons: Products in this category are complete NMS suites with almost all the popular NMS functions. The product itself is charged and the add-ons are charged a premium. Add-ons provide modules with functions that would not be available in the usual NMS products or given as a specialized function product by other vendors. Example of a product that falls in this category is IBM Netcool. The add-on could be a module with a set of standard adapters or a specialized function for managing IP networks.
3. Function specialized NMS: Products in this category are developed to perform a specific NMS functionality. For example, "Granite Inventory" a product of Telcordia is intended to perform inventory management and not the general FCAPS functions. Though there is competition between vendors working on NMS framework and packages, these products don't directly compete with products developed to perform specialized functions. For inventory management, the competition for Telcordia is inventory management products from companies like Cramer, MetaSolv, NetCracker, and so on and not products like IBM Tivoli, AdventNet, or OpenNMS.
4. Domain specialized NMS: Products in this category are for managing networks in a specific domain. For example, "netspan" a product of Airspan is for performing FCAPS functionality on the WiMAX network and not a generic product for multiple domains. OEM (original equipment manufacturers) who want to provide/add a specialized WiMAX NMS as part of their WiMAX product suite for service providers, will opt to buy netspan rather than go with a generic NMS product.
5. Hybrid type: Several combinations of the four NMS product types already discussed is also possible. This type has been termed Hybrid. For example, IBM Tivoli is an NMS framework that is developed without using open source products and hence the product is not free. Another example of the Hybrid again from IBM is Tivoli Netcool IP Multimedia Subsystem (IMS) Manager. Though the name IMS manager shows it is a domain specialized NMS, this product is build over a framework/package product and can also be interpreted as an add-on to the product rather than a separate NMS product.

It should be kept in mind that the NMS product types discussed above are merely to understand the concept and there is no such classification in telecom space for NMS products.

## 13.2 OpenNMS

OpenNMS is a network management solution developed in the open-source model. So the application and its source can be downloaded for free and distributed under terms of the GPL/LGPL license. OpenNMS is a functionality rich NMS and can be used for managing both enterprise and telecom networks. Its features like

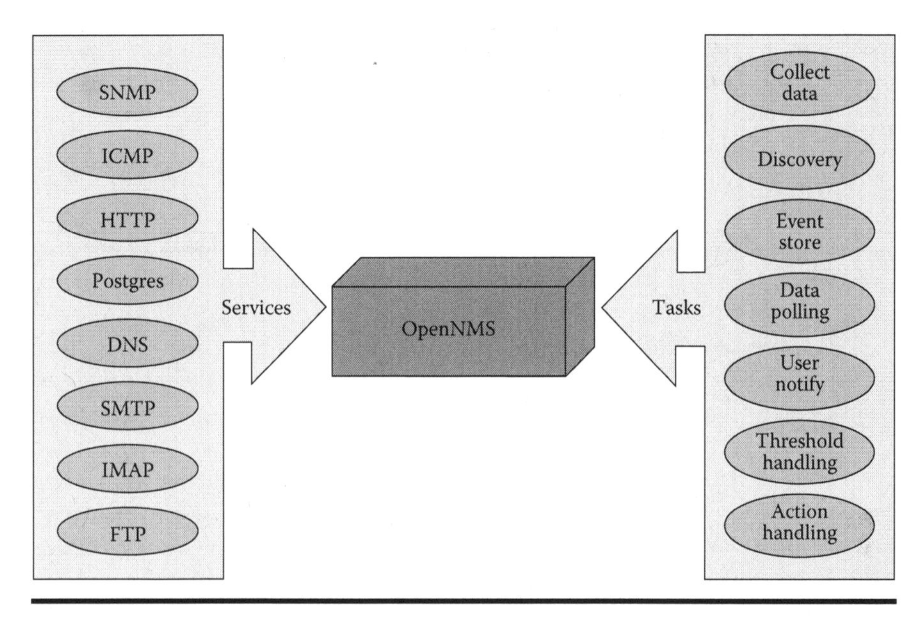


**Figure 13.1 OpenNMS tasks and services.**

automated network discovery, configuration using xml, and support for multiple operating systems, make it one of the most popular NMS products that can be compared with enterprise grade NMS products from leading network equipment manufacturers.

After discovery of elements in the network, OpenNMS discovers the services running on the discovered elements. The status of the services are polled and updated by OpenNMS. Some of the services that can be identified by OpenNMS (see Figure 13.1) are Citrix Service, DHCP Service, DNS Service, FTP Service, HTTP Service, ICMP Service, IMAP Service, LDAP Service, POP3 Service, SMTP Service, SSH Service, and Telnet Service.

OpenNMS runs as a set of daemon processes, each handling a specific task. Each of these daemon processes can spawn multiple threads that work on an event-driven model. This means that the threads associated with the daemon process listens for events that concern them as well as submit events to a virtual bus. Some of these processes are discussed next.

1. Discovery (daemon process: OpenNMS.Discovery: This process uses ICMP (Internet control message protocol) packets to find (discover) devices in the network. The range of addresses to be scanned for response is specified in file discovery-configuration.xml. It is also possible to trigger discovery of a network element from the web-based OpenNMS GUI by specifying the IP address of node to be discovered. A discovery.log file records the discovery process.

2. Polling (daemon process: OpenNMS.Pollerd: This process is used to find the status of a discovered service. An attempt is made to establish connection with the service, and failure to connect leads to retry for connection. The retry is only performed a predefined number of times and a timeout interval determines how much time to wait for a response from the service. Configuration of the poller can be performed by making changes in poller-configuration.xml file. The poller also has special models to handle polling when a service goes down, like a higher polling frequency initially expecting the service to come up and then a lower frequency in polling.
3. Notification (daemon process: OpenNMS.Notifd: As the name suggests, the purpose of this process is to notify the user about an event. There are multiple methods available in OpenNMS to send notifications. E-mail and paging are the most common methods of issuing notifications to a physical user. Mechanisms that employ XMPP and SNMP traps are available for sending notification to another application and HTTP GET/POST can be used to send notifications to a Web site. The notification is usually a text message with details on the event that triggered the notification, along with information on the node that caused the event. When the notification method employed is e-mail, then a subject line is also introduced. Notification can be issued to users (users.xml), groups (groups.xml), or configured roles. Additional capabilities include the capability to add escalations when a notification is not acknowledged and to introduce a delay before sending notification. Notifications have a unique id.
4. Data collection (daemon process: OpenNMS.Collectd: SNMP was the only data collection method supported previously by OpenNMS to collect data from the network elements from southbound interface. A majority of the data processing functions including certain performance reports still rely on SNMP MIB to data from devices being managed. Current versions of OpenNMS also have support for data collection using HTTP and JMX (Java Management Extensions). The HTTP data collector can collect metrics for generating a performance report that gives the application ability to collect data from elements without having to implement the SNMP agent. A dedicated JMX agent should be running on the device to be monitored, for OpenNMS to collect data using JMX. The main configuration file for data collection is collectd-configuration.xml.

There are many more daemon processes like OpenNMS.Eventd that is concerned with writing and receiving all the event information, OpenNMS.Trapd to handle asynchronous trap information, and OpenNMS.Threshd process to manage threshold settings. These processes use numerous open source products like joesnmp for SNMP data collection, log4j for event generation, postgresql as DBMS interface, and (JRobin) RRDTool to store graph data, just to name a few.

OpenNMS got a Gold Award for "Product Excellence in 2007" in the Network Management Platform category, against competing products like OpenView of HP

(Hewlett-Packard) and Tivoli of IBM. This is a major achievement for an open source product when competing with some of the best NMS products available in the telecom industry. OpenNMS application and source can be downloaded from sourceforge.net and the product Web site is www.opennms.org.

## 13.3 AdventNet Web NMS

AdventNet is a framework for building customized network and element management solutions. It provides a variety of fault, configuration, performance, and security functionalities as part of network monitoring and offers inventory management and provisioning services. Being a framework, it supports multiple adapters for interfacing with northbound GUI applications, southbound network elements, and relational databases. Protocols supported for interaction with network elements include TL1, SNMP, CORBA, XML-based messaging, and JMS. In addition to Web servers like Java Web server and Apache Tomcat, Web-based publishing to northbound interface is facilitated using application servers like JBoss and Weblogic. From a database perspective, AdventNet provides data integration with Oracle, Sybase, Mysql, MS Sql, Solid, Timesten, and many other popular databases.

Multiple service providers have adopted AdventNet for unified management of its networks. This would involve AdventNet collecting data from elements of multiple networks and representing it in a common GUI. Centralized configuration of devices and quick launch of element managers is also possible. Distributed mediation servers is a remote management feature of AdventNet in which multiple mediation servers collect data from elements in the network and forward the data to a centralized server.

Some of the functionalities in AdventNet Web NMS are:

- Fault management: The usual capabilities provided in NMS for alarm monitoring, like filtering a particular alarm, suppression of alarms that are not of interest for user and correlating alarms is performed. Segregation of alarms based on severity and the element where the alarm originated can be performed from AdventNet Web GUI. Clearing and acknowledging an alarm is also supported.
- Configuration management: Network elements are automatically discovered and their topology is created. An audit is performed on the discovered elements and their attributes, to get status updates that are not propagated with asynchronous notifications from the devices. Inventory management capability is also in AdventNet with information on all devices in the network.
- Performance management: The performance data collected from the network are aggregated and analyzed to generated graphical reports in AdventNet Web GUI. The functionality of filtering helps to customize performance reports. Thresholding is also supported and AdventNet gives the user, graphical

interfaces to set thresholds on key performance parameters for ease of monitoring and to generate event notifications of trouble scenarios before the actual trouble condition occurs.

- Security management: AdventNet Web NMS has extensive security features that include both functional and architectural features. The functional features include an initial authentication followed by an authorization window that gives authorization to users to work on the NMS based on their login ID, role, or group. Each user who wants to use AdventNet is recognized using a unique login identity (ID). The user can be assigned to a role like "administrator" as there will be a specific set of functionalities permitted for the role. Users can also be grouped under a group name and functionalities permitted for the group can then be defined. Group/role needs to be created first before allocating users to a specific role/group. The architectural feature includes support for Java security model, with use of centralized RMI server for access control and use of SSL for communication.

Compliancy to standards is another key capability of AdventNet that helps in making this solution suitable for multitechnology and multivendor environment. AdventNet is compliant to TMF MTNM standard and implements the NML-EML interface in conformance to standards TMF513/608/814.

AdventNet Web NMS has a layered (tier) architecture (see Figure 13.2) where the functionalities (components that perform specific functionality) are grouped under five layers. They are:

1. Mediation layer: This layer has components that provide southbound interface capabilities. Multiple southbound protocols are supported for dynamic deployment. TCP, HTTP, RMI, and CORBA are some of the transports supported with Java API and XML interface for application development.

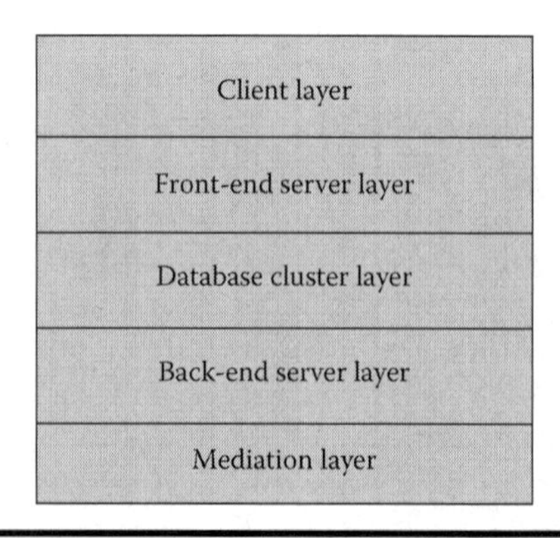


**Figure 13.2 Layered architecture of AdventNet Web NMS.**

2. Back-end server layer: The back-end server performs FCAPS management functions. This is the layer that incorporates the core management logic that works on the data from the mediation layer. There is a primary back-end server and a secondary back-end server in this layer. In case of any issues or failure in the primary server, control is transferred to the secondary server, which acts as a backup for primary server.
3. Database cluster layer: Multiple database servers that support RDMS reside in this layer. Having a separate server for the database leads to high availability. Management data integrity, security, and database synchronization are all handled in this layer.
4. Front-end server layer: The role of the front-end server is to create client views for clients using Web NMS. That is when the client selects an option, the specified operation needs to be performed, and the resulting output needs to be converted to a format that can be accessed by the client. Multiple front-end servers are used so that more client requests can be handled.
5. Client layer: AdventNet supports both Java client and HTML Web access client. The Java client can provide a consistent user experience and supports TCP, RMI, and HTTP transports while the Web access client supports HTTP and HTTPS.

Some of the project tools offered by AdventNet for Web NMS are:

- Web NMS License Administration Tool: To generate license for applications built over Web NMS. License can be implemented for function level granularity (i.e., license for using specific functions in the NMS), and the license can be for a trial or registered version.
- Web NMS Rebranding Tool: OEMs using AdventNet as a framework may opt to rebrand the final application built on AdventNet framework for which this tool can be used.
- Web NMS Test Suite: This test suite provides the test environment and facilitates automated testing of AdventNet Web NMS based applications.
- Web NMS TMF EMS Pack: Interface module that implements TMF 513/814 northbound CORBA support for interaction with EMS.

## 13.4 HP OpenView

OpenView was a product of Hewlett Packard. It can be used for network, system, data, application, and service management. The HP OpenView can be described as a suite of software applications with multiple optional modules/smart plug-ins and third-party applications. HP OpenView was rebranded in 2007. Its open architecture and multiplatform support helps in out-of-the-box integration and rapid deployment that leads to a quick return on investment (ROI).

Some of the products in HP OpenView Suite are:

- OpenView network node manager: It is primarily a network monitoring product. The network node manager (NNM) collects management data and event notifications from servers, storage devices, and clients and displays data on the OpenView NNM console. This centralized management console leads to better management of devices and lesser time to resolve a network issue. NNM can perform an automatic discovery of devices running Insight management agents. The product uses SNMP as the primary network management protocol and has an event browser to display SNMP events that are collected and analyzed by the NNM.
- OpenView operations (OVO): This is an application monitoring product. It is built using NNM as its foundation, though NNM is used for network monitoring. SNMP is the network management protocol used in OVO and it requires a management agent running on the managed nodes for monitoring the nodes. In addition to displaying SNMP trap as alarms, OVO also performs alarm correlations. Users can log onto OVO using a Motif GUI on UNIX host, or a Java based GUI for UNIX or Microsoft Windows host. OVO also supports role-based security and customization features. It supports operation on Unix and Microsoft Windows operating systems.
- OpenView smart plug-ins: The SPIs are optional modules that extend the capabilities of management applications. Capabilities like correlation intelligence can be added to an existing event manager by simply installing the smart plug-in. These SPIs can be used to integrate with business applications running on multiple infrastructure and platforms. Functions like configuration of application by altering policies and threshold alarm setting for configuration changes can be done with SPIs. In most scenarios, OpenView SPIs are intended to expand the capabilities of OpenView Operations and OpenView Network Node Manager to other management areas or third-party products. HP OpenView SPIs are available for IBM DB2, IBM Websphere application server, Oracle application server, Citrix, Microsoft Exchange, Unix OS, OpenVMS, PeopleSoft, SAP, and so on.
- OpenView Performance Insight (OVPI): This OpenView product monitors network performance. Its capability in identifying trouble conditions makes it effective for meeting QoS demands. In addition to displaying performance data in intuitive graphical formats, the Performance Insight also offers generation of customized performance reports and tracking of traffic patterns. It can be integrated with OpenView Network Node Manager and OpenView Service Information Portal. Other performance management products of OpenView include OpenView Performance Agent (OVPA), OpenView Performance Manager (OVPM), OpenView Reporter (OVR), and OpenView GlancePlus
- OpenView Internet Services (OVIS). OVIS is used for monitoring and maintenance of Internet services. Monitoring involves isolating trouble scenarios

and generating reports on the service, while maintenance operations are mainly for installation and troubleshooting issues. Availability of service, response time and compliance to service level agreements can be tracked using this product. OVIS Probes are available for operation on Windows, HP-UX, Sun Solaris, and Red Hat Linux Operating Systems.

HP OpenView supports ITIL (Information Technology Infrastructure Library) standards. Let us discuss some of the OpenView product suite functions in more detail next.

1. Fault management: Events generated by multiple networks, systems, and applications are presented in a single Java-based console. The fault console gives customized service event reports and presents data in a user-friendly format after parsing the fault data collected. OpenView fault management permits configuring of templates and event management in a secure environment. This would map to the ITIL functions like incident management and problem management. The products in the OpenView suite that offer this function include OpenView Network Node Manager, OpenView Operations, OpenView TeMIP, and OpenView Smart Plug-ins.
2. Service management: Management of service in the information technology industry would involve configuring the services and its maintenance, to ensure performance that meets guaranteed service levels and continuous availability of the services. This would lead to aligning the service operation to achieving specific business objectives. HP OpenView products on service management optimize service delivery. Configuration management is an ITIL function that falls under service management. Service management products from OpenView include OpenView Service Reporter, OpenView Service navigator, OpenView Service Desk, and OpenView Operations.
3. Performance management: In an information technology infrastructure there will be services, applications, networks, and storage devices to be managed. Performance management of this infrastructure should ensure a peak performance at all times for all managed entities. OpenView functions monitor performance to identify and isolate issues before they arise. A simple example would be to check the utilization level of the systems to ensure that they are not overloaded at all times. Trend analysis using historical performance records can also done using OpenView products. Performance management maps to change management, availability management, and capacity management functions of ITIL. The OpenView Products that provide performance management functionality include OpenView Internet Services, OpenView Performance Insight, OpenView Performance Manager, OpenView Problem Diagnosis, OpenView Storage Optimizer, and Open View Transaction Analyzer.
4. Accounting management: Based on service and resource consumption, the allocation of cost for maintenance and improvement needs to be allocated.

So usage is critical information that governs optimal return over IT investment and best business results. OpenView tools for accounting management provide usage information by consolidating fault, performance, and capacity data. The accounting management functionality offered by OpenView products map to financial management and capacity management of ITIL. The OpenView products that provide accounting management functionality include OpenView Dynamic Netvalue Analyzer, OpenView Enterprise Usage Manager, and OpenView Storage Area Manager.

5. Provisioning: HP OpenView products provide end-to-end provisioning of the entire IT infrastructure. This includes provisioning of service, network application, and storage devices. The provisioning is performed to align with business objectives and goals. The HP OpenView products make it easy to adapt infrastructure to change in business requirements by editing or reconfiguring the entities using simple business rules. Provisioning function in OpenView maps to Configuration management of ITIL. Service provisioning products from OpenView include OpenView Service Activator and OpenView TeMIP.

Some of the OpenView Configuration Management (OVCM) products are OVCM Application Manager, OVCM Inventory Manager, OVCM OS Manager, and OVCM Patch Manager. Similarly the OpenView Storage (OVS) management products include OVS Area Manager, OVS Data Protector, OVS Mirroring, and OVS Mirroring Exchange Failover Utility. In addition to applications on the network, service, and business layers, OpenView even has customer centric products like OpenView Web Transaction Observer and OpenView Internet Services that can provide to the customer, personalized views of the services enjoyed by the specific customer.

## 13.5 IBM Tivoli

IBM Tivoli consists of a management framework and a set of enterprise products that use the framework as the base infrastructure. Tivoli Management Framework works on multiple operating systems like Windows and different flavors of UNIX. The framework that can manage distributed networks can be customized for specific applications using the Tivoli enterprise products installed on top of the management framework. The Tivoli desktop that is part of the framework, supports both graphical user interface (GUI) and a command line interface (CLI).

In addition to management services and administration capability, Tivoli management framework also offers:

- Application development environment: This development environment is intended for programmers to develop customized applications using the services offered by the framework. These applications form a layer of management functions over the framework.

- Event adaptation facility: The events generated from the managed nodes and events that can be collected from other network managers have a different format from what is expected by Tivoli framework and its enterprise solutions. The event adaptation facility provides the capability to convert the events from external sources into a format compactable with Tivoli.
- Application extension facility: This capability allows users to customize Tivoli products based on requirements at a specific site.

Some of the functions offered by Tivoli Management products are:

- Manage accounts: Tivoli product user and group accounts can be created, modified, or deleted. A field in the user account can be edited and the changes applied to a target system or multiple systems in the network.
- Software maintenance: This involves installing new software or an upgrade of existing software on elements in the network. It is possible to get a list of installed software with the details on software version and upgrade systems with old releases.
- Third-party application management: Applications like "Lotus Notes" can be launched with Tivoli tools along with management data. Monitors and task libraries help to maintain these applications in Tivoli environment.
- Trouble management: While monitoring the network, if Tivoli identifies a system crash then commands to take corrective actions can be performed or a page message can be issued to the operator.
- Inventory management: The assets of an enterprise can be tracked and managed using Tivoli products. It is possible to perform operations, administration, and maintenance on these assets. Centralized control of enterprise assets is made possible with Tivoli products.
- Storage management: Managing the storage devices and the data in it is an integral part of enterprise management. In addition to products for service, network, and application management, Tivoli also has products for storage management.

IBM Tivoli family of products has specialized products for all aspects of management. In storage management alone IBM Tivoli has multiple products like Tivoli Storage Manager, Tivoli Storage Manager FastBack, Tivoli Storage Manager FastBack Center, Tivoli Storage Manager HSM for Windows, Tivoli Storage Optimizer z/OS, and Tivoli Storage Process Manager. The capabilities of some of the network management products in the IBM Tivoli family are discussed next.

1. Tivoli Monitoring. This provides an end-to-end view of the entire distributed IT infrastructure. It monitors performance and availability of the distributed operating system and its applications. It has a user-friendly browser with a customizable workspace. The Monitoring Agents collect data from the managed

node and sends it to the monitoring server. Tivoli Monitoring also has the capability to detect problem scenarios and take corrective action. IBM Eclipse helps the server by providing help pages for the monitoring agents that aid the users of Tivoli Monitor.

2. Tivoli Netcool/OMNIbus: This is used by many service providers as an integration platform acting as a manager-of-managers offering a single console for data from network managers like HP OpenView, NetIQ, and CA Unicenter TNG. It supports operation on Sun Solaris, Windows, IBM AIX, HP-UX, and multiple flavors of Linux platforms. Event handling is the main management function in Netcool/OMNIbus. Its interaction with other applications that facilitates smooth integration is made possible using probes and gateways.

Probes connect to an event source and collect data from it for processing. The different types of Netcool/OMNIbus probes are:

- Device probes: These probes connect to a device like router or switch and collect management data.
- Log file probes: The device to be monitored or target system writes event data to a log file. This log file on the target system is read by Netcool/OMNIbus probe to collect data.
- Database probes: A table in the database is used as the source to provide input. Whenever an event occurs this source table gets updated.
- API probes: These probes acquire data using APIs in the application from which data is collected. The application here is another telecom data management application for which Netcool/OMNIbus is acting as a manger-of-manager.
- CORBA probes: These probes use CORBA interfaces to collect data from a target system. The interfaces for communication are published by vendors in IDL files.
- Miscellaneous probes: In addition to the generic common probes discussed above, there are probes for specific application or to collect data using methods other than the ones discussed. These are called miscellaneous probes and an e-mail probe to collect data from a mail server is an example of a miscellaneous probe.

While probes are used to collect data from a target, the gateway is used when Netcool/OMNIbus itself acts as the source for another application. For example a CRM (customer relationship management) application uses Helpdesk gateway to get alerts from this Tivoli product.

1. Tivoli provisioning manager: Using this Tivoli product, provisioning can be completely automated using work flows. Work flow is a step-by-step definition of the desired provisioning process. This work flow is much simpler compared to scripts in that they can be created by the drag and drop operation

of provisioning blocks using the development kit provided with Tivoli provisioning manager. Multiple elements in the network can be provisioned with a single work flow. The automation work flows can be created across domains. It follows service oriented architecture for more re-use of functionality. It is more of an IT infrastructure provisioning application rather than just a system provisioning manager. It supports operation on AIX, Windows, Linux, and Sun Solaris. The graphical user interface can be customized to generate role based views and a provisioning work flow can be re-used just like a template for future provisioning operations.

2. Tivoli configuration manager: This product handles secure inventory management and software distribution. Software distribution mainly involves identifying missing patches on the client machine and installation of patches. The product even helps in building a patch deployment plan and can do package installation in distributed environment. The inventory module in the product identifies the hardware and software configuration of devices in the IT infrastructure, with functions to change system configuration. The secure environment is implemented using multilevel firewalls.

## 13.6 Telecom Network Management Products

The current network management solutions are expected to be capable of handling multiple domains like enterprise networks, telecom networks, storage networks, and so on. Most enterprise management solutions already discussed in this chapter handle multiple domains and can be used for application and service management in addition to network management. For example IBM Tivoli Network Manager IP Edition from the Tivoli family can be used for layer 2 and 3 data management associated with telecom networks. However there are products for telecom network and element management alone and specialized for specific telecom network and devices. Most of the network equipment manufacturers are the major developers of telecom specific network management products. Cisco's Unified Communications Network Management suite (the suite consists of Cisco Unified Provisioning Manager, Cisco Unified Operations Manager, Cisco Unified Service Monitor and Cisco Unified Service Statistics Manager), Alcatel-Lucent's 1300 Convergent Network Management Center (CMC) for integrated network management, and Nortel's Integrated Element Management System (IEMS) are a few examples of products that fall in this category.

## 13.7 Conclusion

There are many NMS products and frameworks available in the market. The official guide published by the vendors for some of these products itself span 300 to

400 pages. A complete explanation of the features and architecture of any specific NMS product was excluded for this reason and a general overview of the most popular frameworks/products was done in this chapter. It can be seen from the Additional Reading section next that some of the products discussed in this chapter have books written for the sole purpose of understanding a specific capability of that product alone. This chapter gives the reader familiarity with products from the leading telecom management players.

## Additional Reading

1. OpenNMS Documentation www.opennms.org/index.php/Docu-overview and AdventNet Web NMS Technical Documentation www.adventnet.com/products/web-nms/documentation.html
2. Jill Huntington-Lee, Kornel Terplan, and Jeffrey A. Gibson. *HP's OpenView: A Practical Guide*. New York: McGraw-Hill Companies, 1997.
3. Yoichiro Ishii and Hiroshi Kashima. *All About Tivoli Management Agents*. Austin, TX: IBM Redbooks, 1999.
4. IBM Redbooks. *An Introduction to Tivoli Enterprise*. Austin, TX, 1999.
5. Tammy Zitello, Deborah Williams, and Paul Weber. *HP OpenView System Administration Handbook: Network Node Manager, Customer Views, Service Information Portal, OpenView Operations*. Upper Saddle River, NJ, Prentice Hall PTR, 2004.

# Chapter 14

# SNMP

Chapter Six in this book had a section on SNMP that gives a one page overview of this protocol. SNMP is a widely used protocol and even the latest developed NMS solutions provide support for SNMP considering that most of the existing network elements support only management using SNMP. This chapter is intended to give the reader a more detailed coverage on SNMP, including its different versions that show how this management protocol evolved over various versions. SNMP MIB has been excluded from this chapter.

## 14.1 Introduction

The SNMP is one of the most popular management protocols used for managing the Internet. IETF initially suggested Simple Gateway Monitoring Protocol (SGMP) for Internet standardization, but SGMP was later replaced with Simple Network Management Protocol (SNMP). While SGMP was mainly defined for managing internet routers, SNMP was defined to manage a wide variety of network elements. It should be understood that SNMP is not just an enhancement to SGMP, the syntax and semantics of SNMP are different from those of SGMP.

The components in the network element to be managed are referred to as managed objects that are defined in the SNMP management information base (MIB). In order to manage the object with SNMP it must follow a certain set of rules as mentioned in the structure of management information (SMI). The object is defined using predefined syntax and semantics. Each object can have more than one instance and the instance can have a set of values.

SNMP has three versions. They are SNMPv1, SNMPv2, and SNMPv3. This chapter will discuss each of these versions in detail. Please refer to the requests for comments (RFCs) mentioned in the Additional Reading section for more information. SNMP mainly consists of two entities: the management station and the SNMP agent that communicate with each other. The management station is the element or network manager that collects data about the network element while the agent runs on the network element and sends management information about the element to the management station.

The basic features of this protocol that makes it so popular are:

- It is one of the most simple management protocols available.
- It is relatively cheap to implement and maintain compared to other protocols.
- It only has a basic set of management operations that makes it easier for operators to work on this protocol.
- It can capture errors generated by the network element asynchronously using traps.
- It is easy to add managed nodes in the network.
- Using the management station: Agent architecture, the management capabilities, and work load on the network element is reduced.
- It is based on UDP and hence is best suited for intermittent networks that use packets like the internet.

It can be seen that the main design philosophy of SNMP was to keep it SIMPLE.

## 14.2 SNMPv1

The network management framework associated with SNMPv1 is composed of four RFCs from IETF. They are RFC 1155, RFC 1157, RFC 1212, and RFC 1213.

- RFC 1155 deals with the syntax and semantics for defining a managed object and is also referred to as structure of management information (SMI).
- RFC1157 is dedicated to the details of SNMP including its architecture, object types, and operations.
- RFC 1212 gives the format for defining MIB modules.
- RFC 1213 defines the basic set of objects for network management, which is also known as MIB–II.

SNMPv1 supports five PDUs. GetRequest, GetNextRequest, and SetRequest PDU are used for communication from manager to agent and GetResponse and Trap PDU are used for communication from agent to manager (see Figure 14.1). This interaction is based on a simple request–response model where the network management system (SNMP Manager) issues a request, and managed devices (agent running on device) return responses.

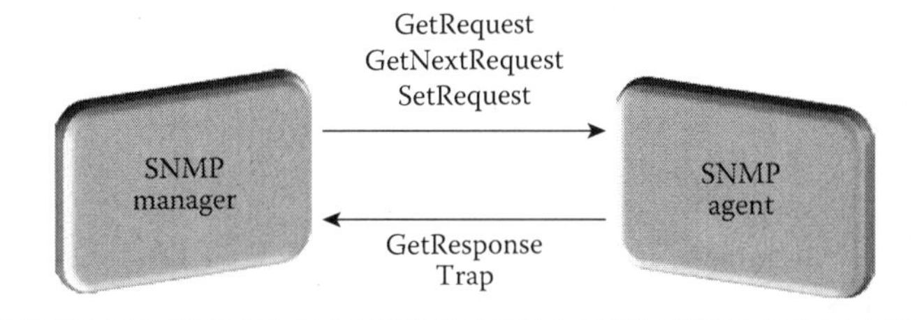


**Figure 14.1 SNMPv1 packet data units (PDUs).**

- The get operation of the GetRequest PDU is used by the manager to retrieve the value of object instances from an agent.
- The agent uses the response operation of the GetResponse PDU to respond to the get operation with the object instances requested. If the agent cannot provide all the object instances in the requested list then it does not provide any value.
- The GetNext operation of the GetNextRequest PDU is used by the manager to retrieve the value of the next object instance in the MIB. This operation assumes that object values in the MIB can be accessed in tabular or sequential form.
- The set operation of the SetRequest PDU is used by the manager to set the values of object instances in the MIB.
- The trap operation (Trap PDU) is used by agents to asynchronously notify the manager on the occurrence of an event.

The PDUs GetRequest, GetNextRequest, SetRequest, and GetResponse have the format shown below:

| Request ID | Error Status | Error Index | Variable Bindings |
|---|---|---|---|

- Request ID is used as an identifier for correlating a response with the request. This way duplicate responses from the agent can be discarded at the manager.
- Error Status indicates the kind of error and can take values like genErr, badValue, tooBig, noSuchName, readOnly, and noError.
- Error Index gives the position of the object instance (variable) associated with the error.
- Variable binding is a list of object instance names and values.

Let us check how error status works taking an example of GetRequest PDU sent to the agent.

- If the agent encounters errors while processing the GetRequest PDU from the manager then error status is "genError."

- When the agent receives a GetRequest PDU from the manager, the agent checks whether the object name is correct and available in MIB present in the network element being managed. When the object name is not correct or when the object name corresponds to an aggregate object type (not supported), then the error status is "noSuchName."
- If the response for the GetRequest PDU exceeds the local capacity of the agent then error status is "tooBig."
- If the agent is able to successfully process the request and can generate a response without errors, then the error status is "noError."

The Trap PDU has the format shown below:

| Enterprise | Agent Address | Generic Trap Type | Specific Trap Code | Time Stamp | Variable Bindings |
|---|---|---|---|---|---|

- Enterprise tag is used to identify the object that generated the trap. It is the object identifier for a group.
- Agent Address is the IP address of the agent that generated the trap.
- Generic Trap Type is a code value used to identify the generic category in which the trap can be associated. Example: coldStart, warmStart, linkup, linkDown, and so on.
- Specific Trap Code provides the implementation specific trap code.
- Timestamp is used for logging purpose to identify when the trap occurred.
- Variable Binding provides a list variable (object instance name)/value pairs.

The SNMP PDU is constructed in ASN.1 (abstract syntax notification) form and the message to be exchanged between the manager and agent is serialized using BER (basic encoding rules).

The SNMP message consists of a message header and PDU. The message header has a version number to specify the SNMP version used and a community name to define the access environment. The community name acts as a weak form of authentication for a group of managers. Figure 14.2 shows the method of creating an SNMP message for sending and how it is handled at the receiving end.

In addition to the agent and manager, another system that is commonly referred in discussions on SNMP is a proxy. Proxy converts messages from one SNMP version to another or converts messages from SNMP to some other protocol and vice versa. Proxy is used when the manager and agent supports two different protocols or different versions of the same protocol.

Security in SNMPv1 is implemented using community strings. The authentication schema shown in Figure 14.3 is a filter module that checks the community string. Community name or community string is a set of octets used to identify a member in the community. An application can have more than one community name and only applications in the same community can interact. Hence community helps to pair two application entities.

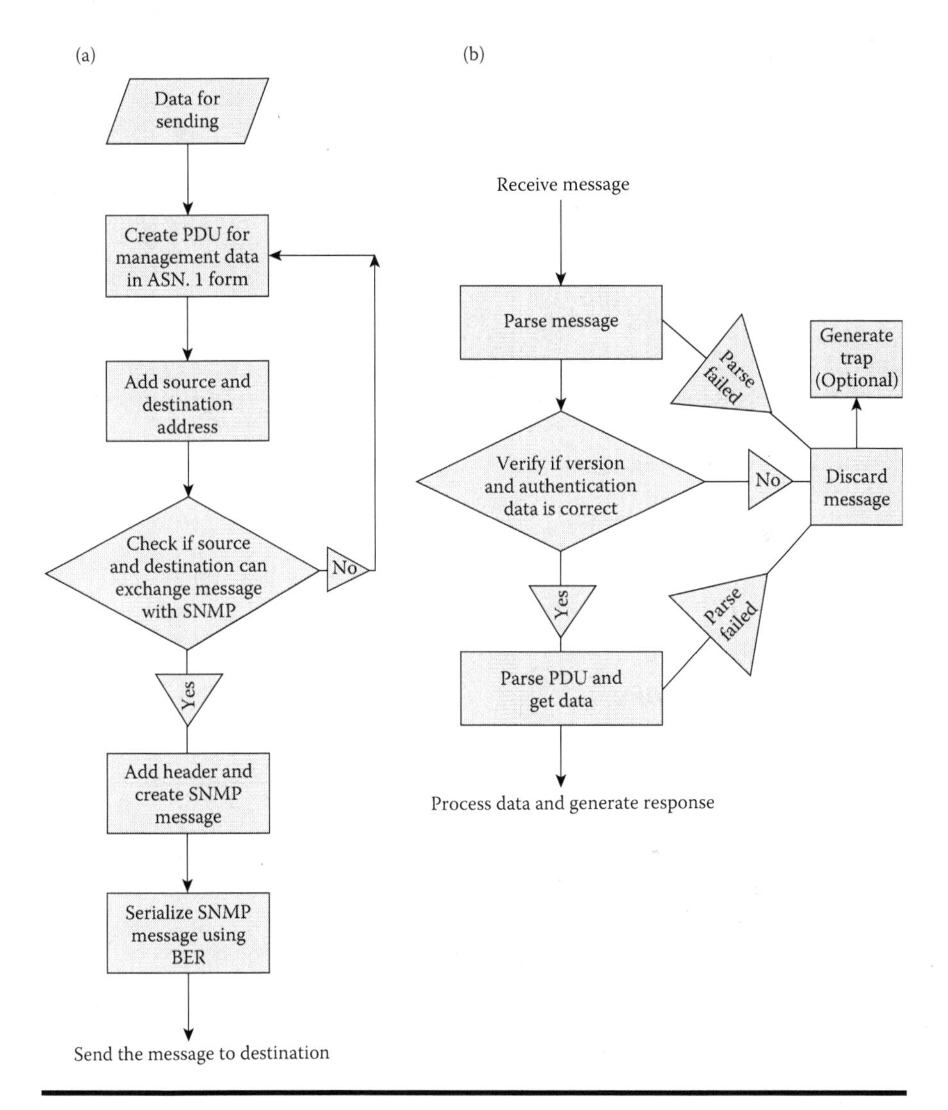


**Figure 14.2 SNMPv1 message handling: (a) sending of SNMP message; (b) receiving of SNMP message.**

The SNMP agent has a community profile that defines the access permissions to the MIB. The agent can view only a subset of managed objects in a network element and the agent can have read or write access to the visible objects. A generic "public" community usually offers read only access to objects. An agent can be associated with the public community to access the read only objects in the community.

Let us describe the access policy shown in Figure 14.4. If "A" corresponds to the agent for managing a Cisco element and "B" corresponds to agent for managing a

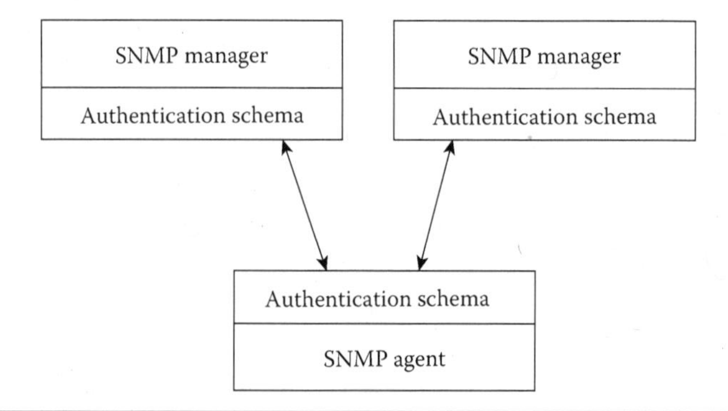


**Figure 14.3 SNMPv1 security.**

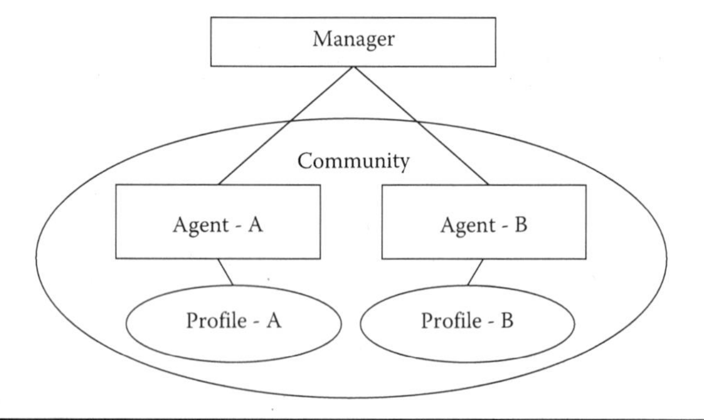


**Figure 14.4 SNMPv1 access policy.**

Nortel element, then the manager has access to the community having both a Cisco and Nortel agent. Cisco agent and Nortel agent have a different community profile, which is "Profile A" and "Profile B." So the Cisco agent will be able to view only the managed objects defined by its profile and not Nortel managed objects that have a different profile, but the manager can view both Cisco and Nortel managed objects. There is also an entity called manager of manager (MoM) according to the OSI model. The MoM is expected to process data that is managed by multiple managers; that is, the MoM must be able to view managed objects from multiple community. That is, if manager-1 can view community-1 and manager-2 can view community-2, then an MoM can view both community-1 and community-2.

## 14.3 SNMPv2

The major weakness in SNMPv1 was the lack of an elaborate security framework. The main intent of having SNMPv2 was to overcome this security issue. While

SNMPv1 and SNMPv3 are IETF standards, SNMPv2 did not get fully standardized and only draft versions are available. In discussions and NMS product specifications on SNMPv2 there are four popular versions that come up. They are:

- SNMPv2P: This was introduced as improvement for SNMPv1 and is based on RFC 1441 and 1452. Unlike community-based security in SNMPv1, the SNMPv2P had a party-based security system. There were many controversies on the administrative and security framework specified in this release and many raised the concern that it was complex and would not be easy for network administrators to adopt.
- SNMPv2C: To address the issues in the security framework offered by SNMPv2P, SNMPv2c was suggested that uses community-based security similar to SNMPv1. It was proposed under RFC 1901 through 1908.
- SNMPv2U: This version was defined as an alternate security model for party-based SNMPv2P. It was proposed in RFC 1909 and has the user-based security model defined in RFC 1910. This version had minimum security and was mainly targeted for small agents.
- SNMPv2*: This is a commercial version that is more refined that SNMPv2U and implements a security system that can be used for large networks. Two important aspects that made this version suitable for commercial use is the consideration for scalability and some level of provisioning capability to perform remote configuration of security.

Though SNMPv2 had enhanced security compared to SNMPv1 and SNMPv2* provided capabilities of remote configuration, it was only in SNMPv3 that a standard that provides commercial grade administration and security framework was introduced. The discussion on SNMPv2 in this chapter deals with generic non-conflicting recommendations and does not relate to a specific version. SNMPv2C endorsed by IETF is the most popular implementation of SNMPv2 that is available in the majority of NMS products.

The advantages offered by SNMPv2 over SNMPv1 are:

- Better event handling using "inform." In SNMPv1 "Trap" message generated by the agent for event notification is simply collected by the manager, while SNMPv2 "Inform" message from the agent is acknowledged by the manager with a response. "Inform" message for event notification generated by the agent will be resent to the manager, if the manager does not send a response message to acknowledge the reception of the message.
- Improved error handling with new error and exception definitions.
- Expanded data types like BITS, Unsigned32, Counter64 in addition to many others.
- Standardized multiprotocol support.
- Enhanced security.

- Large data retrieval possible using Get-Bulk operator, leading to better efficiency and performance.
- Improved data definition with additions to SMI, new MIB objects, and changes to syntax and semantics of existing objects.

The Get, GetNext, and Set operations used in SNMPv1 are exactly the same as those used in SNMPv2. They perform the action of getting and setting of managed objects. Even the message format associated with these operations has not changed in SNMPv2.

Though Trap operation in SNMPv2, serves the same function as that used in SNMPv1, a different message format is used for trap operation in SNMPv2.

SNMPv2 defines two new protocol operations:

- GetBulk: This operation is used to retrieve a large bulk of data. This operation was introduced to fix the deficiency in SNMPv1 for handling blocks of data like multiple rows in a table. The response to a GetBulk request will try to fill as much of the requested data as will fit in the response message. This means that when a scenario arises where the response message cannot give all the requested data, then it will provide partial results.
- Inform: This operation was introduced to facilitate interaction between network managers. In a distributed environment where one manager wants to send a message to another manager, then the Inform operation can be used. It is also used for sending notifications similar to a trap. Inform is a "confirmed trap" operation.

The PDUs GetRequest, GetNextRequest, SetRequest, Trap, Inform, and GetResponse have the format shown below:

| PDU Type | Request ID | Error Status | Error Index | Variable Bindings |
|---|---|---|---|---|

- PDU type is used to identify the type of PDU, whether it is a GetRequest, SetRequest, GetResponse, and so on.
- Request ID will associate an SNMP response with its corresponding request.
- Error Status and Error Index is used to describe the error scenario for an operation similar to SNMPv1 PDU.
- Variable bindings serves the same purpose as in SNMPv1 to give variable-value pairs.

The GetBulk PDU in SNMPv2 has a different format to handle bulks of data:

| PDU Type | Request ID | Nonrepeaters | Max Repetitions | Variable Bindings |
|---|---|---|---|---|

- PDU Type always corresponds to GetBulk PDU.
- Request ID is to associate a bulk response with the correct request.
- Nonrepeaters (N) and Max Repetitions (M): Before we explain these two fields it should be clear that the GetBulk operation is similar to performing the GetNext operation multiple times. While GetNext operation just returns the next object, the GetBulk performs the GetNext multiple times based on N and M values to get a bulk of data. The first N elements of the variable binding list are treated as a normal GetNext operation and the next elements are considered as consisting of M number of repeated GetNext operation.
- Variable bindings serves the same purpose as in SNMPv1 to give variable-value pairs.

SNMPv2 also introduced "Report PDU." The intent of the report PDU is to account for administrative requirements. A report PDU could be used for faster recovery by sending information for time synchronization. It is not defined in SNMPv2 PDU definitions listed in RFC 1905. Implementers can define their own usage and semantics to work with "Report" PDU in SNMPv2.

SNMPv2 mainly supports five transport services for transferring management information. This association of SNMPv2 with a transport service is referred to as snmpDomains and is described in RFC 1906. The domains are:

| *snmpDomain* | *Description* |
|---|---|
| snmpUDPDomain | This corresponds to SNMPv2 over UDP (User Datagram protocol). For telecom management purpose UDP is the most popular transport service used with SNMP. |
| snmpCLNSDomain | This corresponds to SNMPv2 over OSI (open systems interconnection). This domain uses a ConnectionLess Network Service (CLNS). |
| snmpCONSDomain | This corresponds to SNMPv2 over OSI, similar to snmpCLNSDomain. This domain uses a Connection Oriented Network Service (CONS). |
| snmpDDPDomain | This domain corresponds to SNMPv2 over DDP (datagram delivery protocol). DDP is AppleTalk equivalent of UDP (and IP). |
| snmpIPXDomain | This domain corresponds to SNMPv2 over IPX (Internetwork Packet Exchange) protocol. Mapping of SNMP over IPX is used in Novell NetWare environments. |

In-spite of several advantages of the SNMPv2 over its predecessor SNMPv1, there still were few unmet goals on security and administrative framework in its recommendations that resulted in SNMPv3. The major drawback of SNMPv2 was the lack of confidentiality. The request message from the origin and the response message from the destination were both not encrypted before transmission. Any third-party application could intercept the message and decode it. Introduction of encryption and several other security features is discussed in the next section.

## 14.4 SNMPv3

The five most important RFCs of SNMP Version 3 are:

1. RFC 3411: An architecture for describing SNMP management frameworks: This document is intended to provide details of the SNMPv3 management framework. It clarifies most of the terms that are used in the RFC documents for SNMPv3. It provides an introduction to multiple logical contexts and defines the user as an identity. The RFC also includes the SNMP entity architecture having the SNMP engine that runs on the agent and the manager.
2. RFC 3412: Message processing and dispatching: There are multiple message processing models with regard to the SNMP versions. This RFC defines how the SNMP message corresponding to a particular version is dispatched for processing to the appropriate model. The SNMP engine defined in Version 3 can support multiple message processing models. For example, the SNMP management system can support SNMPv1, v2, and v3 protocols at the same time. The RFC also includes description of SNMPv3 Message Processing Model.
3. RFC 3413. SNMP Applications: This RFC describes five types of SNMP applications.
4. The five SNMP applications are:
   - Command generators: An application that initiates an SNMP read or write request falls in this category.
   - Command responders: An application that responds to SNMP read or write request falls in this category.
   - Notification originators: An application that generates an SNMP notification falls in this category.
   - Notification receivers: An application that receives the SNMP notification falls in this category.
   - Proxy forwarders: An application that forwards SNMP messages falls in this category.
5. RFC 3414: User-Based Security Model (USM): This RFC handles message level security involving authentication and privacy implemented using USM. Along with authentication using security parameters in the SNMPv3

message, the privacy is obtained by encrypting the message at the sender and decrypting the message at the receiver.

USM ensures that data security is achieved by:

- Preventing unauthorized access to data on transit from sender to receiver.
- Ensuring that the sender/receiver is valid and authorized to send/receive a particular message.
- Ensuring that the message arrives in a timely manner and no re-ordering is done by an unauthorized entity.

Secret keys are used by the user for authentication and privacy. The authentication protocols suggested for use are MD5 (Message Digest 5) and SHA-1 (Secure Hash Algorithm 1) and the privacy protocol suggested for encryption and decryption is DES (Data Encryption Standard). Other authentication and privacy protocols can also be used.

6. RFC 3415: View-Based Access Control Model (VACM): VACM details administrative capabilities involving access to objects in the MIB. The community string that was discussed in SNMPvl and SNMPv2 is also used in SNMPv3. Here the community string is used to identify the data requester and its location, and to determine access control and MIB view information. This is achieved in SNMPv3 using a mode dynamic access control model that is easy to administer.

The Architecture of SNMPv3 is explained using an SNMP entity which is an implementation feature. In SNMPv3, the protocol entities and application entities involved in an SNMP interaction are clearly marked under SNMP engine and SNMP application, respectively. SNMP manager and SNMP agent are two SNMP entities. The architecture of SNMPv3 is shown in Figure 14.5.

The protocol engine performs actions like message processing, dispatch of processed message, and security handling. It does not deal with application specific interactions, which are done by "SNMP Applications."

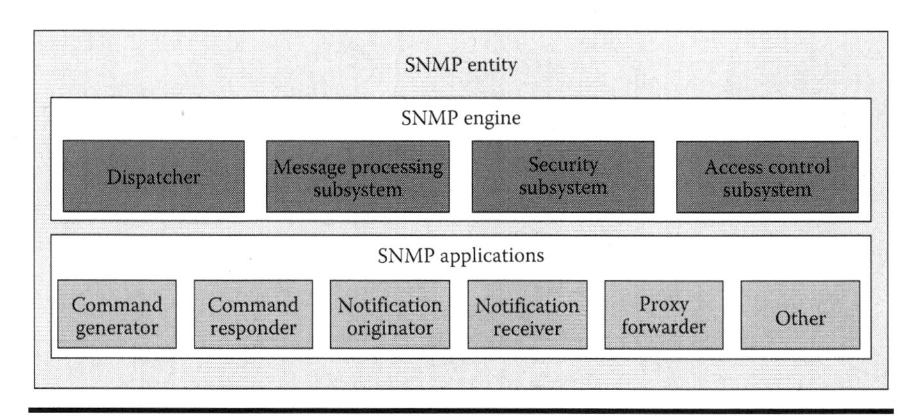


**Figure 14.5 SNMPv3 architecture.**

The components in the SNMP engine are:

- Dispatcher: This module handles sending and receiving of messages. This module does the transport mapping, interacts with the message processing module to get messages and dispatches the PDUs to command and notification handlers.
- Message processing subsystem: The message process subsystem supports one or more message processing models. This subsystem has one or more modules corresponding to the versions of SNMP supported. The dispatcher module identifies the version of the SNMP message and interacts with the appropriate module in the message processing subsystem.
- Security subsystem: It handles message level authentication and privacy. User-based security model (USM) is usually used when SNMPv3 module existing in the message processing subsystem. When the messaging supported includes other versions of SNMP then community-based security model or other security model would also be supported in the security subsystem.
- Access control subsystem: Authorization of access to specific objects is handled in this component. View-based access control model (VACM) is supported in the access control subsystem of SNMPv3.

The components of the SNMP applications are:

- Command generator and command responder: The generator component performs the action of command generation and the responder will respond to commands. The generator is linked with commands like get, get-next, and get-bulk while the responder is linked with commands like get-response. The SNMP manager will contain a command generator and the SNMP agent contains the command responder.
- Notification receiver and notification originator: These components handle notifications like trap and inform. The SNMP manager usually contains a notification receiver and the SNMP agent contains the notification originator.
- Proxy forwarder: This component forwards the request without consideration on the object being handled.
- Other: The possibility of adding extensions in SNMP applications entity is accounted for using a component named Other. The Other Application box acts as a placeholder for applications that may be developed in the future.

SNMPv3 message structure is shown below:

| Message Version | Msg ID | Message Maximum Size | Message Flags | Message Security Model | Message Security Parameters | Context Engine ID | Context Name | PDU |
|---|---|---|---|---|---|---|---|---|

- The message version is used by the message processing unit to identify the SNMP version used in the request or response message. This ID is critical for the dispatcher in identifying the version of message.
- The message ID, message maximum size, message flags, and message security model is used by the message processing system to define unique properties of the message. This set of tags are specific to the SNMPv3 message structure and would not be required when working with other versions of SNMPv3.
- The message security parameters are used by the security subsystem. There is a separate tag to specify the security model that is USM for SNMPv3.
- The context engine ID, Context name, and PDUs are encapsulated to a collection set called scoped PDU. Next let us discuss the concept of "Context" in SNMPv3. A collection of objects is called an SNMP context. Each of these collections is accessible by an SNMP entity. A context can be a collection based on a network element, a logical element, multiple elements, or a subset element(s). Each context has a unique identifier called the contextName, which is unique within an SNMP entity. A contextEngineID represents an instance of a context within an administrative domain. Figures 14.6 and 14.7 shows sequence diagramon command and notification. Each SNMP entity and administrative domain can have multiple contexts, so the combination of contextEngineID and contextName uniquely identifies a context within an administrative domain.

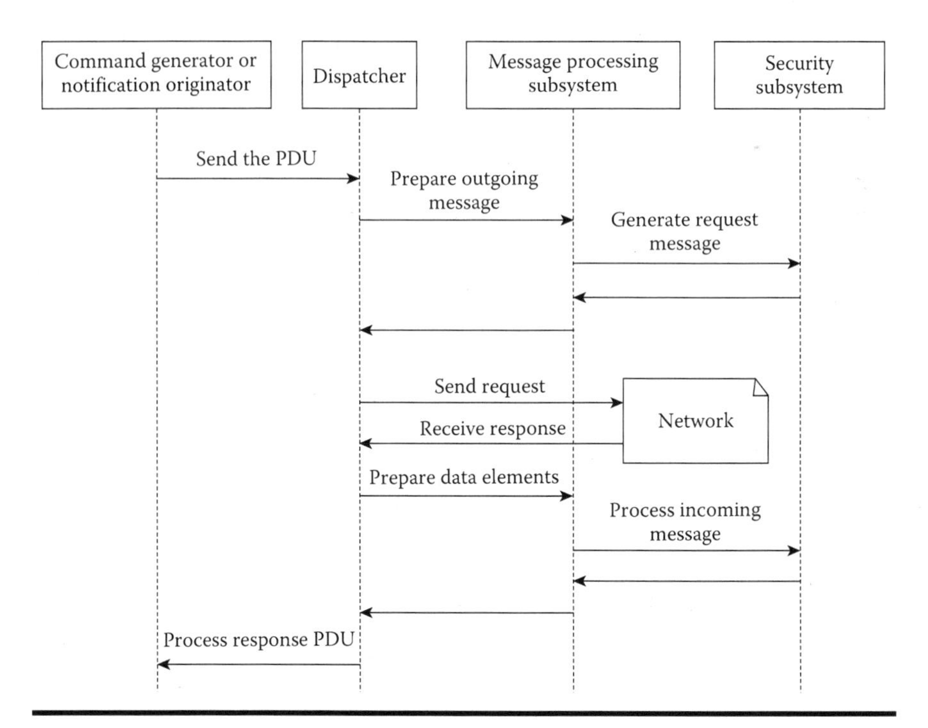


**Figure 14.6 Sequence diagram for command generation/notification origination.**

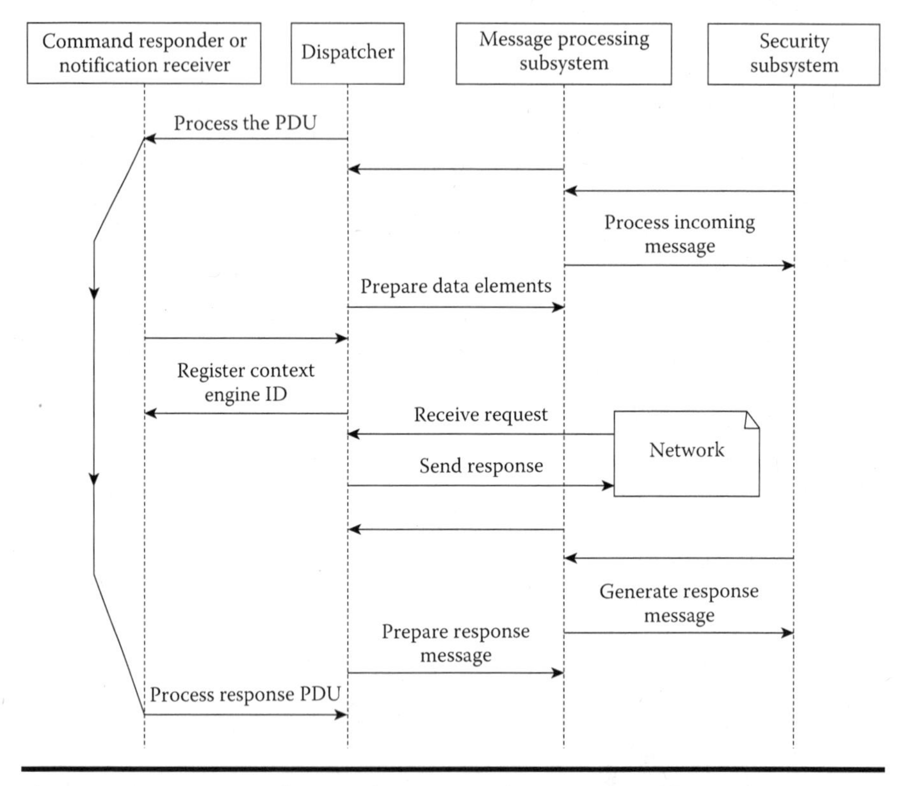


**Figure 14.7 Sequence diagram for command response/notification receive.**

## 14.5 Conclusion

SNMP basic framework is suited for small networks where most of the functionality is implemented at the manager with minimum capabilities in the agent. For managing large heterogeneous networks RMON (remote network MONitoring) is used. The RMON collects data from the elements and processes the data locally and sends processed data to the manager using SNMP. This takes care of scalability and decentralized management at the SNMP manager implementation.

For distributed and client–server environments, CORBA (common object request broker architecture) is popularly used. To use SNMP in this domain interoperability between CORBA and SNMP needs to be worked out. Performance and security play a critical role in working in distributed management domain. There is a separate Internet group called, the distributed management (DISMAN) group working toward defining standards in this area.

New protocols based on XML are now becoming popular for network management and most of the legacy NMS solutions are based on SNMP. Adapters for conversion between protocols are getting popular and some of the NMS solutions support multiple protocols in addition to different versions of SNMP.

The simplicity of SNMP made it one of the most popular network management protocols. Though other management protocols were introduced, SNMP is still the protocol that has the highest number of implementations. To support new devices or extend existing devices only change in the MIB is required when using SNMP. SNMP still needs changes and new specifications to make it a complete management protocol that supports all forms of FCAPS functionality, but that would probably take out the biggest power of SNMP, which is simplicity as at stands for simple network management protocol.

## Additional Reading

1. Marshall T. Rose. *The Simple Book: An Introduction to Internet Management.* New York: Prentice Hall-Gale, 1993.
2. William Stallings. *SNMP, SNMPv2, SNMPv3, and RMON 1 and 2.* 3rd ed. Upper Saddle River, NJ: Addison-Wesley, 1999.
3. Allan Leinwand and Karen Fang. *Network Management: A Practical Perspective.* 2nd ed. Upper Saddle River, NJ: Addison-Wesley, 1995.
4. Douglas Mauro and Kevin Schmidt. *Essential SNMP.* 2nd ed. Sebastopol, CA: O'Reilly Media, Inc, 2005.
5. Sidnie M. Feit. *SNMP: A Guide to Network Management.* New York: McGraw-Hill, 1993.
6. Sean J. Harnedy. *Total SNMP: Exploring the Simple Network Management Protocol.* 2nd ed. Upper Saddle River, NJ: Prentice Hall PTR, 1997.
7. Paul Simonaü. *SNMP Network Management* (McGraw-Hill Computer Communications Series). New York: McGraw-Hill, 1999.

# Chapter 15

# Information Handling

This chapter is about management information handling. The concepts that are discussed include the abstract syntax for application entity using ASN.1 (Abstract Syntax Notation One), the transfer syntax for presentation entity using BER (Basic Encoding Rules), and the management information model using SMI (Structure of Management Information).

## 15.1 Introduction

Information handling consists of two components. One is storing transfer of data and the other part is storing the data. When two systems exchange data, the syntax and semantics of the message containing the data needs to be agreed upon by the systems for effective communication. The way the data is arranged should be defined in an abstract form to understand the meaning of each data component in the predefined arrangement.

This abstract form defined for application interaction is called abstract syntax and it is an application entity. ASN.1 is one specific example of abstract syntax. The set of rules for transforming or encoding the data from a local syntax to what is defined by abstract syntax is called the transfer syntax. It deals with the transfer of data from presentation to application and hence it is a presentation entity. BER is one specific example of transfer syntax.

Managed object for which management data is collected needs to have a structured form of representation and storage. The management data is maintained in a logical collection called the MIB. MIB objects are objects in the MIB and a set of MIB objects make up a MIB module. Though MIB can be implemented as a software database, it is not necessarily a database and it is best defined as a description of objects. Information management model SMI, performs the activity of defining the general characteristics and types of MIB objects.

Some of the standards that are relevant from information handling perspective are:

| *Standard* | *Description* |
|---|---|
| ISO/IEC 8824-1 (X.680) | Basic ASN.1 Notation |
| ISO/IEC 8824-2 (X.681) | Information Objects Specification |
| ISO/IEC 8825-1 (X.690) | Basic, Canonical, and Distinguished Encoding Rules |
| RFC 1155 | Defines the Structure of Management Information (SMI) |
| RFC 1213 | MIB for Network Management of TCP/IP-based Internets |
| RFC 3418 | Management Information Base (MIB) for the Simple Network Management Protocol (SNMP) |

## 15.2 ASN.1

The ASN.1 has the capability to represent most types of data involving variable, complex, and structures that need to be extended. It was developed as a joint effort from ISO and ITU-T and is mainly defined in ITU-T X.680/ISO 8824-1.

### 15.2.1 ASN.1 Simple Types

The simple types can be classified into basic types, character string types, object types, and miscellaneous types. A few examples for each of the simple types are given below:

| *Basic Types* | *Character String Types* | *Object Types* | *Miscellaneous Types* |
|---|---|---|---|
| • BOOLEAN<br>• INTEGER<br>• ENUMERATED<br>• REAL | • NumericString<br>• PrintableString<br>• GraphicString<br>• TeletexString | Has an "OBJECT IDENTIFIER" and "ObjectDescriptor" | • NULL<br>• UTCTime |

The basic type is similar to data types in most programming languages. The character string presents different flavors of string for specific applications. An object type is for defining an object. It has an identifier called OBJECT IDENTIFIER and description of the object given in ObjectDescriptor. Miscellaneous types are to include generic types like GeneralizedTime (e.g., 2010), UTCTime (e.g., yy = 80..99 : 1980–1999), and so forth.

Syntax for type definition: <type name> ::= <type>
Examples of type definition:

1. TIMER ::= INTEGER
2. MacAddress ::= OCTET STRING
3. Days ::= ENUMERATED { Monday (1), Tuesday (2), Wednesday (3) }

Syntax for value assignments: <value name> <type> ::= <value>
Examples of value assignments:

1. ipAdd MacAddress ::= 'FFEEFF00'H
2. firstName VisibleString ::= "Jithesh"
3. currentDay Days ::= Monday
4. currentTime UTCTime ::= "000204062010+0100"

Subtypes as the name suggests, are derived from existing types.
A subtype is used when:

- Inheriting characteristics from a reusable type. That is, if multiple types have common characteristics then these common characteristics can be moved to a parent type and specialized subtypes can be created by deriving from the parent type. This way the subtype will include both the parent characteristics and its own individual characteristics.
- Subtypes are also used to create subsets from a parent set; that is, when the parent set is large or for some other reason the contents of a parent type needs to be put in two subtypes.

A subtype must have a value when it is derived from the parent type.
Syntax for subtype definition: <subtype name> ::= <type> ( <constraint> )
Examples of subtype definition:

1. TIMER ::= INTEGER ( 0..100 )
2. MacAddress ::= OCTET STRING ( SIZE(4) )
3. Today ::= Days ( Monday | Tuesday )

Before we move from simple types to structured types, let us look into object types. Object type discussed under simple type is a very important format for information handling, as management information in a MIB has management entities handled as objects. ASN.1 object identifiers are organized in a hierarchical tree structure. This makes it possible to give any object a unique identifier. This tree is shown in Figure 15.1.

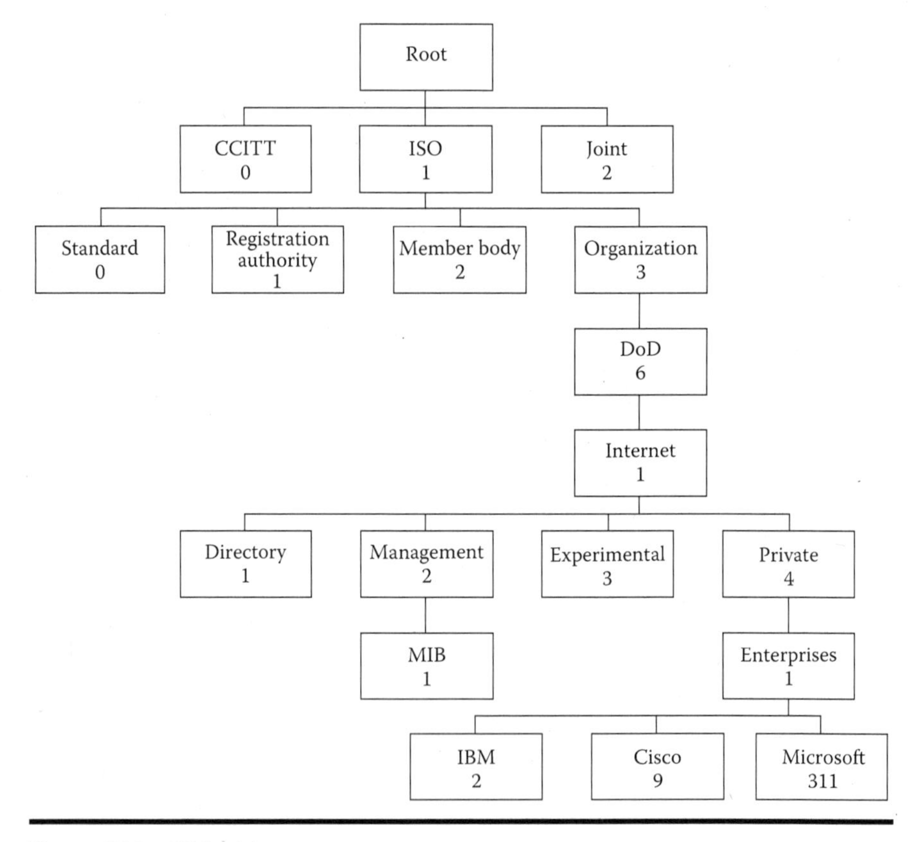


**Figure 15.1 Object tree.**

Each entity to be managed needs to be uniquely identified using a name, number, or a combination of the two, known as an Object Identifier. The object is a part of the tree, its identity is defined using the traversal path starting at the Root and ending at the object position. The starting point in an object tree is always the Root. Below the root are identifiers for standardizing bodies CCITT and ISO, along with these is an identifier for joint definitions of CCITT and ISO. ISO has four children namely standard, registration authority, member body, and organization with identifier values 0, 1, 2, and 3, respectively.

Example of value assignments for an object identifier:

1. management OBJECT IDENTIFIER ::= { internet 2 }
   // name-number combination
2. internet OBJECT IDENTIFIER ::= { iso(1) org(3) dod(6) 1 }
   // name-number combination
3. internet OBJECT IDENTIFIER ::= { 1 3 6 1 }
   // identification using number alone

There can be scenarios where the object descriptor will not be unique, but the combination of object descriptor and object identifier will always be unique globally.

## *15.2.2 ASN.1 Structured Type*

In this section we will look into four ASN.1 structured types. They are:

- SEQUENCE: Used for collection of variables of different type (can have a few variables of the same type but all variables are not of same type) where order is significant.
  Type Definition:

  ```
  NodeInfo ::= SEQUENCE {
  nodeName    VisibleString,
  nodeState   VisibleString,
  BladeNo     INTEGER
  }
  ```

  Value Assignment:

  ```
  MediaGateway NodeInfo ::= {
  nodeName    "MGW-10",
  nodeState   "Offline",
  BladeNo     4
  }
  ```

- SEQUENCE OF: Used for collection of variables of same type where order is significant.
  Type Definition:

  ```
  UserList ::= SEQUENCE OF VisibleString
  ```

  Value Assignment:

  ```
  operators UserList ::= {
  "Joe"        "Patrick",
  "Dan"        "Peter",
  }
  ```

- SET: Used for collection of variables of different type (can have a few variables of same type but all variables are not of same type) where order is insignificant. Tagging of items in the set is required to uniquely identify them. If the tags are not specified explicitly then ASN.1 will automatically tag the items in a SET.

Type Definition:

```
NodeInfo ::= SET {
nodeName    VisibleString,
nodeState   VisibleString,
BladeNo     INTEGER
}
```

Value Assignment:

```
MediaGateway NodeInfo ::= {
nodeName    "MGW-10",
nodeState   "Offline",
BladeNo  .  4
}
```

- SET OF: Used for collection of variables of same type where order is insignificant.
  Type Definition:

```
UserList ::= SET OF VisibleString
```

Value Assignment:

```
operators UserList ::= {
"Joe"       "Patrick",
"Dan"       "Peter",
}
```

ASN.1 supports explicit and implicit tagging. An advanced feature of ASN.1 that is not discussed here is the concept of modules. Module definitions are used for grouping ASN.1 definitions.

## 15.3 BER

BER (basic encoding rules) is a transfer syntax discussed in this section. Encoding involves converting ASN.1 data to a sequence of octets for transfer. Though other methods of representing ASN.1 exist like packed encoding rules (PER), BER is suggested for OSI data transfer. BER has two subset forms called canonical encoding rules (CER) and distinguished encoding rules (DER). DER is suited for transfer of small encoded data values while CER is commonly used for large amount of data transfer.

There are three methods of encoding ASN.1 data to BER:

- Primitive, definite-length encoding: This method is employed with simple types and types derived from simple types using implicit tagging, excluding simple string type. The length of the value to be encoded needs to be known to perform this method of encoding.
- Constructed, definite-length encoding: This method is employed with simple string type, structured type, types derived from simple string, and structured type using implicit tagging and type derived from any type using explicit tagging. The length of the value to be encoded needs to be known to perform this method of encoding.
- Constructed, indefinite-length encoding: This method is employed with simple string type, structured type, types derived from simple string, and structured type using implicit tagging and type derived from any type using explicit tagging. The length of the value to be encoded does not have to be known in advance to perform this method of encoding.

Structure of BER primitive and constructed, definite length encoding:

| Identifier<br>Octets | Length<br>Octets | Contents<br>Octets |
|---|---|---|

Structure of BER constructed, indefinite length encoding:

| Identifier Octets | Length<br>Octets | Contents<br>Octets | End-of-Contents<br>Octets |
|---|---|---|---|

The fields in BER data element are:

- Identifier octets: This field is used to provide information on the characteristics of the data in the contents field (see Figure 15.2). It has three components:
  - Class: Used for classifying the data element.
  - P/C: To specify if the BER is primitive or constructed.
  - Tag number: Describes the data element.
- Length octets: In BER structure using fixed length encoding, this field gives the number of contents octets. When nondefinite length encoding is used, this field indicates that length is indefinite. The length is indicated in one of three forms:
  - Short form: Used for definite length encoding when content length is smaller than 128 octets.
  - Long form: Used for definite length encoding when content length is greater than 128 octets.
  - Indefinite form: Used for indefinite length encoding.

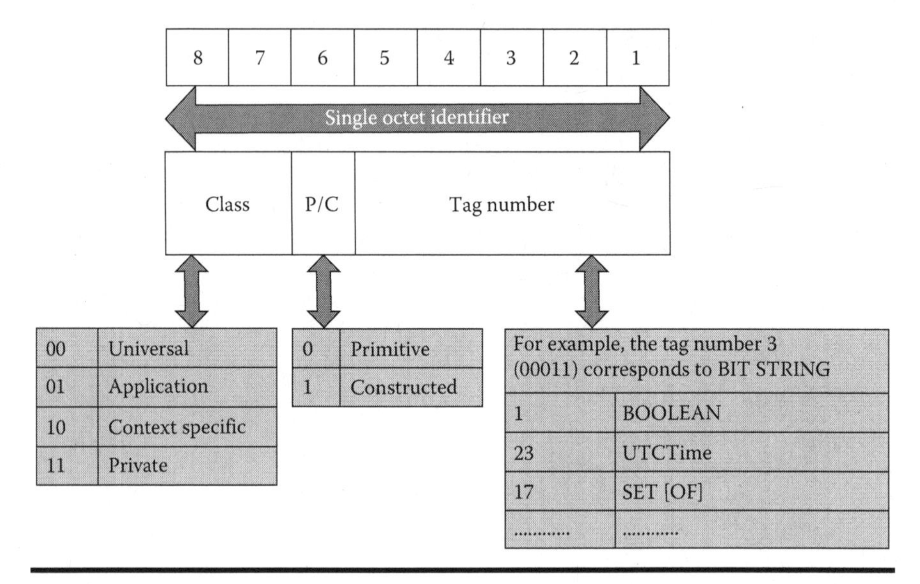


**Figure 15.2 Identifier octet.**

- Contents octets: The ASN.1 data in encoded format is present in this field.
- End-of-contents octets: When the encoding is using a constructed, indefinite-length method this field denotes end of contents. This field is not present when other encoding methods are used.

Let us close the discussion on BER with an example of encoding with BER as shown in Figure 15.3. For encoding zero, the identifier corresponds to a universal private integer with length one and the contents will have two's complement of zero, which is zero itself.

## 15.4 SMI

Structure of management information (SMI) is a subset of ASN.1 used to define the modules, objects, and notifications associated with management information. Any resource that has to be managed is referred to as managed object and is represented by a class called the managed object class. Defining generic guidelines for information representations helps to create specific subclasses from the generic parent classes in SMI.

The main ITU-T documents on SMI are:

- X.720 Management information model: It defines the concept of managed object and managed object class. The principles for naming managed objects, the attributes in managed objects, and the relationships like inheritance,

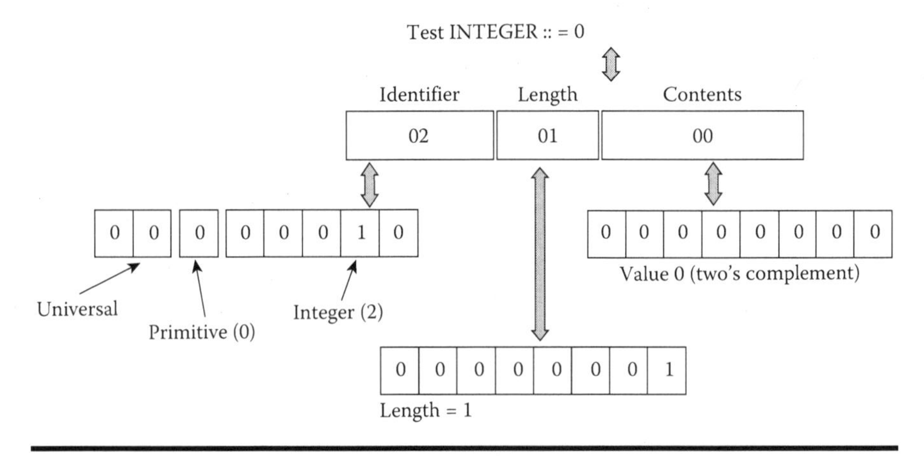


**Figure 15.3 Example of BER encoding.**

specialization, and containment associated with managed objects are also discussed in this document.

- X.721 Definition of management information: It defines the managed object classes that are required for systems management functions.
- X.722 Guidelines for the definition of managed objects (GDMO): It specifies the syntax and semantics for the templates in SMI-like managed objects, attributes, notifications, actions, and so on.
- X.723 Generic management information: It defines the managed object classes at different layers or more specifically at resource level like ports.
- X.724 Requirements and guidelines for implementation conformance statement proformas associated with OSI management: It relates to conformance testing of managed objects and systems management functions.

The management information is stored in a management information base (MIB). The data can take two forms. It can be a "Scalar" like integer, character, and so forth or it can be a "Table" that is a two-dimensional array of scalars. While scalar can be used to store the value of a variable in the managed object, the table can be used to capture different instances of the managed object.

As already mentioned, SMI is defined using ASN.1 constructs. The data types for SMI can be classified into three groups. They are:

- Simple types: These are taken from ASN.1 and comprised of types like INTEGER and OCTET STRING that were discussed in the preceding sections.
- APPLICATION-WIDE types: These are data types that are specific to SMI. It includes scalars like Integer32, IpAddress, and Counter32. Note that integer represented for the simple types of ASN.1 are written in full capital

INTEGER, while the integer specific to SMI only has the first alphabet capitalized Integer32.

- PSEUDO types: This type has the look and feel of an ASN.1 type but is not defined in ASN.1. An example for this is BITS, which is written in full capital similar to ASN.1 but is not part of the ASN definitions.

The above discussion on types is based on SMIv2. The initial version of SMIv1 did not have data types like Counter32 and Counter64 defined, instead a single data type called Counter represented both 32 and 64 bit types. In addition to giving more meaningful definition for existing types of SMIv1, the second version also introduced new data types like "Unsigned" and made some of the data types like "NetworkAddress" defined in version one obsolete.

Next let us use an example to discuss the definition of objects. Let us consider switch as an entity for our example. It has a unique "address" of type IpAddress as its identifier and its attributes include a "status" showing the usability state of the switch of type OCTET STRING, number of "connection" of type Integer32 and a table named "Routing" having a collection of data of type INTEGER showing the switch routing logic. This can be represented in the tree diagram (see Figure 15.4).

The address, status, connection are all objects of the switch entity. If "status" has its value as "offline" and to get the status object, the query is to 1.2.1, and to get the value of the object the query is for 1.2.1.0. So to get an instance of an object, "0" is added to the query. Similarly when the value of the address is 101.101.101.101, the query 1.1 returns an address and to get the value in an address the query should be 1.1.0. For the example in Figure 15.4, the request/query 1.2 or 1.2.0 will lead to an error as the response. For SNMP MIB query/request a GET command can be used.

In the routing table, for the sake of simplicity let us consider that "dest" column is a unique identifier for that table and the elements in "dest" do not repeat and hence can be used as an index for query. A query on the index 1.3.1.5 will get a response of "5" while a query on 1.3.2.5 will give the value "3" corresponding to index value of "5."

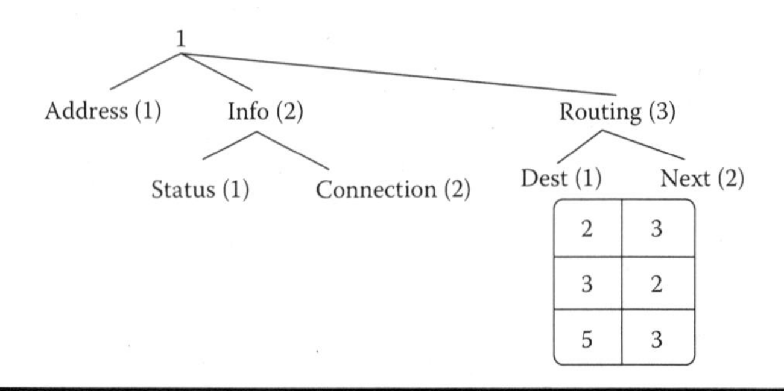


**Figure 15.4 Tree diagram for switch entity data in a MIB.**

Some of the important fields in scalar and table definition are discussed next. It should be understood that this list does not handle all the fields defined in SMI. The explanation is also supplemented with examples on usage.

1. The fields that appear in object definition of leaf objects are:
   - OBJECT-TYPE: This field is used after the name of the object and encapsulates the remaining structure of the object. In short this tag is for a composite object type definition.
   - SYNTAX: The primary type associated with the definition is specified with SYNTAX. It can take values like INTEGER, BITS, IpAddress, Intger32, and so on. Example, the SYNTAX of "address" will be IpAddress.
   - MAX-ACCESS: This tag is not to signify the access rights. As the name suggests it gives the maximum access operation that can be performed. It can take values read-only, read-write, read-create, nonaccessible, and accessible-for-notify. Example "Info" in Figure 15.4 will be nonaccessible, while "status" updated dynamically is read-only.
   - STATUS: This tag is used to show the current status of the object. It can take values current, deprecated, and obsolete. Current should always be implemented, deprecated should be implemented but it will become outdated, and obsolete is used when the object is outdated.
   - DESCRIPTION: A description on the object can be given after this tag. Example:

```
address        OBJECT-TYPE
SYNTAX         IpAddress
MAX-ACCESS     read-write
STATUS         current
DESCRIPTION    "Address of the switch"
::= { MIB 1}
```

2. The fields that appear in the object definition of leaf objects are:
   - OBJECT IDENTIFIER: It is used to define a nonleaf object. In the example we discussed, "Info" object falls in this category. Example:

```
Info OBJECT IDENTIFIER: = { MIB 1 }
```

   - OBJECT-IDENTITY: This tag provides an alternate method to define nonleaf objects. It uses some of the tags when defining leaf objects in creating the definition. Example:

```
Info           OBJECT-IDENTITY
STATUS         current
```

```
DESCRIPTION  "Scalar Objects for switch are registered
             under this object"
::= { MIB 1 }
```

3. The fields that appear in table definition are:
   - INDEX: This tag is used to specify the index of the table or the column/group of columns that will act as the unique identifier for rows in the table.
   - SEQUENCE: The columns in the table can be grouped into a sequence using this tag.
     Example of table definition using "Routing" shown in Figure 15.4:

```
RowEntries ::=
     SEQUENCE { dest INTEGER, next INTEGER
     }
```

*Comment*: Now the rows in the table are defined

```
Routing      OBJECT-TYPE
SYNTAX       SEQUENCE OF RowEntries
MAX-ACCESS   nonaccessible
STATUS       current
DESCRIPTION  "Routing table for the switch"
::= { MIB 1}
```

*Comment:* Now routing table has been defined

```
REntry       OBJECT-TYPE
SYNTAX       RowEntries
MAX-ACCESS   nonaccessible
STATUS       current
DESCRIPTION  "An entry specifying a route"
INDEX        { dest }
:: = { Routing 1 }
```

*Comment:* Now individual entries and index for table is defined

*Comment:* Finally dest and next needs to be defined like scalar objects except that these scalars are associated to the routing table and not directly to MIB.

4. Other fields include:
   - NOTIFICATION-TYPE: This tag is used to define notifications. In SNMP context a notification type is used for defining traps.
     Example:

```
switchDown NOTIFICATION-TYPE
OBJECTS     { ifIndex }
STATUS      current
```

```
DESCRIPTION "This trap signifies that the switch is
            down"
::=         { snmpTraps 2 }
```

- TEXTUAL CONVENTION: This tag is used to refine the semantics of existing types. For defining the set of states the switch can have, we could use the TEXTUAL CONVENTION tag as shown in the example. In this example the type INTEGER has been refined.
  Example:

```
State ::=     TEXTUAL CONVENTION
STATUS        current
DESCRIPTION   "Run Time States"
SYNTAX        INTEGER { Offline (0), Busy (1), Active
              (2) }
```

## 15.5 Conclusion

Another important item to be discussed before we conclude is GDMO. The GDMO, a part of the SMI document, is a structured description language that provides a way of specifying the class for objects, object behaviors, attributes, and class hierarchy. GDMO uses a set of templates for information modeling. Some of the templates include definitions for managed object, packages, attributes, and notifications. Packages are used to group characteristics of a managed object and they can be included using a CHARACTERIZED BY or CONDITIONAL PACKAGES tag. Inheritance property is achieved using a DERIVED FROM tag and the object class is registered with an identifier using a REGISTERED AS tag. Refer X.722 document for more details about GDMO and tags like ALLOMORPHIC, PARAMETERS, and so on that are not discussed here.

A sample managed object class template is shown below:

```
<class> MANAGED OBJECT CLASS
       [DERIVED FROM <class> [,<class>]*;]
       [ALLOMORPHIC SET <class> [,<class>]*;]
       [CHARACTERIZED BY <package> [,<package>]*;]
       [CONDITIONAL PACKAGES
              <package> PRESENT IF <condition>
              [,<package> PRESENT IF <condition>]*;]
       [PARAMETERS <parameter> [,<parameter>]*;]
       REGISTERED AS <object-identifier>;
```

This chapter gives the reader an overview on ASN.1, BER, and SMI. These concepts form an important part of coding the network management information framework.

## Additional Reading

1. ISO/IEC/JTC 1/SC 6. ISO/IEC 8825-1:2002, *Information technology: ASN.1 encoding rules*. Washington, DC: ANSI, 2007.
2. John Larmouth and Morgan Kaufmann. *ASN.1 Complete*. San Francisco, CA: Morgan Kaufmann Publishers, 1999.
3. Olivier Dubuisson and Morgan Kaufmann. *ASN.1 Communication between Heterogeneous Systems*. San Diego, CA: Academic Press, 2000.
4. ISO/IEC/JTC 1/SC 6. ISO/IEC 10165-1:1993, *Information Technology—Open Systems Interconnection—Management Information Services—Structure of management Information: Management Information Model*. Washington, DC: ANSI, 2007.

# Chapter 16

# Management Information Base (MIB)

This chapter is intended to give the reader an overview of the Management Information Base. After going through this chapter, the reader will have a basic understanding and the different types of MIB. A detailed explanation of MIB-II and SNMP MIB is also available in this chapter.

## 16.1 Introduction

A management information base is a collection of managed objects. This is a virtual database that stores management data variables (managed objects) that correspond to resources to be managed in a network. The management information is monitored by updating the information in a MIB. The network manager will request information from the MIB through an agent (see Figure 16.1). The agent fetches the management information and responds to the manager. The network manager can use the MIB to monitor the network and it can also update the MIB.

The actual MIB definition is just a file giving the format information and the MIB instance is the actual variables associated with the managed object. So the MIB is instantiated within the agent and an actual copy of the values associated with the MIB variables are available in the network element that also houses the agent. The manager usually keeps a local copy of the MIB variables for ease of access in performing and rendering the management information. A suitable synchronization mechanism is usually in place to ensure that the copy available with the manager is updated periodically based on the actual information collected using the agent.

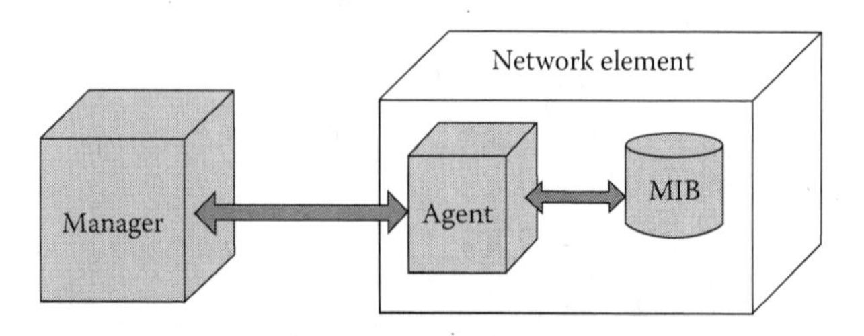


**Figure 16.1 Manager-MIB interaction using agent.**

Practical implementations with MIB do not define all objects of a managed element in a single definition. From a product development perspective this makes sense as the architect having expertise in ethernet may not have expertise in storage or switching. Also the management functionality in a router will be different from that of a Web server. So having separate MIB modules in defining a managed network is required. This leads to multiple MIB modules. From a usage perspective, there are vast varieties of MIB modules to work with—IETF has more than 140 MIB modules most of which are draft versions, enterprise, or vendor specific MIB modules account to figures in the thousands and other participants like universities also have defined MIB modules.

In this chapter, MIBs are grouped into a set of categories. Like most grouping in technical space, a miscellaneous category is used here to incorporate MIBs that do not fall in the common categories that were identified for discussion.

## 16.2 Types of MIB

MIB can be broadly classified into the following types:

- Protocol MIBs: These MIBs relate to protocols at different layers in the OSI model (see Figure 16.2). It can be seen from the diagram that SNMP MIB falls in the application layer of protocol MIB. Similar to SNMP protocol for network management, there are protocols defined for various layers in the communication stack. Protocol MIBs are a general grouping of MIBs associated with these protocol layers. Between the transmission and network layers there are interfaces for which MIBs are defined. Both the physical and data link layer is handled as a single transmission layer in the protocol MIB grouping.

  Some of the RFCs associated with protocol MIBs are:

  a. Transmission MIBs
     - RFC 2515: Definitions of managed objects for ATM management
     - RFC 2558: Definitions of managed objects for the SONET/SDH interface type
     - RFC1512: FDDI management information base

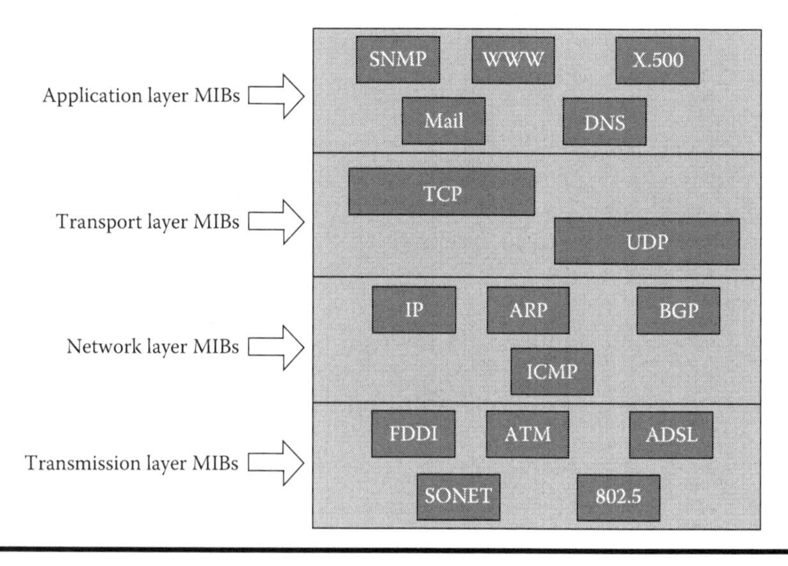


**Figure 16.2 Protocol MIBs.**

b. Network MIBs
   - RFC 1657: Definitions of managed objects for the fourth version of the border gateway protocol (BGP-4) using SMIv2
   - RFC 2667: IP tunnel MIB

c. Transport MIBs
   - RFC 2012: SNMPv2 management information base for the transmission control protocol using SMIv2
   - RFC 2013: SNMPv2 management information base for the user datagram protocol using SMIv2.

d. Application MIBs
   - RFC 2789: Mail monitoring MIB
   - RFC 1611: DNS server MIB extensions

- Remote monitoring MIBs: This category is mainly to group MIBs used for defining remote network monitoring (RMON). RFC 2819 is the main standard in this category. Other RFCs associated with remote monitoring include:
  - RFC 2021: RMON version 2
  - RFC 1513: Token ring extension to RFC
  - RFC 2613: RMON MIB extensions for switched networks version 1.0
- Hardware specific MIBs: Data management for equipment like a printer, modem, and UPS is done using hardware specific MIBs. There are separate RFCs to define the objects that are associated and need to be managed in hardware. Some of the RFC documents that fall in this category are:
  - RFC 1696: Modem management information base using SMIv2
  - RFC 1759: This document details printer MIB

  - RFC 1660: It has definitions of managed objects for parallel printer-like hardware devices using SMIv2
  - RFC 1659: Definitions of managed objects for RS-232-like hardware devices using SMIv2
  - RFC1628: UPS management information base
- Distributed management MIBs: Data about events, notifications, trace, and scheduling operations in a distributed environment are handled using the distributed management MIBs. Some of the RFC documents that fall in this category are:
  - RFC 2981: This is event MIB and handles four areas associated with event. They are trigger, object, event, and notification. Trigger is what causes the event on an object that could lead to a notification
  - RFC 2925: Definitions of managed objects for remote ping, trace route, and lookup operations
  - RFC 2591: Definitions of managed objects for scheduling management operations
- Vendor specific MIBs: There would be scenarios when the standard and experimental MIBS provided by IETF does not provide access to all statistical, state, configurations, and control information of the managed object of the vendor. To solve this issue vendors develop MIB extensions called enterprise or vendor specific MIB that should satisfy a specific set of requirements. Companies like Cisco and Juniper have an extensive list of enterprise specific MIBs. Some of the RFC documents that fall in this category are:
  - RFC 1742: MIB for Appletalk (this is for Apple)
  - RFC1559: DECnet phase IV MIB extensions (this is for Digital Equipment Corporation)
- Miscellaneous MIBs: This is a general catch all category to capture the MIBs that does not fall in the master groups discussed. Some RFCs that fall in this category include:
  - RFC 2758: Definitions of managed objects for service level agreements performance monitoring
  - RFC 2513: This document deals with definition of managed objects for controlling the collection and storage of accounting information for connection-oriented networks

## 16.3 MIB-II

This section details MIB-II, which is one of the most frequently updated modules in the MIB tree. SNMP is covered under MIB-II and is handled in more detail in subsequent sections. There was a MIB-I document defined in RFC1156, which is obsolete now and was replaced with MIB-II defined in RFC 1213. This RFC is about management information base for network management of TCP/IP-based internets. So it defines the variables to manage TCP/IP protocol stack. Only a limited number of these variables can be modified as most of them are read-only. There are several groups in MIB-II that are detailed in separate RFCs. Some of the groups include TCP (RFC 2012), UDP (RFC 2013), IP (RFC 2011), SNMP (RFC 1907), and so on. A significant difference between MIB-I and MIB-II, is the presence of CMOT (value 9) in MIB-I that was removed in MIB-II, and the introduction of SNMP (value 11) in MIB-II.

Some of the design criteria that were used in defining MIB-II are:

- Every variable on fault and configuration management needs to be implemented unless it is not applicable to that specific device. For example, a device that does not have a TCP implementation will not require these variables. In most cases it will not contain variables on items like security and accounting.
- The number of control objects that can set the behavior of a system should be minimal and these objects should not make significant changes in the behavior of the system. This is mainly because of security concerns with protocols like initial versions of SNMP.
- Have a small number of objects that are divided between groups.
- If a variable can be derived from some other variable, then avoid redundancy by creating a new variable. For example if there is a variable for "packets received" and another variable for "packets lost," then a separate variable is not required for capturing "total packets sent."
- Only the variables everyone agreed upon were used while others were eliminated. This ensures that every variable that is defined will serve an actual business purpose for management data collection.
- Since the maintenance of the variable will be an issue when the number of variables increases, it was decided to focus more on variables to capture abnormal behavior rather than defining variables that are associated with normal operation.
- Another major criterion was to ensure that the MIB is not for a specific implementation and is more of a functional MIB. Host resource MIB handles device specific variables.

There are multiple sections under MIB-II, like System (1), Interface (2), AT (3), and so on (see Figure 16.3). Each of these sections has a further subsection attribute to describe it. Let us look into these objects in a little more detail.

a. System (1)

   This group has the following variables under it:
   - sysDescr: This is the system description.
   - sysObjectID: This is the object identifier that points to enterprise specific MIB.
   - sysUpTime: Used to identify how long the system has been running. More specifically the time since last reinitialization of the system.

- sysContact: The contact person for the system; usually the e-mail address is specified.
- sysName: The name of the system to identify it in a network.
- sysLocation: The physical location of the system.
- sysServices: This is the services offered by the system. A number is used to indicate the functionalities performed (e.g., bridge functions, repeater functions, router functions, etc.).

Not all these objects are writable. In the above example sysContact, sysName, and sysLocation are writable while others are read-only.

b. Interface Group (2)
   This group consists of the ifNumber and ifTable, where ifNumber corresponds to the number of rows in the ifTable (interface table). Thus ifNumber is a direct indication of the number of entities or interfaces of the system. The ifTable has a set of attributes (columns) for each interface associated to the system. Some of these attributes are:
   - ifIndex: This is the index column of the ifTable.
   - ifDescr: This field is used to give a description for the interface. Usually the vendor name and number for the interface is listed in this field.
   - ifType: This is the interface type. It can take values like 15 for FDDI interface, 23 for PPP interface, 32 for Frame Relay interface, and so on. There are more than 150 different types of interfaces and the type list is maintained by IANA (Internet assigned numbers authority).
   - ifPhysAddress: It corresponds to the interface address and is usually the MAC address.
   - ifAdminStatus: This shows the admin status and is a writable field.
   - ifInErrors/ifOutErrors: This field maintains a count of input (ifInErrors)/output (ifOutErrors) errors for the interface.
   - ifInDiscards/ifOutDiscards: This field maintains a count of packets discarded during input (ifInDiscards)/output (ifOutDiscards).
   - ifInOctets/ifOutOctets: This field gives the number of octets received (ifInOctets)/transmitted (ifOutOctets) over the interface.

Not all columns in the ifTable are discussed here, but for collecting all possible information about the interface, a complete understanding of ifTable is required.

c. AT (Address Translation) Group (3)
   The address translation group only has one object, the atTable (1) associated to it. The table is used to map the network address to a physical address or subnetwork specific address. Some of the fields in this table are:
   - atIfIndex: This column is used to capture the interface table index.
   - atPhysAddress: The physical address that needs translation and is media dependent is captured here.

- atNetAddress: Network address that corresponds to the media dependent physical address is captured here. This is the usual IP address.

The access permissions for the above fields allow both read and write operation.

d. IP Group (4)

This group has two writable variables, few tables, and a set of read-only variables. The writable variables are ipForwarding (1) and ipDefaultTTL (2). ipForwarding is used to indicate whether the system is a router. When the system is a router, IP forwarding will be enabled. ipDefaultTTL gives the Time to Live (TTL) field of packets generated by the system. The read-only variables are:

- ipInReceives: Input packets received.
- ipInHdrErrors: This gives the errors generated due to version mismatch, error in check sum, and so on that are issued due to packet errors.
- ipInAddrErors: This error is specifically related to packets received for the wrong address. When a packet is received that is not intended for the specific system and the error is registered in this variable.
- ipForwDatagrams: When the system is a router, there will be forwarding packets and this variable is used to keep track of packets forwarded.
- ipReasmReqds: For some packets the fragmentation flag will be set. This is an indication that reassembling will be required. This variable gives a count of number of packets with a fragmentation flag set.
- ipReasmFails/ipReasmOKs: Reassembling can be either a success or it will fail. The count of fails in reassembling is handled in the ipReasmFails variable and the count of reassembling success is maintained in ipReasmOKs. The failures are common when a specific fragment does not arrive in the input. A missing or lost packet can cause reassembling failure.
- ipInDiscards: This shows the number of packets discarded. Packets can be discarded due to internal problems in the system.
- ipOutDiscards: This variable gives a count of the output packets discarded.
- ipFragCreates: This variable keeps a count of the output packets with the fragmentation flag set.
- ipFragFails/ipFragOKs: Fragmentation can be either a success or it can fail. The count of fails in fragmentation is handled in ipFragFails variable and the count of fragmentation success is maintained in ipFragOKs.

There are few more read-only variables like ipInUnknownProtos, ipInDelivers, ipOutRequests, and ipOutNoRoutes that have not been discussed in this section. Now let us look into the tables in an IP Group. The tables are:

- ipAddrTable: This is the IP address table. It has an index field (ifAdEntIfIndex) and information like NetMask address that is used for classless interdomain routing.
- ipNetToMediaTable: This table uses ipNetToMediaIfIndex as the index field. It is used to map the physical address to a network address. A field ipNetToMediaType in this table can take values "static" when the address mapping is configured by an administrator, "dynamic" when it is automatically done using a protocol, or "in-valid."
- ipRouteTable: This table is used to give routing information including the route type and next hop or jump for a packet routing. The problem with this table is that its field ipRouteDest takes unique values, so multiple routes to a destination cannot be configured using this table. Since there can be multiple routes to a destination in real-life scenarios, this table was depreciated and a new table named ipForwardTable was introduced.
- ipForwardTable: ipForward (24) was introduced to ensure that multiple routes can be set to a particular destination. This object has ipForwardNumber (1) and ipForwardTable (2). The ipForwardNumber is an indication of the number of entries in the ipForwardTable. ipForwardDest field in this table is not unique, making it possible to enter duplicate values corresponding to different routes to the same destination. Other key fields include ipForwardNextHop for the next jump information, ipForwardIfIndex corresponding to interface index, ipForwardPolicy to specify the routing policy, and ipForwardProto to specify the routing protocol used. An example entry to ipForwardProto field is OSPF corresponding to Open Shortest Path First, which is a dynamic routing protocol.

e. ICMP Group (5)

This group has variables for storing information on Internet control message protocol (ICMP). Some of the variables in the ICMP group are:

- icmpInMsgs: This variable keeps a count of ICMP input messages.
- icmpInErrors: This variable keeps a count if input errors occur when the system uses ICMP protocol to receive data.
- icmpOutMsgs: This variable keeps a count of the number of output ICMP messages.
- icmpOutErrors: This variable keeps a count of output errors occurring when the system uses ICMP protocol to send data.
- icmpOutRedirects: This variable keeps a count of output message redirects that occurred with ICMP messaging.

f. TCP Group (6)

The transport control protocol (TCP) group has variables on the TCP connections. Some of the variables in this group are:

- tcpMaxConn: This is the maximum number of TCP connections permitted.

- tcpActiveOpens: This variable gives the number of active open TCP connections.
- tcpPassiveOpens: This variable gives the number of passive open TCP connections.
- tcpAttemptFails: This variable keeps a count of the TCP connection attempts failed.
- tcpEstabResets: This variable has information on number of resets performed on an established connection.
- tcpCurrEstab: It is an indication of the current established connection.
- tcpInSegs/tcpOutSegs: The number of input and output TCP segments is stored in tcpInSegs and tcpOutSegs, respectively.
- tcpRetranSegs: It keeps a count of segments retransmitted.
- tcpInErrs: The input errors for TCP connection is maintained in this variable.

This group also has a "tcpConnTable" connection table. This table has fields to capture the connection state, local address, local port, remote address, and remote port. The connection state is a writable field that can be changed and can take values like sync, abort, listen, and closed.

g. UDP Group (7)
   The user datagram protocol (UDP) group has a few variables and a table. Some of the variables are:
   - udpInDatagrams: This variable keeps a count of input datagrams.
   - udpNoPorts: The number of UDP ports at any point of time can be obtained using this variable.
   - udpInErrors: The number of UDP error messages can be kept in this counter.
   - udpOutDatagrams: This variable keeps a count of UDP output datagrams.

The udpTable is used for collecting information on the status of UDP connection.

h. EGP Group (8)
   The exterior gateway protocol (EGP) group has variables similar to the UDP group except that it is for the EGP protocol. Some of the variables that keep a count are egpInMsgs, egpOutMsgs, egpInErrors, and egpOutErrors.
i. Transmission Group (10):
   This group does not have any variables but is a place holder for MIB modules. Some of the modules covered in the transmission group are FDDI (15), RS-232 (33), LDAP (16), X.25 (38), and SONET (39).

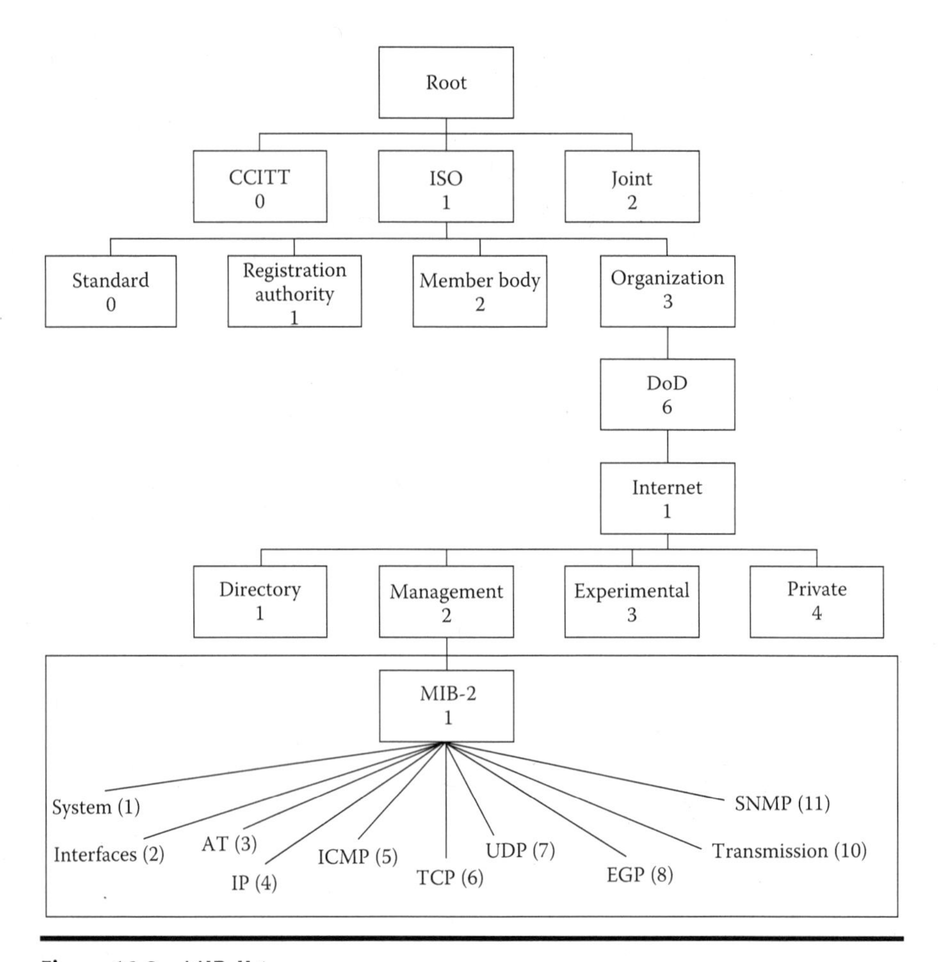


**Figure 16.3 MIB-II tree.**

j. SNMP Group (11)
This group is for the simple network management protocol (SNMP). It has a lot of counters to capture information on the PDUs (packet data units). The various kinds of PDUs, the number of PDUs received, the number of PDUs transmitted, and the PDU errors are all captured using counters. It is also possible to enable and disable authentication traps. Being the MIB for the most popular network management protocol, SNMP MIB is handled in more detail in the next section. SNMP has three popular version and the MIB changes associated with each of these versions are discussed under SNMPv1 MIB, SNMPv2 MIB, and SNMPv3 MIB.

## 16.4 SNMPv1 MIB

SNMP uses ASN.1 for defining objects and PDUs exchanged by the management protocol. Only a subset of ASN.1 data types is used. They are INTEGER, OCTET STRING, OBJECT IDENTIFIER, SEQUENCE, SEQUENCE OF, and NULL. RFC 1155 states how objects are defined. In MIB, object types have an OBJECT DESCRIPTOR that is a textual name and an OBJECT IDENTIFIER. A representation bladeLocation {blade 10}, has bladeLocation as the OBJECT DESCRIPTOR, and {blade 6} as the OBJECT IDENTIFIER.

The OBJECT IDENTIFIER for the SNMP subtree is 1.3.6.1.2.1.11. Objects can be added to the tree for data collection. For example a count of SNMP input messages can be maintained in snmpInMsgs using the OBJECT IDENTIFIER is 1.3.6.1.2.1.11.8, where 8 is the subidentifier for this new object. Each scalar object of the input message object has only one instance and it is identified by concatenating 0 to the OBJECT IDENTIFIER of snmpInMsgs. This instance will be represented as 1.3.6.1.2.1.11.8.0.

Tables in SNMP are two-dimensional in nature and only the objects in a table, known as columnar objects, can be manipulated by SNMP. Instances of objects in tables are identified by the INDEX clause in SNMPvl where this clause refers to a row in a table. Instances of objects in tables can also be identified using the AUGMENTS clause, which is an addition in SNMPv2. They retrieve values from a table and the length of each column are traversed completely before moving to the next column. To add the value of an object instance to the table, the value is entered in a row using the SetRequest operation and to delete an entry the value is set to invalid using SetRequest operation.

A specific set of rules is followed in defining new MIBs. This is required because multiple versions of MIB can coexist. Some of the rules are:

- Old object types are only deprecated, they are not deleted.
- The semantics of old object types are not changed between versions.
- Instead of changing the semantics of an old object, a new object type is formed.
- Only essential objects are defined to keep SNMP simple.
- There is enough flexibility to define implementation specific objects.

## 16.5 SNMPv2 MIB

The structure of management information (SMI) for SNMPv2–RFC 1902 describes rules for defining modules, objects, and traps. Some of the definitions articulated in this document are:

- The macro MODULE-IDENTITY is used to provide the syntax and semantics for defining a module. Information module is an example of a module identified by a MODULE-IDENTITY macro. There are three types of information modules and it is not mandatory that an information module should have all these three modules. The three types are:
  - MIB modules that contain information on objects.
  - Compliance statements that provide compliance requirements for MIB.
  - Capability statements that contain information on the capabilities of the agent implementation.
- The macro OBJECT-TYPE is used to provide the syntax and semantics for defining an object.
- The macro NOTIFICATION-TYPE is used to describe a notification. A notification is issued as confirmation for an event.

Some new terms that have been introduced in RFC 1902 for SNMPv2. These include:

- BITS: Enumeration of named bits
- Unsigned32: Unsigned integer
- MAX-ACCESS: Maximum level of access allowed for an object
- Counter64: For keeping count
- UNITS: Textual units of measurements associated with an object
- AUGMENTS: Gives the object instance identification in a table. It is an alternative to the index clause.
- IMPLIED: For objects with a variable-length syntax

In the hierarchy tree SnmpV2 has many subtrees. Some of these subtrees are:

- snmpDomains(1): It is for the domains used in transport mapping
- snmpProxys(2): It is used for giving the proxies used in SNMP transport
- snmpModules(3): It is used for defining the module identities. Under the snmpModules there is again three subtrees:
  - snmpMIB(1): This is the agent-to-manager interaction MIB
  - snmpM2M(2): This is the manager-to-manager MIB
  - partyMIB(3): This has been removed from the SNMPv2 MIB in RFC 1907.

There are subtrees below the above levels also and further levels below these subtrees. "snmpMIB" under snmpModules, has snmpMIBObjects and snmpMIBConformance subtrees. Below the snmpMIBObjects, we have snmpTrap(4), snmpTraps(5), and snmpSet(6).

## 16.6 SNMPv3 MIB

SNMPv3 has many additional objects introduced to account for the new security features that were introduced like the user-based security model and view-based access control model. Some of the MIB modules present in the object hierarchy of SNMPv3 are:

- snmpFrameworkMIB: This is used to define the MIB module for SNMPv3 architecture. The subtrees under it in addition to snmpFrameworkConformance are:
  - snmpFrameworkAdmin: This is used for the registration of SNMPv3 authentication and privacy protocol objects.
  - snmpFrameworkMIBObjects: This is used to define SNMP engine-related objects like:
    - snmpEngineID
    - snmpEngineTime
    - snmpEngineBoots
    - snmpEngineMaxMessageSize
- snmpMPDMIB: This has the MIB module for message processing and dispatching. The subtrees under it in addition to snmpMDPConformance are:
  - snmpMPDAdmin: This module is used for the registration of message processing and dispatcher objects.
  - snmpMPDMIBObjects: This provides a collection of objects for collecting information like number of packets dropped.
- snmpProxyMIB: This module has objects used to remotely configure the parameters in a proxy forwarding application. The proxy objects in snmpProxyMIB are defined under snmpProxyObjects.
- snmpTargetMIB: This module has objects that are used for defining the addresses, destinations, and transport domains of the targets. In addition to snmpTargetConformance, there is snmpTargetObjects. The snmpTargetObjects in snmpTargetMIB module has tables that capture information about the transport domains and security information for sending messages to specific transport domains.
- snmpNotificationMIB: This module has objects for configuring parameters to generate notifications. The snmpNotifyObjects under this module is used to set the notification types and targets that should receive notifications. It also has snmpNotifyConformance.
- snmpUsmMIB: This module has objects used in the user-based security. The usmMIBObjects perform authentication and ensure privacy is provided for management operations of users. This is along with usmMIBConformance for the conformance statements.

- snmpVacmMIB: This module has objects for view-based access control. The vacmMIBObjects can provide details of the access. This module also has vacmMIBConformance.

The RFCs under which these modules are described are:

- RFC 2271 for snmpFrameWorkMIB
- RFC 2272 for snmpMDPMIB
- RFC 2273 for snmpProxyMIB, snmpTargetMIB, and snmpNotificationMIB
- RFC 2274 for snmpUsmMIB
- RFC 2275 for snmpVacmMIB

## 16.7 Conclusion

This chapter is dedicated to the discussion of management information base and SNMP MIBs are also discussed briefly in this chapter. The MIB is a vast topic and there is much to be included to give a detailed understanding of MIB. There are books dedicated to understanding MIBs, to the extent that some books only detail a specific MIB group in the tree. The concept of information handling was handled in Chapter 15 to give a strong base before giving an overview of MIB. For professionals implementing a specific MIB, it is suggested to refer to the books suggested under this next section and also the relevant RFC.

## Additional Reading

1. Alexander Clemm. *Network Management Fundamentals.* Indianapolis, IN: Cisco Press, 2006.
2. Larry Walsh. *SNMP MIB Handbook.* England: Wyndham Press, 2008.
3. Stephen B. Morris. *Network Management, MIBs and MPLS: Principles, Design and Implementation.* Upper Saddle River, NJ: Prentice Hall PTR, 2003.
4. M. T. Rose. *The Simple Book.* Upper Saddle River, NJ: Prentice Hall, 1996.

# Chapter 17

# Next Generation Network Management (NGNM)

Networks are evolving rapidly to meet current demands of unification and convergence. This new holistic view of the network also requires significant change in the management of the network. Many standardizing bodies are working in this regard for coming up with the requirements for Next Generation Network Management (NGNM). This chapter is intended to give the reader an overview of NGNM. Some of the most popular documents on NGNM, M.3060 from ITU-T, and TR133 from TMF are discussed to provide information on the direction in which standards are being defined for managing next generation networks (NGN).

## 17.1 Introduction

Current network management solutions are targeted at a variety of independent networks. The wide spread popularity of IP multimedia subsystem (IMS) was a clear indication that all of the independent networks will be integrated into a single IP-based infrastructure referred to as next generation networks (NGN). The services, network architectures, and traffic patterns in NGN will dramatically differ from the current networks. The heterogeneity and complexity in NGN, including concepts like fixed mobile convergence, will bring a number of challenges to network management. Various standardizing bodies are coming up with specifications for NGNM to address these challenges. This chapter is intended to shift the focus from simple network management concepts to next generation network management.

The high degree of complexity accompanying the network technology necessitates network management systems (NMS) that can utilize this technology to provide more service interfaces while hiding the inherent complexity. As operators begin to add new networks and expand existing networks to support new technologies and products, the necessity of scalable, flexible, self-manageable, and functionally rich NMS systems arises. Another important factor influencing network management is mergers and acquisitions among the key vendors. Ease of integration is a key impediment in the traditional hierarchical NMS framework. These requirements trigger the need for a framework that will address the NGNM issues.

This chapter will discuss the NGNM concepts in the TR133 TMF document and M.3060 ITU-T document. TeleManagement Forum (TMF) has published a member evaluation version of NGNM strategies (TR133) that considers the challenges involved in the development of an NMS system for a multiservice, heterogeneous network environment. TMF has created an NGNM lifecycle comprising of business, system/architecture, implementation, and deployment stages to explain the concepts and ITU-T document M.3060 covers NGNM architecture as four different architecture views, namely the business process view, management functional view, management information view, and management physical view. Other standardizing bodies like 3GPP, OASIS, and ETSI/TISPAN are also participating in defining standards for NGN and management of NGN. Next generation management solutions are intended for multitechnology (CDMA, IMS, WiMax, etc.), multivendor (common information and interface standards defined), and multidomain (enterprise, telecom, etc.) kind of environment, bringing in the concept of a unified solution that meets all requirements.

## 17.2 NGNM Basics

Multiple standardizing bodies are coming up with standards for NGNM. The intent of TMF with the release of TR133 was to come up with details of the progress so far by other bodies, how liaison can be formed between bodies working on developing standards in the same space, and how to realize business benefits by bringing up standards for NGNM. The holistic view of telecom space is achieved in TR133 by splitting the NGNM space across business, system/architecture, implementation, and deployment. It is in line with the TMF (TeleManagement Forum) concept of NGOSS (Next Generation Operation Systems and Software), which demarcates the service provider space from service developer space and technology specific implementation from a technology neutral implementation.

NGNM originates from the need for managing the next generation network (NGN). NGN according to ITU-T standards is a packet-based network that can support a wide variety of services. Some of the key buzzwords around NGN are its ability to offer services anytime and anywhere. It is multiservice, multiprotocol, and multiaccess-based network. NGN also fosters a multitechnology and multivendor environment. Most of all it provides a secure and reliable communication channel. Now let us explore these terminologies in more details. Anytime and anywhere means that it is not tied to a specific kind of network or equipment to access the services offered on NGN. The services are always on and moving, change in underlying infrastructure or access technology does not affect the customer in terms of accessing the service. Multiservice means that multiple services can be hosted on the network, multiaccess means that the access layer can be fixed, mobile terminals, or any other kind of access technology, and multiprotocol means that published and well-defined protocol interfaces make it possible to use different types of protocols for interaction. The next generation network fosters multitechnology that is support for technologies like GSM, UMTS, CDMA, and multivendor that is open standards for ease of integration leading to different competing vendors.

NGN has the following characteristics:

- Packet backbone for communication
- Separation of service, access, and core network
- Unified service experience to end user
- User access to different service providers leading to service mesh up
- Convergence of fixed and mobile space
- QoS (Quality of Service) determined on an end-to-end basis
- Open interfaces for interaction
- Wide range of services and applications

Managing a network with the above characteristics ensures that everything from the end customer to the lower most nodes in a network element is managed, with seamless information flow from end-to-end. This makes minimal manual intervention a key feature to be available in next generation network management. Traditionally there were different standardizing bodies working different technologies in telecom. With convergence in the telecom domain, these standardizing bodies need to work together in developing standards that can be used across technologies. Convergence is also happening at the domain level. All telecom companies have an enterprise aspect. That is in addition to the telecom infrastructure there is also an enterprise network that needs to be managed. Convergence of enterprise and telecom domain leads to the need for standards that makes it possible for a single management solution that can manage both the telecom and enterprise network in a company.

Let us next look into an outline of NGN management requirements and what is usually defined in this space, which is not currently limited to just a network. A general set of capability map for NGN management was captured as part of the OSS vision document of ETIS TISPAN. Many other standardizing bodies also contributed in this space and came up with an extensive set of NGNM requirements. The major ones that came out with an initial set of NGNM requirements in addition to ETSI TISPAN are ATIS and ITU-T. TMF has compiled these requirements in TR133 document.

Given below are some of the main categories in which NGNM requirements are defined.

1. Customer centric requirements: Customer needs and end-user experience is a key capability that drives current industry. This has led to specific requirements centered on the customer. In a competing telecom market the best vendor is the one that can satisfy customer demands. So choice of service provider is a default requirement of NGNM. Personalization and mobility is one key customer requirement in next generation network management. Personalization would involve each customer getting the experience he/she wants, this way the service capabilities are customized to suit the needs of the specific user. Personalization is coupled with mobility to ensure that the NGN infrastructure has features for the personalized services to always follow the customer. Personalization and mobility are usually based on the user role. Another customer centric requirement is flexibility of service. Flexibility ensures service anytime and where, using any access mechanism chosen by the customer using the terminal device of choice. The customer should also be able to aggregate his services and understand the different billing models that are used. This means that the services are bundled and the billing model is simplified to improve customer experience. Two more high level requirements would be self-care strategies and security. As part of NGNM most service providers are giving capabilities to the end customer to handle the issues that they usually encounter when using a service. This would include functionality like calling a hot line where a software program will try to understand the problem that a user is facing and give the user steps to fix the problem. This way the end customer does not have to wait for an operator and in most cases can get an issue fixed within a few minutes. Security from a customer perspective would be to ensure that the content is secure and reliable and not misused in anyway by the service provider.
2. Service requirements: The main requirement in service space is to increase the number of services offered by the service provider. Killer application or the single application that can capture the market and build a customer base now mostly being a myth, the sure strategy to get customers is to increase the number of services. One of the adopted methods to achieve this without burden on capital expenditure is to mesh up existing services to create a new service. Diversity in the services that are offered is another service level requirement. Service management spans multiple areas like fraud management, billing management, provisioning, service fulfillment, and so on. NGNM service level requirement is to bring in agility to service management by mass customization. Similar to simplicity in customer experience, the service management should be made simple using the customization.

Context and location-based service can be specified as a requirement in a service level under NGNM, though this has already been implemented in most systems. Context aware services would ensure that the customer would be offered a service based on the current context. An example would be that the customer takes some pictures using his/her user equipment from a place that does not offer internet access due to reasons like low signal strength. Now the pictures get stored in the user equipment and when the customer moves to a location where the signal strength is restored, he/she is prompted with a message on whether the service of uploading the pictures to a personal album on the web needs to be issued as a service request. This message to offer the service is based on the user context and understands what the user might want. Location-based services identifies the user location and sends appropriate service requests that might be of interest to the user when in a specific location.

3. Operation requirements: Minimal manual intervention is the key operation requirement in next generation network management. Automatic provisioning, as well as automatic handling of issues, is expected from new management capabilities. A single click to provision a service should enable all the service capabilities in configuring the service parameters, setting up the underlying network to provisioning individual elements in the network to offer the service and finally to provision billing of the customer for the service. Event driven management system should parse events to identify issues, raise alarms in the case of SLA jeopardy, and take corrective action based on preprogramed rules to avoid SLA breach. The operational environment comprising of the service, network, and enabling technology needs to provide quick service delivery and high levels of service assurance. Having an underlying process in place that supports optimal operations is also a requirement in NGNM. The process should be lean in that with minimal effort more operational capabilities must be possible.
4. Business/industry direction requirements: Mergers and acquisitions are rising and the NGN management capabilities should allow ease in the process of consolidating process and products. Interoperability is a key requirement and seamless integration should be available for products to compete in the market. Most management products are expected to follow open information models based on SID (shared information and data)/CIM (common information model). The management interfaces are also published and based on standards like MTOSI (Multi-Technology Operations System Interface) and OSS/J (OSS through Java Initiative). With management products that are based on open standards building adapters for communication to integrate with other products is easy. Also when the business process is aligned with telecom process frameworks like eTOM, it is easier to acquire or merge the business process for a particular service offering. The FAB component (fulfillment, assurance, and billing) needs to align to best meet the interests of

the customer. The business should be focused on meeting the SLA set with the customer. Along with quality of service from the service provider space, the quality of end-user experience from customer space needs to be parameterized and measured. A key industry direction is toward service oriented architecture and COTS (commercial off-the-shelf) based products. This means that reuse should be maximized. Service abstraction, service discovery, and service reusability features of SOA are suitable for providing services that can be extended to legacy and next generation at the same time. New NMS solutions especially for next generation network and services are based on loosely coupled, service discovery architecture. COTS-based implementation also leads to reuse when the product is developed for integration with a wide variety of components. The main intent of splitting to components is to achieve better management at the component level and at the same time having the entire system controlled by managing specific components, thus reducing the complexity.

5. Other requirements associated with NGNM: Compliance to standards is a key requirement for NGNM. This includes standards given by telecom and enterprise standardizing organizations, compliance specified by regulatory bodies and legal obligations. Regulatory requirements include emergency services and legal obligations can include security features like lawful interception.

With multiple NGNM requirements from different standardizing bodies, service providers and developers have difficulty in determining what standards to use. So liaison of standardizing bodies to do collaborative work and combining standards and refining them to suit business needs is also being looked into as an outcome of NGNM initiatives.

## 17.3 TR133

TeleManagement Forum (TMF) has published a member evaluation version of NGNM strategies (TR133), which suggests a lifecycle approach to give holistic view to NGNM. The lifecycle model has four blocks (see Figure 17.1) that correspond to intersection of stake holder and technology parts. Stake holder part has the service provider and service developer space. Technology part has the technology neutral and technology specific space. The four blocks are:

- Business: This block is the intersection of technology neutral space and service provider space. The service provider defines requirements in a technology agnostic manner, which is the business block.
- System/architecture: This block is the intersection of technology neutral space and service developer space. Based on the requirements defined in a business

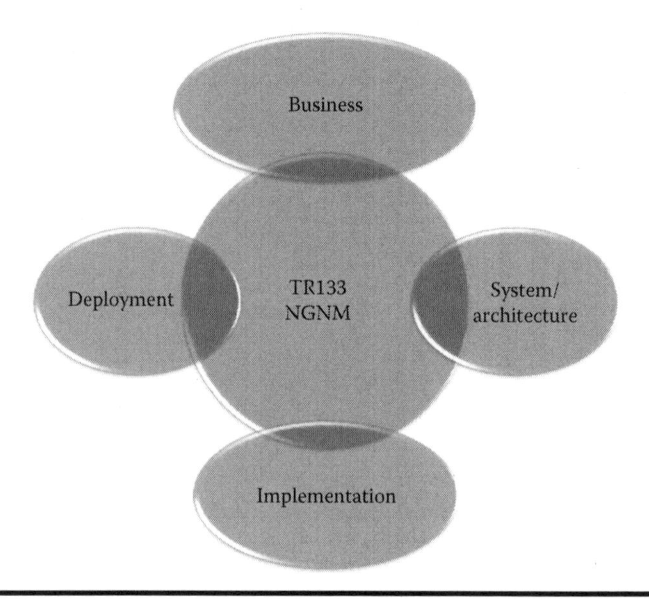


**Figure 17.1 NGNM lifecycle. (From TMF TR133 NGNM Strategy document).**

block by service provider, the service developers create an architecture that is not tied to a specific technology or vendor product and is a technology neutral architecture (TNA) that satisfies the requirements.

- Implementation: This block is the intersection of technology specific space and service developer space. A technology specific implementation of the TNA is performed in this block. The implementation is performed by the service provider.
- Deployment: This block is the intersection of technology specific space and service provider space. Once the implementation is completed by the service developers, the service providers need to deploy the solution. This activity is captured in the deployment block.

These four blocks constitute the telecom lifecycle and is used to give holistic vision in all forms of development activities. The lifecycle is discussed in detail in the chapter on NGOSS lifecycle. NGNM focus is also split into these four blocks by TMF in the TR133 document. Rather than activities in the block, TR133 shows what should be the contribution from the key stake holders of individual blocks toward NGNM.

1. Business analyst dealing with business block in NGNM
   - The operational parameters in a service provider space needs to be clearly defined for which the business analyst needs to define key performance indicators (KPIs) for operations that can be measured and monitored. For

proper monitoring of these operational KPIs, there should be tools that can capture the required information and represent it in a user-friendly format like plots/graphs or events that can be tagged. An end-to-end process needs to be defined for the same.

- Revenue assurance is an important part of business and has key challenges to be addressed in the NGN environment. Research and resulting white papers to address the issue in revenue assurance can be brought out by a business analyst. Another key area where white papers are required from a business analyst is in fraud management space. The NGN is an IP-based network and the chances of security breach and fraud that existed in the IP world is valid in the case of NGN. Fraud management is a key area of focus for NGNM and the business analyst can contribute with white papers in this space.
- OSS development needs input from the business analyst and so regulatory guidelines are required for telecom space. The changes in regulations have an impact on the business guidelines on the company and the regulatory implications in an NGN environment can be identified by the business analyst.

2. System analyst/solution architect dealing with system/architecture block in NGNM
   - System analyst/architect can look into technical marketing that involves promoting NGOSS components including eTOM, SID, TAM, and TNA as the enabler for managing heterogeneous networks with common models. Various NMS solutions are coming up in the market that have capabilities to better manage NGN and are built using NGOSS models. Legacy solution needs are best aligned with NGOSS to integrate these applications. Adoption of eTOM as a standard by ITU-T shows the wide spread use of NGOSS concepts. Legacy System owners can start by preparing a business map, information map, application map, and architecture map of their telecom system. The next step involves use of NGOSS methodology of SANRR (scope-analyze-normalize-rationalize-rectify). The end result of multiple iterations with SANRR is a set of level 4 eTOM business models, well-defined interfaces, and contracts with SID businesses and system models, technology neutral architecture wherever possible in implementation, and a detailed application map.
   - Architecting based on SOA (service-oriented architecture) is another area of focus for architects. Benefits of using SOA in NMS, have driven NMS developers to come up with new SOA-based NMS suites. NGOSS does provide guidelines for developing SOA, but what is required is an architecture that can convert the functionality rich legacy NMS solutions to SOA, addressing the capabilities of NGNM.
   - Architects need to look at SID as the ideal abstract common model that provides an excellent foundation for OSS/BSS integration. In addition to

using SID business and system views to provide well-defined contracts of the existing system, SID can also be used for mapping application interfaces and attributes for reuse. Applications that can enable content-based transformation logic to be captured on the SID without writing low-level code and capture validation rules on the SID are already available in the market. Enterprise common information model (CIM) and telecom shared information and data (SID) mapping is an area TMF and DMTF are jointly working on based on the convergence in management of telecom and enterprise space.

- COTS integration is one of the major drivers in developing meta-models by NGOSS and architects needs to play a key role in promoting the use of meta-models. In a multivendor heterogeneous network environment, integration is the key word. From an NMS perspective, this would involve adding management functionality supplied by multivendors for various network elements to an NMS. Having a holistic telecom view and designing systems that can be easily integrated would involve following common models at various stages of development. NGOSS defines a set a meta-models that would ease integration.

3. Implementation provider dealing with implementation block in NGNM
   - The main goal of the implementers is to churn out NGOSS solutions to meet current market requirements in a timely manner. The activities should target deliverables that are ready to use and can be plugged into the telecom management solutions currently available. An example of this would be an MTOSI adapter that will convert messages from legacy calls in the solution to interactions that are compliant with MTOSI.
   - Building the technology specific implantation over the technology neutral architecture of TMF and working on a feedback loop between technology neutral and technology specific implementation is a key activity that will help in enhancing the TNA. Software vendors need to make implementations based on the guidelines of holistic telecom vision and ensure that seamless information flows happen between different blocks to make final deployment a success with minimal faults.
   - Prosspero is an activity group in TMF that is aimed at providing "ready-to-go" standards. Based on the standards recommended, vendors participating in this activity group create solution packages having reference code, guides, and test information. The implementers are the major contributors for this activity of creating solutions that will help companies to easily adopt the standards.
   - Telecom application map (TAM) of TMF is an application map in telecom space similar to the eTOM process map. The intent of an application map is to identify the areas in telecom space where applications are developed to meet specific requirements in resource, service, and business management. The implementation provider needs to ensure that TAM is

defined to the lowest level of granularity process to make it beneficial for adoption.
- The development of procurable NGOSS contacts based on TAM also needs to be driven by the implementation provider. The implementation of a contract library including strategy for developing contract registry and repository vests on the implementation provider.
- In addition to solutions based on standards done as part of Prosspero, work should also go into development of tools that will aid the lifecycle, methodologies, interfaces, models, and maps defined by TMF.

4. Operations /deployment team dealing with the deployment block
   - The operations center should be implementing component-based management that uses the capabilities provided by NGOSS system management architecture and reference implementations.
   - Another activity of the operations center is to do business and service performance tooling. The performance tooling will include QoS, business activity, and equipment performance measurements to meet the SLA agreed upon with the customer.

## 17.4 M.3060

According to this document, NGNM is not a new generation of management systems, but a management system that supports monitoring and control of the NGN resources and services. The resources of NGN has both a service component and transport component. The role of NGNM is to have seamless flow of data across interfaces from the network element layer to the business management layer. Having both network and services in the scope of management, the suggestions of ITU-T addresses both network operators and service providers.

The NGNM architecture can be divided into four views as shown in the diagram (see Figure 17.2). They are:

- Business process view (architecture): This view spans across the three other views and is concerned with the business process framework in the service provider space. eTOM is suggested as the reference framework for process view.
- Management functional view (architecture): This view is to capture the management functions to be implemented in NGNM. The management functions are categorized under a set of blocks like the operations systems function block (OSF) and network element function block (NEF).
- Management information view (architecture): In order for entities to perform functions that are defined in a functional view, these entities need to communicate management information. The characteristics of this management information are defined in this view.

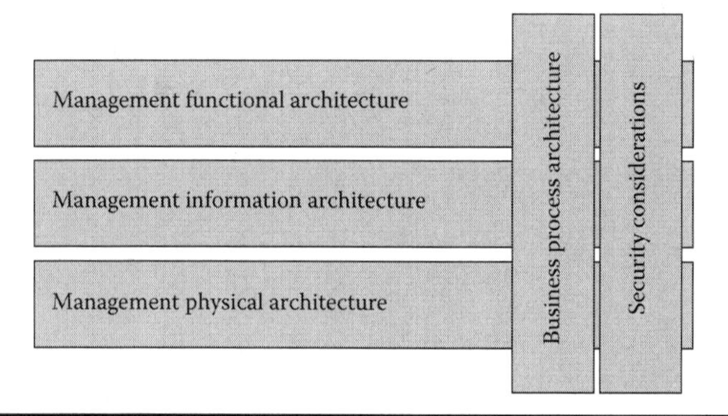


**Figure 17.2 NGNM management architecture. (Based on ITU-T documentation.)**

- Management physical view (architecture): The deployment and physical configuration for mapping the functions is a critical part of network management and is handled in this view.

Security considerations are also shown as spanning across the physical, information, and functional view. Security in the next generation network and the NGN management infrastructure is part of NGNM and is given under a separate set of definitions from ITU-T.

Another important aspect in the M.3060 document is the logical layered architecture (see Figure 17.3) that brings back the conventional concept of dividing the management pane into a business management layer (BML), service management layer (SML), network management layer (NML), element management layer (NML), and network element layer (NEL). A set of functional blocks are defined across each of these layers to list the entities and management functions that span across the different layers in NGNM as defined in M.3060.

The M3060 identifies NGNM functionality to support NGN by:

- End-to-end management of physical and logical resources. The networks to be managed would include core network, access network, and customer networks. The interconnect components and the supporting terminals also need to be managed.
- Management platform that ensures secure access to customer and management information.
- Have management segregation between the service and transport layers, ensuring there will not be any dependency for service management on a specific form of transport.
- Management capability available to authorized users any time and any place.
- Management capability to ensure that customers can personalize their service as well as create new services using the services offered on NGN.

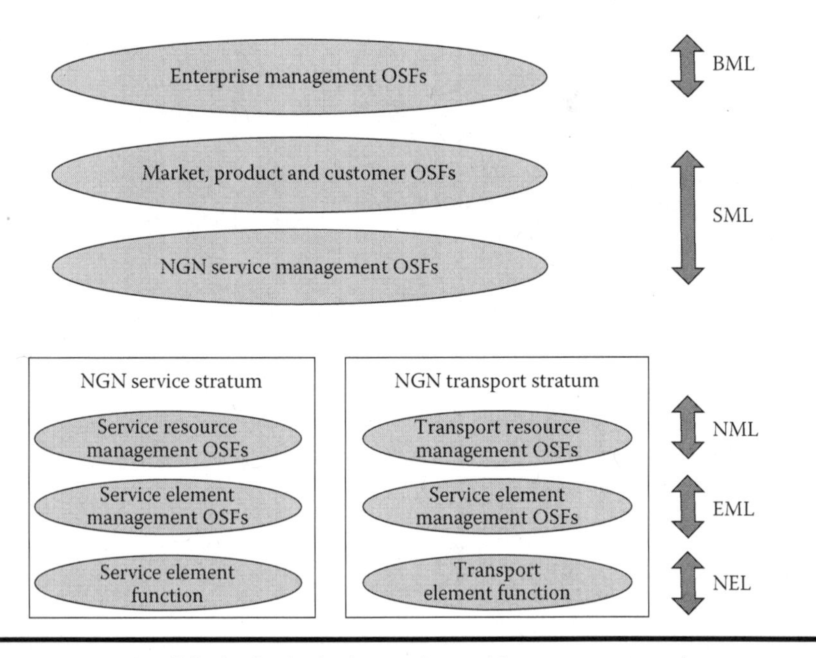


**Figure 17.3 Simplified logical layered architecture. (Based on ITU-T documentation.)**

- Management capability that can offer end-user service improvements and ability to implement customer self-care strategies.
- Management solution to support eBusiness value networks based in concept to roles like customer, service provider, supplier, and so on.
- Management layer should form an abstraction layer in hiding the complexity of technologies and domains in the resource layer made up of network, computing resources, and application.
- Have support for different forms of billing process, where the accounting layer in the management solution can do both offline and online charging.
- The management solution should be an autonomic computing environment where bug fixing can be handled with minimal manual intervention.
- The management solution should be able to do dynamic end-to-end provisioning. This means that when the operators request for activation of service, the management solution should provision the service, provision the network, and configure the individual network elements to offer the specific service to be activated and start billing for the activated service.
- A standard-based management platform that is not tied to a specific organization, technology, or domain.
- An event-based management platform that does trend monitoring to take actions in the event of an error and maintains the quality of service.

- Well-defined interface and information points between the transport and network stratum, the management and control plane, and finally between administrative boundaries.
- Provide an integrated view of management across various planes from element to business management.
- NGN management should have an architecture and supporting business process that will go along with seamless introduction of new services.

M.3060 was one of the first and most comprehensive documents from a standardizing body. Many other standardizing bodies followed the lead of ITU-T and came up with specifications on next generation network management to satisfy the requirements in various technology and domain segments.

## 17.5 Conclusion

ITU-T and TMF came up with many more documents for NGNM, other than M.3060 and TR133. Some of the notable ones include ITU-T M.3016 series on security for the management plane, ITU-T M.3050 series on enhanced telecom operations map that is based on the TMF document GB921, ITU-T M.3341 for QoS/SLA management service, ITU-T M.3350 on ETS management service, TMF shared information and data model GB922, and TMF513 on multitechnology network management (MTNM).

Other standardizing bodies that have published specifications for NGNM include:

- 3GPP: The specifications were mainly on charging management and integration reference points (IRP) for alarm, configuration, notification, and state management.
- ATIS: Specification on usage data management for packet-based services.
- DMTF: Their main contributions include the common information model (CIM) and web-based enterprise management (WBEM).
- ETSI: Main documents were TS 188 001 on NGN OSS architecture, TS 188 003 on NGN OSS requirements, and TS 188 004 on NGN OSS vision.
- IETF: Contributions include the NETCONF protocol that is mainly quoted as NGNM protocol, SNMPv3, and SMIv2 (structure of management information version 2).
- MEF: Main document was titled EMS-NMS information model, which was to detail interface standards for Metro Ethernet Network.

There were many more documents on NGNM from various standardizing bodies and this chapter is just to provide a basic understanding of NGNM to the reader.

## Additional Reading

1. M.3060 document that can be downloaded from www.itu.int
2. TR133 document from TMF that can be downloaded from www.tmforum.org
3. Guy Pujolle. *Management, Control and Evolution of IP Networks.* New York: WileyBlackwell, 2008.
4. Yan Ma, Deokjai Choi, and Shingo Ata. *Challenges for Next Generation Network Operations and Service Management.* New York: Springer, 2008.

# Chapter 18

# XML-Based Protocols

XML (Extensible Markup Language) based protocols are rapidly replacing traditional management protocols like SNMP and are now the defector standard for use in web-based network management. Protocols like SOAP, which are XML-based, can be viewed in management standards like WBEM for enterprise management defined by DMTF as well as MTOSI for telecom management defined by TMF. This chapter is intended to give the reader a basic understanding of XML protocol. SOAP the most popular XML protocol and NETCONF hailed as the next generation network management protocol is discussed as specific examples.

## 18.1 Introduction

Traditionally SNMP has been the preferred protocol for network management of packet networks like Ethernet, Token-ring, and so forth. Efficient and low bandwidth transport mechanisms using UDP over an IP) as well as an optimized, but limited, data model were the two main reasons for its popularity. The simplistic data model architecture used by SNMP (a flat MIB definition) was well suited for low complexity legacy packet nodes.

With the convergence of traditional circuit (voice) and packet networks (data) onto a common IP-based packetized core as in next generation networks like IMS, the complexity of the network elements involved, and the services supported have undergone a huge change. With convergence in networks space, a flat and simple model like SNMP does not scale up to meeting the next generation network management requirements. A typical example of convergence at a network element level is an L3-L7 switch in a converged network. This switch performs multiple other functionalities in addition to switching including, but not limited to, authentication, encryption, and application routing. As a result, the information base to be managed is much more complex. Attempts to serve this complex information base using SNMP has resulted in a force-fitting of this information into the limited 2-dimensional SNMP MIB model. This is clearly indicated in the proliferation of "helper tables" in any modern network element MIB. Another characteristic difference between legacy and next-generation packet networks is the number of network elements involved. Next generation networks have witnessed a mushrooming of both numbers and types of network elements to support a variety of services. This effective collection and analysis of events from all these elements is another area where a trap-based notification model of SNMP is showing signs of strain.

The discussion so far was on why SNMP is not suitable for NGNM and the need for a new protocol. XML as a language very much suited for network management because it helps to define the structure of management information in an easy and flexible manner. XML is able to reliably transfer management data over widely deployed protocols like HTTP and TCP. Using the defined structure, XML ensures interoperability among management applications from different vendors and facilitates better integration of management data from disparate sources by flexibly linking between a managed object and management application. XML-based protocols bring in the advantages of XML for information exchange and also address the challenges in current network management making it the preferred protocol for web-based management across multiple domains like enterprise, telecom, and storage.

## 18.2 XMLP Overview

Let us start with the XML protocol (XMLP) requirements from the World Wide Web Consortium (W3C). XMLP according to W3C is a lightweight protocol that is used for exchanging structured management information in a distributed environment.

The key advantages of using XML protocol are:

- Usage of XML for information exchange, which brings with it a wealth of readily available convertors, parsers, and other development and implementation toolsets.
- XML protocol ensures that management information can be defined in an easy, flexible, and powerful manner.
- XMLP provides a message construct that can be exchanged over a variety of underlying protocols. This makes it independent of the underlying transport protocol and suited for situations where an extensible lean transport protocol is required. One of the focuses of the XML protocol work group (WG) was to ensure the transport protocol independence.

- XML is well suited for distributed environments where there are inbuilt mechanisms for usage scenarios on involving multiple data readers, multiple data writers, and other control entities.
- XMLP provides flexible linkages between managed object and management application. This is again a projection of the distributed nature of XMLP.
- XML protocol provides a strong security paradigm with capabilities for automatic and centralized validation.

The intent of defining requirements on XML protocol (XMLP) was to create a foundation protocol whose definition will remain simple and stable over time. The design based on this foundation that explicitly used modularity and layering will assure longevity and subsequent extension of the design keeping the foundation intact.

XMLP inherently supports the following:

- Message encryption: This means that the message header and the associated payload can be encrypted when using XML protocol.
- Multiple intermediaries: The protocol support third-party intermediary communication where the message can be collected, parsed, and worked with. There would be transport and processing intermediaries.
- Non-XML data handling: This means the protocol can handle sending of non-XML data, which may be a required functionality when working with certain legacy systems.
- Asynchronous messaging and caching with expiration rules that will improve efficiency due to latency or bandwidth.

Before we get into the details of message encapsulation, let us quickly check some of the usage scenarios for use of XML protocol.

- Fire-and-forget to single receiver and multiple receivers: The sender wants to send an acknowledgment message to a single receiver or broadcast it to multiple receivers.
- Request-response: This involves one entity requesting information and the receiver processing the request and sending a response.
- Request with acknowledgment: Here the sender expects an acknowledgment from the receiver on the status of data delivery.
- Request with encrypted: Here an encryption methodology is agreed between sender and receiver and this is followed in the message exchange where the payload, and in some cases the header also, will be encrypted.
- Single and multiple intermediaries: Here an intermediary is present between the sender and ultimately the receiver.
- Asynchronous messaging: Here the sender does not expect an immediate response from the receiver. The sender will tag the request with an identifier to correlate with the response when it arrives.

## 18.3 XML Protocol Message Envelope

The XMLP message envelope is the data encapsulation component of XMLP protocol. The message envelope is a key component of the XMLP protocol that allows it to be extremely versatile and adaptable for use in carrying management information.

The envelope is a syntactical construct that contains:

- XMLP header block: There is only one header block for multiple body blocks. The header block consists of:
  - Routing information for the message envelope.
  - Data describing the content of the body blocks.
  - Authentication and transaction information.
  - Flow control and other information that will help in message management.
  - XMP namespace field for grouping. It can also be used for versioning to ensure a transparent conversion of the message for different versions supported by the various XMLP processors.
- XMLP body blocks: This is the actual payload of the message and there will be multiple body blocks in a single message envelope. The body blocks do not have any information about message transfer and is only processed at the final destination.

## 18.4 XML Protocol Binding

The binding defines the mapping of an XMLP message, also called an XML infoset into a specific "on-the-wire" representation. This on-the-wire representation is specific to the transport and application protocol that is used to actually send the message across. So the advantage of making message definition in XML protocol independent of the transport is achieved using the concept of XMLP binding. For example, the same XML infoset could have:

- An XML document representation that could be sent across a vanilla TCP/IP session.
- A DOM object representation that could be used to send it in a distributed computing environment.
- An HTML object representation that could be used to send it as a regular Web page.

The XML infoset (short for information set) is an abstract data model to describe the information available from an XML document based on specification from W3C. This makes it easier to work on an XML document without the need to analyze and interpret the XML syntax. There are different ways by which information in the XML infoset can be accessed with DOM object and HTML being some of the ways that have specific APIs to satisfy the requirements of the programming language. XML protocol binding with HTTP is the most popular transport for most specifications considering that the XML protocol can be used for Web-based network and operations management. For packet-based intermittent traffic, even UDP over the XML protocol can be looked into inline with what is used conventionally with SNMP.

## 18.5 Mapping with SNMP

Since the reader is already familiar with SNMP, which was handled in much detail in some of the preceding chapters, the capability of XML protocol is analyzed in this chapter by showing how we can achieve the functions in SNMP using the XML protocol. This is followed by looking at some of the XML-based protocols like SOAP and NETCONF in more detail.

To start with mapping of functions for SNMP, the first requirement would be binding with UDP (user datagram protocol). This will not be a problem as XML protocol can bind with UDP and is not coupled to a specific transport protocol (see Figure 18.1). The main difference here is that XML schema is used instead of the ASN.1 structure and the SNMP PDU is replaced with XMLP message envelope.

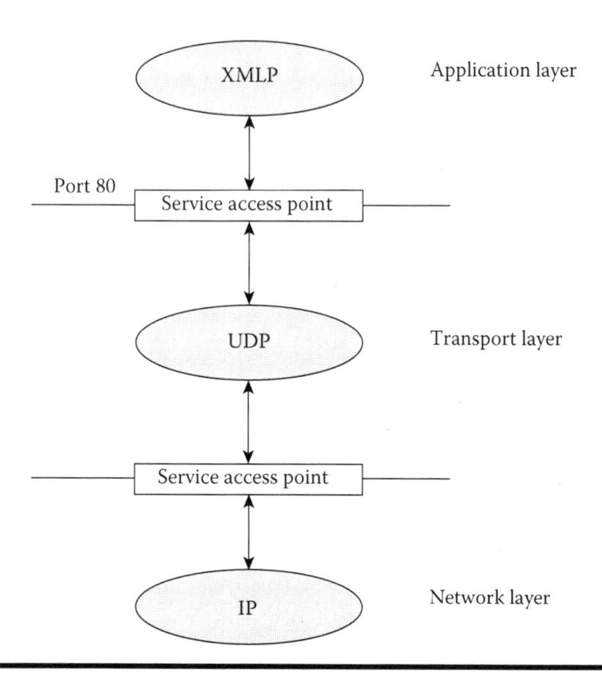


**Figure 18.1 Protocol binding with UDP.**

### 18.5.1 Messaging

The five basic messages that are used in SNMP and needs to be available in a replacement protocol are GET, GET-NEXT, GET-RESPONSE, SET, and TRAP. The example here is for binding with UDP, so the connection details needs to be explicitly provided in the namespace of XMLP envelope.

1. GET and GET-NEXT message: In SNMP these messages are used by the manager to request specific information from the agent. The header, which is optional, gives the identification of the sender. The "host" in the header corresponds to where the information needs to be sent. The URI will be used while sending the envelope to the packet network. Header can also be used for authentication and encoding. The body contains the actual request to be performed.

   In the specific example used here, the "Body" of the envelope is used to query the status of a card in a specific shelf. It can be seen that the XML protocol message is easier to read and interpret compared to the SNMP message. The tags like <getenvp:Header> and <getenvp:Body> in the message make it possible for third-party parsers to be used in processing the XML protocol message.

GET Message with XML:

```
<getenvp:getenvp to='host' xmlns='reqinfo:client'
xmlns:getenvp='---URI---'>
  <getenvp:Header>
    <getenvp:getenvp from='host' id='stream-id'
    xmlns='reqinfo:client' xmlns:getenvp='---URI---' />
  </getenvp:Header>

  <getenvp:Body>
    <iq type='get' to='mib@host' id='query-id'>
      <query xmlns='reqinfo:iq:shelf-6.card-2.status'/>
    </iq>
  </getenvp:Body>
</ getenvp:getenvp >
```

GET-NEXT message with XML:

```
<getenvp:getenvp to='host' xmlns='reqinfo:client'
xmlns:getenvp='---URI---'>
  <getenvp:Header>
    <getenvp:getenvp from='host' id='stream-id'
    xmlns='reqinfo:client' xmlns:getenvp='---URI---' />
  </getenvp:Header>

  <getenvp:Body>
    <iq type='get-next' to='mib@host' id='query-id'>
      <query xmlns='reqinfo:iq: shelf-6.card-3.status'/>
```

```
    </iq>
  </getenvp:Body>
</ getenvp:getenvp >
```

2. GET-RESPONSE message: The agent responds to the GET or GET-NEXT message from the manager with a GET-RESPONSE message that contains either the information requested or an error indication generated while processing the request from manager. The example below shows the response message with status information as "In-Service" for card-2 in shelf-6.

   In the GET-RESPONSE message, the Body will contain details of the request for which the response is being sent and the response. The query ID is populated in the message response Body along with specific information on the value given as response to the query. This makes it easy to tag the response against the request.

GET-RESPONSE message with XML:

```
<getenvp:getenvp to='host' xmlns='reqinfo:client'
xmlns:getenvp='---URI---'>
  <getenvp:Header>
    <getenvp:getenvp from='host' id='stream-id'
    xmlns='reqinfo:client' xmlns:getenvp='---URI---' />
  </getenvp:Header>

  <getenvp:Body>
    <iq type='get-response' to='mgr@host' id='query-id'>
      <query xmlns='reqinfo:iq: shelf-6.card-2.status'>
        <var>In-Service</var>
      </query>
    </iq>
  </getenvp:Body>
</ getenvp:getenvp >
```

3. SET Message: This message allows the manager to request a change to be made in the value of a specific variable. The agent based on design could respond to this message with a GET-RESPONSE message indicating that a change request was successful or an error message indicating the reason for change request failure. In the example below, the SET request is to change the status of link-1 for card-6 in shelf-2 to "Offline." It can be seen that the SET message body will have information about what needs to be set and where the information needs to be set.

SET message with XML:

```
<setenvp:setenvp to='host' xmlns='setinfo:client'
xmlns:setenvp='---URI---'>
  <setenvp:Header>
```

```
    <setenvp:setenvp from='host' id='stream-id'
    xmlns='setinfo:client' xmlns:setenvp='---URI---' />
  </setenvp:Header>

  <setenvp:Body>
    <iq type='set' to='mib@host' id='query-id'>
      <query xmlns='setinfo:iq: 'shelf-2.card-6.link-1.
      status'/>
        <var>Offline</var>
      </query>
    </iq>
  </setenvp:Body>
</setenvp:setenvp>
```

4. TRAP message: This message allows the agent to spontaneously inform the manager of an important event. Traps are very effective in sending fault information. The body of the trap message sent by the agent will contain an identifier and the information details on the reason for the generated trap. These XML traps can be collected by the manager and parsed for details that can be used for debugging the error scenario captured by the TRAP message.

TRAP message with XML:

```
<trapenvp:trapenvp to='host' xmlns='trapinfo:client'
xmlns:trapenvp='---URI---'>
  <trapenvp:Header>
    <trapenvp:trapenvp from='host' id='stream-id'
    xmlns='trapinfo:client' xmlns:trapenvp='---URI---' />
  </trapenvp:Header>

  <trapenvp:Body>
    <iq type='trap' to='mgr@host' id='query-id'>
      <query xmlns='trap:iq:Trunk.ABC.FaultDetected'>
        <trapenvp:Fault>
          <faultcode>Fault Code here</faultcode>
          <faultstring>Fault String here</faultstring>
        </trapenvp:Fault>
      </query>
    </iq>
  </getenvp:Body>
</ getenvp:getenvp >
```

### 18.5.2 Transfer of Message

The XMLP envelope needs to be packaged with the UDP header with details of source and destination port. Figure 18.2 shows the XMLP envelope with the UDP

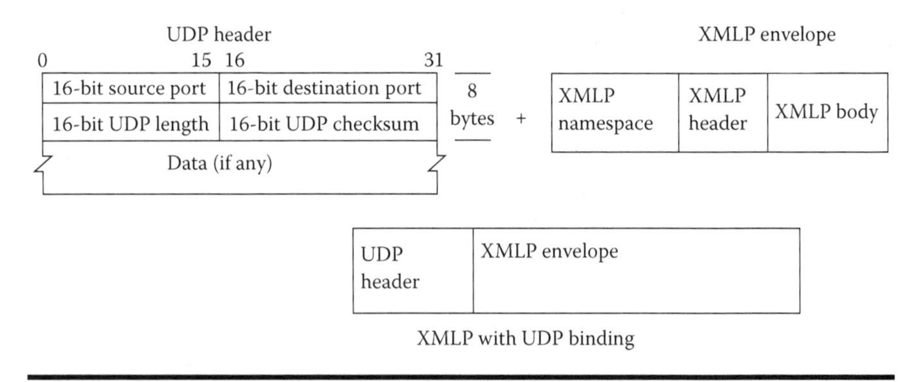


**Figure 18.2 Message for transfer with transport binding.**

header. This is not enough for transmission over the network. Before transmission to a packet network, the message needs to be appended with an IP header. The IP frame that is to be transmitted will contain an Ethernet header and IP header on the UDP datagram with the XMLP envelope.

Transmitting the host information in the IP frame helps to identify the host IP address. Port information should also be available for proper transfer to the specified application in the host machine otherwise default port is assumed. Multicast is also possible where the transmission is to a group of machines. This happens when an arbitrary group of receivers expresses interest in receiving a particular data stream. The destination in this case for the XMLP–UDP message will be the multicast group. The source address is specified in the UDP packet and needs to be the IPv4 or IPv6 address of the sender. The receiver needs to be programed to reject a packaged XMLP message that has inappropriate values for the source address. The XMLP envelope needs to fit within a single datagram with inherent encryption capability.

## 18.5.3 XMLP-Based Management Framework

Let us start with the management data originating from the network elements. There is usually a single manager for multiple network elements. So there will be multiple agents running on the different network elements, collecting management data and feeding data to a common XMLP manager. The management information obtained from the agents is first passed through a data conversion/XML parser module, where the body of the XMLP envelope is parsed to extract useful information. This information is stored in a database and it will consist of fault, configuration, accounting, performance, and security (FCAPS) information. The purpose of the FCAPS management code is to organize the data collected from the agents after parsing to appropriate containers in the database (DB). This FCAPS management module is not mandatory but is usually present in most XMLP management frameworks (see Figure 18.3).

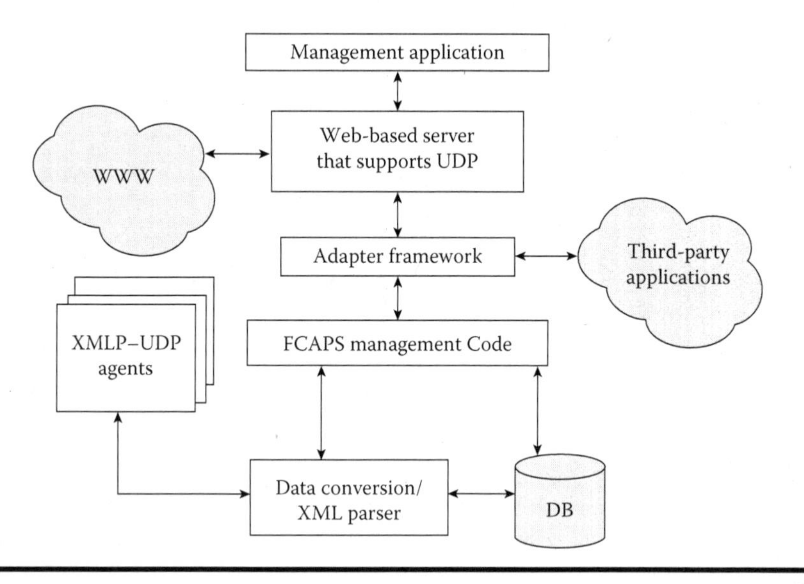


**Figure 18.3 XMLP management framework.**

The data derived can be sent out via an adapter framework to feed third-party management applications. An example of this would be CORBA IDL, which is supported by most third-party management applications. Before XML became popular, CORBA was the interface of choice for most management applications. This is quite evident even in standards defined by bodies like TMF, where the MTNM (multitechnology network management) definitions mapped to CORBA. Later the working group of MTNM merged with MTOSI (multitechnology operations system interface) where the mapping is using SOAP, which is an XML based protocol.

Using a Web server this information can also be passed to the www pool. Any queries from the www pool can also be managed with the framework. An example of such a Web server that can support UDP is the Windows Media Server that does data streaming. There could also be a stand-alone XMLP management application that takes the FCAPS information, processes the information, and represents the information in a GUI (graphical user interface).

The XMLP agent that runs on the network element has a set of distinct modules (see Figure 18.4). Consider the scenario in which XML protocol needs to be used on a network element where SNMP was previously used. This means that an SNMP MIB was used to store management data. An SNMP MIB to XMLP schema converter should then be available in the agent to make the initial conversion to XML and make the protocol stack a drop in for SNMP. This would not be present in an XML-based agent for which support for SNMP is not required. The data upon conversion is stored in a XML data structure, which can then spread from a flat MIB to an object-oriented framework. The movement from flat MIB to object-oriented framework is a major requirement for making the system suitable for managing complex next generation networks and its elements.

The SNMP MIB to XML schema conversion is a two-step process. In the first step the MIB definition is converted into a simple set of flat object definitions. This involves collecting all the scalar MIB variables under a module into a single object, and then modeling all the tables as contained objects. Each table would result in a separate object type or class. In the second step, the converter would undergo a process of "reverse normalization," where it would examine the table(s) indices in a MIB group and form parent–child (containment) relationships between the object classes formed in the first step. This step helps to ensure that a rich hierarchical information model is created as part of the transformation that can be examined and manipulated by management applications.

The XMLP processor performs the operation of data creation, processing of data, and transfer and reception of XML data. The embedded network element applications is just a fancy term used to mean nonmanagement functional blocks in the network element (NE) performing the core functionality of the element. The NE specific applications can use the XMLP processor for creating management information and storing it into the XML database using a management interface. The NE core applications, for instance, can create performance records in the XML format and provide it to the processor. Once the message is ready the transport binding is performed in the UDP binder, followed by an engine that initiates transfer and reception. This section on mapping to SNMP is intended to make it comfortable for the reader to switch from the popular SNMP-based management

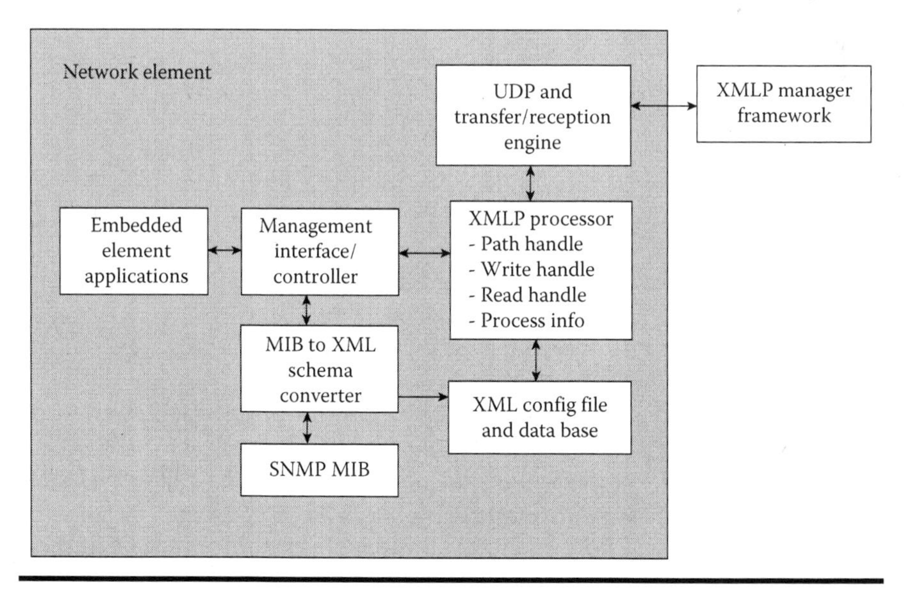


**Figure 18.4 Block diagram of XMLP agent.**

to XML-based protocol for the network management. XML-based protocol is mostly linked with HTTP, but most of the concepts discussed here with UDP holds good.

## 18.6 SOAP

SOAP (simple object access protocol) is a lightweight protocol that uses XML to define the messaging framework to generate messages that can be exchanged over a variety of protocols. SOAP provides a structured information exchange in a decentralized and distributed environment using all the capabilities offered by XML technology. It should be understood that SOAP is just a management protocol and does not require development of client or server in a specific programing language, though most open source, SOAP-based, client-server developments are using Java.

SOAP is stateless and does not enforce a specific schematic for the management data it has to exchange. SOAP was a recommendation from W3C (World Wide Web Consortium).. The SOAP 1.2 recommendation has two parts. The first part defines the SOAP message envelope while the second part defines the data model and protocol binding for SOAP.

Following with the XMLP discussion, the SOAP message envelope has a SOAP header and SOAP body. The header is identified using the tag <env:Header> and the body is identified using the tag <env:Body>. The header contains the transaction and management information as in most XML protocols. Standardizing bodies for Web services are working toward defining standard entries for the header in specific domains.

The SOAP header block is not mandatory for message transmission. Specific information that can be used by SOAP intermediaries or destination nodes can be put in the header block. The SOAP body blocks/entries are mandatory for any message envelope to be valid and this block is the key information that is processed at the receiver. It can contain either the application data, the remote procedure call with parameters, or SOAP fault information.

Some of the attributes in the SOAP header block are:

- env:role: The "role" attribute is used to identify the role of the specific target associated with the header block. This role can be standard roles or custom roles, and the absence of a role indicates that the role will be the final destination.
- env:mustUnderstand: The "mustUnderstand" attribute is used to specify the header block needing to be processed. It is more of a Boolean variable and can have a value of either "true" or "false." If the value of "mustUnderstand" is true, then destination node *must* process the header block. If there is an issue during the processing or the processing of the header block fails, then a fault event has to be generated. If the value of "mustUnderstand" is

false then the destination node may or may not process the header block. If this attribute is not present in the header block attribute then it is equivalent to setting the attribute to false. As discussed earlier the header block is not mandatory, but there could be critical information at times in the header specific for an application. In these scenarios, setting the attribute value as true will ensure that the header block is processed as required for the application.

- env:relay: Control of header blocks during a relay operation can be done with the "relay" attribute. Not all header blocks will be processed by intermediary nodes and default behavior is to remove the header before a relay operation. So unnecessary propagation of header blocks can be avoided and maintaining of specific headers can be done using the relay attribute. It is a feature that is newly introduced in SOAP 1.2.

SOAP fault as defined with the tag <env:Fault> and can be used to carry error and status information. The fault information will contain the identifier or group to which the fault message belongs (tag: <faultcode>), the short description of the fault (tag: <faultstring>), the node or actor associated with the fault (tag: <faultactor>), and details on the fault that can help in debugging and fixing the fault (tag: <detail>). An example of a predefined SOAP fault code value is "VersionMismatch" corresponding to invalid namespace in the envelope. Namespace is an XML construct (xmlns) used for grouping of elements in a SOAP application. It can also be used for version controlling. Some of the namespaces that are defined for SOAP 1.2 include envelope, serialization, and upgrade.

Example for using header attributes in SOAP message

```
<env:Envelope xmlns:env="http://www.w3.org/2003/05/soap-
envelope">

<!-- Header with two blocks (BlockOne and BlockTwo) -->
  <env:Header>
    <B1:BlockOne xmlns:B1= http://example.com env:role=
    http://example.com/Log env:mustUnderstand="true">
    <!-- URI example.com is reserved for documentation -->
    <!-- role is set to Logger so the header is to be
    processed -->
    <!-- To ensure header processing "mustUnderstand is set
    to "true" -->
    ...
    </B1:BlockOne>

    <B2:BlockTwo xmlns:q= http://example.com env:role=
    http://www.w3.org/2003/05/soap-envelope/role/next
    env:relay = "true">
```

```
    <!-- "mustUnderstand attribute takes value "false" here
    by default -->
    ...
    </B2:BlockTwo>

  </env:Header>

<!-- Sample Body with no contents -->
  <env:Body >
  ...
  </env:Body>

</env:Envelope>
```

SOAP capabilities are not restricted to conventional XML document exchange between the sender and receiver nodes. SOAP can also be used for remote procedure calls (RPC). The RPC is popular for message exchange in heterogeneous environment. To execute RPC with SOAP, the address of the destination node and procedure name with arguments is required. So for normal Web-based services involving simple message exchange, the conventional document-based technique is used and when a specific behavior is expected to be performed that needs mapping to a procedure, then SOAP-based remote procedures are used. There are specific request and response rules that need to be followed when working with RPC with SOAP messaging. One of them is to ensure encoding of the RPC request and response in a structure (struct). A SOAP request struct has the same name and type as the method/procedure being invoked. It should be noted that XML-RPC and SOAP are not the same and have a separate specification associated with use of RPC in XML. The XML-RPC has specifications for basic request-response RPC in XML with HTTP as transport, while SOAP is a more enhanced version of XML-RPC having user-defined data types, the ability to specify the destination node, message specific processing, and other features that are absent in XML-RPC.

When a SOAP message is received at the destination node the message is processed. The processing would involve parsing the message, checking if the message is syntactically correct and identify the action to be performed. Intermediate nodes (nodes between source and destination nodes in a message path) can be added to the message path without changing the parsing and associated message processing logic. Even removing the header, changing the header, reinserting the header, or adding a new header can be performed.

There are two types of intermediaries in SOAP. They are the forwarding intermediaries and active intermediaries. Forward intermediary performs the basic capability of forwarding the SOAP message to another SOAP node that can be another intermediary or the final destination node. The actions performed by forwarding intermediary is limited to removing of header blocks that are processed and that which cannot be relayed. Active intermediary performs additional processing that involves modifying the SOAP message before the basic forwarding action. This processing can be related to content manipulation or security related service offering. This additional processing performed by an active SOAP intermediary should be marked by inserting a header block to ensure that the changes are detectable by other nodes downstream, which includes intermediaries and the final destination node.

SOAP infoset should be serialized and concrete to ensure proper transmission and reconstruction without loss of information. This is achieved using protocol binding of the XML document. An encrypted structure can also provide serialization and is supported in SOAP. The most popular transport binding with SOAP is HTTP. The SOAP-HTTP pair is used extensively in Web-based applications. HTTP binding with SOAP provides request message and response message correlation. HTTP implicitly correlates its request message with the response message that can be leveraged as part of binding. So a SOAP application can choose to infer a correlation between a SOAP message sent in the body of a HTTP request message and a SOAP message returned in the HTTP response. In SOAP if a specific feature that is required for the application is not available through the binding, then it may be implemented within the SOAP envelope, using SOAP header blocks.

SOAP binding mainly supports two types of message exchange patterns. They are request–response exchange pattern and response exchange pattern. In SOAP request–response message exchange pattern, SOAP messages can be exchanged in both directions between two adjacent SOAP nodes. The HTTP POST method can be used for conveying SOAP messages in the bodies of the HTTP request and response message. SOAP response message exchange pattern is used when non-SOAP message exchange is involved and is done when a non-SOAP message as an HTTP request is followed by a SOAP message included as part of the response. The HTTP GET method in a HTTP request is used to return a SOAP message in the body of a HTTP response. It is suggested to use SOAP request–response message exchange pattern when information resource is manipulated and SOAP response message exchange pattern when an application is assured that the message exchange is for information retrieval where the information resource is "untouched" as a result of the interaction.

SAAJ (SOAP with attachments API for Java) gives Java APIs for SOAP message creation and manipulation. When it comes to SOAP programing with Java, JAX-RPC (Java API for XML-based RPC) is more popular. JAX-RPC has Java APIs for SOAP that uses SAAJ as an underlying layer. SOAP 1.2, which is most commonly used and based on XML infosets, has been used for the discussion in this chapter, compared to SOAP 1.1 that is based on serializations. SOAP 1.1 does not have a rich feature set in RPC and messaging and missing some of the functionalities defined in SOAP 1.2.

## 18.7 NETCONF

A wide range of techniques were in use for configuring network elements and IETF formed the NETCONF Working Group to standardize network configuration to fix the disparity in current configuration techniques, each having its own specifications for session establishment and data exchange. NETCONF protocol, as the name suggests, mainly deals with network configuration and provides the specifications for configuration functions like installing, querying of information, editing, and deleting the configuration of network devices. NETCONF also offers trouble shooting functionality by viewing network information as faults and taking corrective action in the configuration.

NETCONF is an XML-based protocol. Since it carries configuration information NETCONF inherently supports a secure infrastructure for communication. The NETCONF requests and responses are always sent over a persistent, secure, authenticated transport protocol like SSH. The use of encryption guarantees that the requests and responses are confidential and tamper-proof.

Another security feature in NETCONF is the capability of network elements to track the client identities and enforce permissions associated with identities. Hence security is mostly implemented in the underlying transport layer and even the identities are managed by SSH and then reported to the NETCONF agent to enforce the permissions imposed by the node.

Two important concepts associated with NETCONF are its commands and datastores.

- Commands: NETCONF provides a "Base" set of commands to perform network configuration. This basic set can be extended using "Capabilities." The main purpose of capability is to augment the base operations with new commands. These new commands can have a unique set of parameters, values, and entities that need not be limited by the base operations.
- Datastores: The device statistics and configuration data associated with a device and required to get the device from its initial state to a desired operational state is termed as datastores of the device. NETCONF commands operate on the data stores that constitute different versions of a device state. The <running> configuration datastore represents the current configuration of the device. This is the only default data store for the NETCONF device. Datastores like <startup> and <candidate> are added by capabilities. <startup> represents the device startup state. This configuration datastore can be used to represent the configuration that will be used during the next startup. Temporary states are handled by <candidate>. The <candidate> configuration can be changed to a <running> configuration. While the configuration data can be changed, the state data is read-only.

NETCONF requests are XML documents encapsulated in the <rpc> tag. There is a unique message identifier associated with each request along with the XML namespace.

Example NETCONF Request

```
<rpc message-id="412" xmlns="urn:ietf:params:xml:ns:netconf
:base:1.0">
  <get-config>
    <source>
      <running/>
    </source>
  </get-config>
</rpc>
```

The request here is using the "get-config" command to get the current (running) configuration from the device. The response can be the whole configuration record as specific fields are not specified, or an error with details why the NETCONF request could not get the desired response from the agent.

Example NETCONF response without error

```
<rpc-reply message-id="412" xmlns="urn:ietf:params:xml:ns:n
etconf:base:1.0">
  <data>
    <!-- Configuration Record -->
  </data>
</rpc-reply>
```

Example NETCONF error response

```
<rpc-reply message-id="412" xmlns="urn:ietf:params:xml:ns:n
etconf:base:1.0">
  <rpc-error>
    <error-type>RPC Connection</error-type>
    <error-tag>Device not found</error-tag>
    <error-severity>Major</error-severity>
    <error-info>
      <!-- Details of the error -->
    </error-info>
  </rpc-error>
</rpc-reply>
```

The request and reply both include the same message-id number. In case of an error, the rpc-error is encapsulated in the rpc-reply. The NETCONF reply can indicate one or more errors of varying severity. The error message is mapped to a specific type, along with severity to identify how quick a corrective action is required, and other error-specific details that will help debugging the error. It can be seen from the example that the error response can be easily parsed and valuable information displayed in any format that is required for a GUI or Web-based application. In an event driven network management system, the contents of the error could be programmed to trigger events that involves corrective actions.

In addition to the usual configuration management protocols that provide create, edit, and delete capability, NETCONF has many other commands that help the operations team to easily manage configuration. Some of these include the merge command used to merge the configuration data at the corresponding level in the configuration datastore and replace command used to replace the configuration data in the datastore with the parameter specified. Another very useful command in NETCONF is lock, which can be used by the management system to ensure exclusive access to a collection of devices by locking and reconfiguring each device. Only when required will the changes be committed. This provides an environment where multiple system administrators and their tools can work concurrently on the same network without making conflicting changes to configuration data.

NETCONF offers sophisticated configuration capabilities where a device configuration can be changed without affecting its current configuration. The configuration changes can be made active only when required. This is made possible using the <candidate> capability, which provides a temporary data store. A <commit> operation makes the candidate store permanent by copying the <candidate> to the <running> data store. The <candidate> data store needs to be locked while the store before modified, this is because <candidate> is not private by default. If the client that locked the candidate datastore for changing the configuration finally decides not to commit the change and then <discard-changes> can be used to ensure that the running configuration is not altered. For test-based scenarios, where the operator wants to change the configuration and test if it is working properly and in the case of issues wants to revert back to the previous configuration, then the confirmed-commit capability in NETCONF can be used. This capability extends the usual <commit> operation. A deep dive into NETCONF reveals many features that will make it the configuration management protocol of choice offering capabilities that were not available previously.

Before we close the overview on SOAP let us also look into some of the transport mappings offered by NETCONF. Some of the transport mappings are:

- SSH: A secure transport binding is offered to NETCONF with SSH. Functions like server authentication and confidentiality of the content can be provided with this transport mapping. The client-side user can also be authenticated to the server. SSH also offers the capability of multiplexing the encrypted tunnel or the transport layer into several logical channels. SSH mapping is usually a mandatory feature in most NETCONF implementations.

- BEEP: The feature of connection-oriented asynchronous interactions can be achieved with BEEP. It supports multiple logical channels similar to SSH, with the capability to both frame as well as fragment. BEEP is an optional transport with NETCONF.
- SOAP with HTTP[s]: This combination of SOAP with HTPP as already discussed is one of the most popular transport mappings in the Web service development. There are many Web service standards associated with SOAP with HTTP that could be reused when NETCONF is intended to be used with SOAP over HTTP. This leads to quicker integration with components that are developed to align with Web standards. However the transport mapping of NETCONF with SOAP does not map NETCONF operations to SOAP operations. So NETCONF should not be considered as a replacement for SOAP.

## 18.8 Conclusion

This chapter gives the reader an understanding of XML-based protocols. Rather than jumping into details about XML-based protocol, this chapter uses a more methodic approach of first shifting the focus from SNMP to XML by explaining the XML-based protocol using SNMP concepts that were already discussed in previous chapters. Then two of the most popular XML protocols, the SOAP and NETCONF are discussed. The intent was to build a strong foundation on XML protocol for use in network management.

## Additional Reading

1. Anura Guruge. *Corporate Portals Empowered with XML and Web Services.* Bedford, Massachusetts: Digital Press, 2002.
2. Don Box, Aaron Skonnard, and John Lam. *Essential XML: Protocols and Component Software.* Upper Saddle River, NJ: Addison Wesley, 2000.
3. Eric Newcomer. *Understanding Web Services: XML, WSDL, SOAP and UDDI.* Upper Saddle River, NJ: Addison Wesley, 2002.

# OPERATION/ BUSINESS SUPPORT SYSTEMS (OSS/BSS)

# Chapter 19

# What Is OSS and BSS?

This chapter is intended to give the reader a basic overview on OSS and BSS, which will help in building a solid foundation for understanding the chapters that follow on the process and applications associated with OSS and BSS. The ambiguity in terminologies used and a brief history on the OSS/BSS evolution is also handled in this chapter.

## 19.1 Introduction

The communication service and network industry is moving toward a converged world where communication services like data, voice, and value-added services is available anytime and anywhere. The services are "always on," without the hassle of waiting or moving between locations to be connected or for continuity in services. The facilities offer easy and effortless communications, based on mobility and personalized services that increases quality-of-life and leads to more customer satisfaction. Service providers will get much more effective channels to reach the customer base with new services and applications. With change in services and underlying network, the ease in managing the operations becomes critically important. The major challenge is the changing business logic and appropriate support systems for service delivery, assurance, and billing. Even after much maturity of the IMS technology, there was a lag in adoption because of the absence of a well-defined OSS/BSS stack that could manage the IMS and proper billing system to get business value to service providers when moving to IMS.

Operations support systems (OSS) as a whole includes the systems used to support the daily operations of the service provider. These include business support systems (BSS) like billing and customer management, service operations like service provisioning and management, element management, and network management applications. In the layered management approach, BSS corresponds to business management and OSS corresponds to service management, while the other management layers include network management and element management.

Let us start the discussion with the complete picture where OSS corresponds to support systems that will reduce service provider operating expenses while increasing system performance, productivity and availability. The OSS is both the hardware and software that service providers use to manage their network infrastructure and services.

There are multiple activities in the service provider space associated with offering a service that requires an underlying OSS to manage. For a service starting up, the service needs to be provisioned first, the underlying network including connectivity and the elements needs to be provisioned, then the service and network needs to be configured, the billing system and SLA management system needs to be configured, the customer records need to be updated and when the service is activated the billing system also needs to start working on records. This is just a basic line up of activities and there are many supporting operations associated with just getting a service ready for the customer. The OSS brings in automation and reduces the complexity in managing the services. After activation of service there needs to be a service assurance operation, customer service operation, service monitoring operations, and many more operations that fall under the service and business management scope of an OSS.

## 19.2 Service Providers

The communication service provider that uses the OSS can be:

1. Local exchange carrier (LEC): The LEC is the term used for a service provider company that provides services to a local calling area. There are two types of LECs, the ILEC (incumbent local exchange carrier) and competitive local exchange carriers (CLECs). Congress passed the Telecommunications Act of 1996 that forced the ILECs to offer the use of the local loop or last mile in order to facilitate competition. So CLECs attempted to compete with preexisting LECs or ILEC by using their own switches and networks. The CLECs either resell ILEC services or use their own facilities to offer value-added services.
2. Internet service provider (ISP): As the name suggests, these service providers offer varying levels of internet connectivity to the end user. An ISP can either have its own backbone connection to the Internet or it can be a reseller offering services bought from a service provider that has high bandwidth access to the Internet.

3. Managed service provider (MSP): A managed service provider deals with delivery and management of network-based services, applications, and equipment. These service providers can be hosting companies or access providers that offer services like IP telephony, network management, managed firewalls, and messaging service.
4. Long distance reseller: This is a service provider company that purchases long-distance telephone service in bulk at a reduced price and then resells the long-distance service as blocks to consumers. The long distance reseller benefits from the bulk purchase and the consumers also can get the service from a long distance reseller at a price lower than what is normally required.
5. Interexchange carrier (IXC): An interexchange carrier offers long distance services. These carriers complete a long distance call by routing the call from its originating incumbent local exchange carrier to the destination in a local service provider domain.
6. Application service provider (ASP): An application service provider traditionally offers application as a service over the network. It can be on-demand software or SaaS (software as a service) based application. Application, systems, and network management can be combined as a single bundled offering. The application is expected to follow a service level agreement and a complete business application can be offered by an ASP.
7. Wireless service provider (WSP): As the name suggests, these service providers correspond to carriers who provides cellular, personal, and mobile communication services over a wireless platform to the end user.
8. Content service provider: Content providers were mostly popular in offering Web content on Web sites. In telecom space the content service provider has a wider scope in providing value-added content in eCommerce and mCommerce environments. Since content service providers mainly offer value-added service (VAS) over an existing service offering, they work closely with ISPs, ASPs, and WSPs who provide the basic service on which content can be added.
9. Network service provider (NSP): The NSP offers networking infrastructure as a service. There will be specific network access points (NAP) through which the equipment and facilities in the network can be accessed. AT&T in the United States and BT in the United Kingdom are some of the top players in NSP space.
10. Master managed service provider (MMSP): These service providers offer one or more managed services for resale as "point solutions" generally on a pay-as-you-go model.

In general any organization/company that offers some form of service can be called a service provider. The listings in this section are just some of the popular terminologies used in association with a service provider in the communication industry.

## 19.3 Drivers for Support Systems

There are both technology and business drivers influencing the market of support systems. Some of them are:

- Multiplatform environments: The technology focus is now toward multiplatform-based environments. This increases the complexity on the infrastructure and business process to be managed. The net result is the requirement for a support system that can manage this environment.
- Emphasis on system integration: In telecom space there is an increased focus on interoperability and quick integration of products. The release of the product package in the minimum time (time-to-market) is a key factor for success of the product with multiple competitors and quick change in technology. Easy integration is possible only when the support systems managing the package are flexible to adopt the change.
- Mergers and acquisitions: In the current industry there are lot of mergers and acquisitions happening. Merger and acquisition is facilitated only when the new company and its product can easily be adapted to the business process and products of the master company. For example, the company being acquired has a product on order management and the master company wants to integrate this project with its existing OSS solution. If the order management product and the OSS solution are both eTOM compliant and both have a standard set of interfaces, then very little effort is required in adopting the order management solution.
- Convergence in telecom space: The convergence has a major impact on OSS. With convergence in network, there is a drive to have a single OSS solution that can manage a variety of networks used in different domains. Most OSS solutions now support both wire-line and wireless networks. Also when new network and management standards are defined for convergence, the OSS solution has to easily adopt these standards.
- Off-the-shelf products: COTS (commercial off-the-shelf) products are becoming increasingly popular in telecom and so are standards around the same. The aTCA (advanced telecommunications computing architecture) is one such standard in telecom space. Most OSS solutions are expected to have a standard set of interfaces that make it easy to use it, such as a COTS product. The move to comply with standards is also a driver in changes to OSS space.
- Increasing network complexity: In addition to convergence at domain and technology level, there is also convergence in functionality at network element level that adds complexity to the network. A typical example of this is an L3-L7 switch in a converged network. In addition to the basic switching operation, the switch would be performing multiple other functionalities including authentication, encryption, and application routing. The support

systems need to be able to handle the changed network with complex elements having an aggregate of functionality and a complex information base.

- Emerging standards for service providers: The deregulations and defining of standards for interoperability is a big influence in changing the OSS industry from legacy modules to interoperating standards compliant modules written by different vendors. Open OSS solutions are also available for download, changing the competition landscape in OSS space.
- Customer oriented solutions: Customer focused modules are also becoming popular in support systems. Customer management and assurance of service is becoming an integral part of business support systems. This change in focus can be seen even in management standardizing forums with work groups specifically for customer centric management.

## 19.4 What Do Support Systems Offer?

The support systems handle the customer, the services that are offered to the customer and the resources that offer the services (see Figure 19.1). Managing the customer is a part of the business support while billing is a business operation of the service management that finally applies to the customer. So in most scenarios the three entities that influence the support systems are the resource/infrastructure, service, and customer.

### *19.4.1 Support Systems to Manage the Customers*

The main part of customer management is to manage the customer account. This account has details on the customer contact, information on the services the customer has prescribed, and the contracts between the service provider and customer. Tracking on customer raised issues and the usage of the service is all mapped to

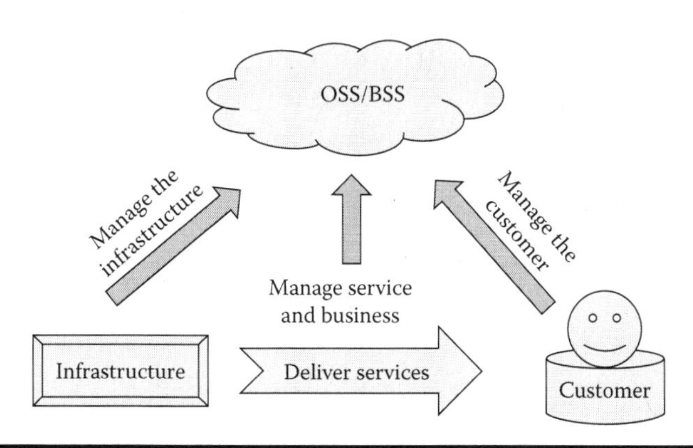


**Figure 19.1 Holistic view of support systems.**

this account. In most cases a unique identification number is associated to each customer as part of the account and used as reference in all transactions associated to the customer across the different modules in support systems. Another aspect in customer management that needs to have a defined process is the sales process. First the sales process should ensure that the customer requirements are satisfied with the specific service offering. The different phases in the sales lifecycle that affect the customer like the ordering process, the change of service or add-ons to the service, and even termination of service needs to be managed. It could be a bundled OSS/BSS solution that takes care of customer management or stand alone modules like an order management system.

Customer billing is also a key aspect of customer management that forms the mainstream activity in business support system. This will include determining how much the customer owes, preparing the customer invoice and also applying any payments made along with adjustments based on discounts or breach in service level agreements. Managing the customer expectations is also becoming an important aspect in customer management. Parameters like quality of end user experience and providing features for customers to manage their service is now part of most customer management systems. Other aspects of managing the customer expectations include communicating the service performance to customer, informing about any scheduled outage well in advance, resolution of any failure in shortest time possible, and obtaining feedback on how much satisfaction the customer has with the resolution, giving offers and discounts, and also having a good customer support desk.

### 19.4.2 Support Systems to Manage Service

The support systems play a major role in managing the services that are offered to the customer. Each service is associated with a set of legal and contractual specifications that needs to be tracked and managed. Automated support systems keep track of the key performance indicators (KPIs) that are linked to the SLA and take corrective action whenever the possibility of SLA breach is flagged. In addition to the contracts for the service, the product offerings in the service, the pricing, promotions, and discounts also need to be planned and tracked. While fulfillment, assurance, and billing of the service are the day-to-day activities, there is also a platform and product planning to be done in the backend. This includes deciding when the services should be made available, what features should be offered, and decision on the possible quote for offering the service.

The order management process also needs to be managed by the support system. This solution is closely integrated with the service provisioning module. So once the order is confirmed, the provisioning is triggered, which includes configuring the network to deliver the ordered service for the contractual specifications agreed upon. Service management does not end with provisioning and configuring of the service. This is followed by active monitoring on the service and an infrastructure to determine the quality of service (QoS) actually delivered by the network. Based on the events generated during monitoring, the network configuration is fine tuned to reconcile the delivered QoS with customer expectations. These events also help in network planning, so that next time a similar service is configured, better results can be obtained from initial settings of the network. Service level events are handled by the support systems. The customer, as well as service providers, also require reports on the service. The customer would be interested in reports like billing, value-added services used, and when a service outage occurred. The service provider would be working on reports related to resource capacity, utilization, bugs raised, and the SLA breaches to be compensated. The generation of reports is a functionality of the support systems.

### 19.4.3 Support Systems to Manage Resources

The infrastructure management is traditionally a part of the support systems. The support system is expected to ensure the proper operation of the network and elements that make the infrastructure for hosting the services. The installation of software and updates in the elements needs to be managed. When service configuration is triggered, the network and its elements need to be configured before setting the service level configuration. Once the elements are configured with necessary hardware and software, the system needs to be tested. This full cycle of network provisioning including configurations needs to be managed by the support systems.

Inventory management is another important module in support systems. Based on requirements, the inventory needs to be supplied for service fulfillment. Faulty inventory needs to be removed and free resources after use needs to be reallocated. Also included in the management of resource is monitoring and maintenance of the resources. This includes detecting faults in the resources and resolving the issues. Collecting usage records from the network and ensuring better utilization of the infrastructure is a part of monitoring. Security of the infrastructure and fraud detection is also an activity that is performed by the support systems. The services should be offered on a secure and reliable infrastructure.

## 19.5 Defining OSS and BSS

On a broad level, OSS is expected to mean any type of support system related to operations in service provider space. In current management space, the scope of OSS is more limited. OSS can be divided into three types of management solutions. They are:

1. BSS (business support systems): These are solutions that handle business operations. It is more customer-centric and a key part of running the business rather than the service or network. Some of the high level functions in BSS include billing, CRM (customer relationship management), marketing support, partner management, sales support, and so forth. These high level

functions encompass several functions each of which can be a specialized solution. For example, billing will have activities like invoicing, payment processing, discounts and adjustments, account management, and credit management. Similarly, CRM will have activities like order management, trouble ticket management, contract management, and SLA breach management.

2. OSS (operation support systems): These are solutions that handle service management. It includes functions like service assurance, manage services reporting, and so on. In most discussions where OSS is mentioned, it is the service management that is the key focus and not any type of support systems. So OSS can be used in the context of any type of support system or to specifically mean service oriented solutions. It is also referred to as "service OSS" to avoid the ambiguity in usage of OSS as the holistic support system.
3. NMS (network management system): These are solutions that handle management of network resources. From support system perspective the element managers are components of the master network manager and the element management layer is embedded in NMS functionality. Network OSSs (NMS) manage specific elements in the network deployed by the operator. NMS solutions are optimized to manage as specific service equipment or a server infrastructure. For example, an NMS for broadband services will be optimized for broadband infrastructure like ATM, Frame relay, or DSL and an NMS for a specific server infrastructure would be optimized for servers like Sun Solaris, HP Unix, or Apache Web Server.

The service OSS is the binding glue between the BSS and NMS domains (see Figure 19.2). There needs to be seamless flow of data between business solutions, service solutions, and network solutions in support systems for a service provider.

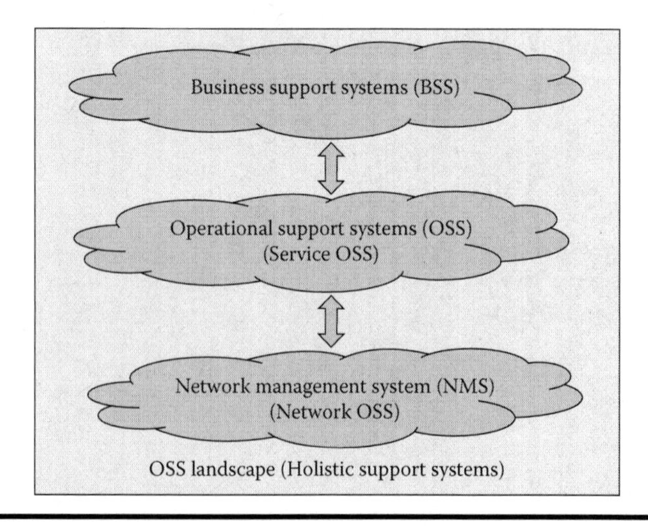


**Figure 19.2 OSS landscape.**

Together these modules form the support system in the service provider space. Most service providers do an end-to-end (E2E) between these modules to bring down the operational expenditure (OPEX).

## 19.6 From TMN to eTOM

In the early 1980s most of the equipment required by local exchange carriers (LECs) were supplied by one or two suppliers in most parts of the world. One of the major changes to this approach was the breakup of the Bell System toward the mid-1980s after with LECs were encouraged to procure equipment from any vendor of choice. As a result the network for offering basic service now had components from multiple vendors. While there were standard protocols for interaction between network elements from different vendors, there were no popular and adopted standards for telecom data and interface management in a multivendor environment. This led to an increase in CAPEX for having new support systems, increase in OPEX with usage of more OSS solutions from different vendors, reduced interoperability between OSS solutions, and lack of flexibility in bringing in new OSS functionality. The first popular attempt to define the management stack that was widely accepted was the TMN (telecommunications management network) model.

The TMN model divided the management stack into five layers. The layers are: business management layer (BML), service management layer (SML), network management layer (NML), element management layer (EML), and the network element (NE) layer. There are management agents running on the network elements in the network element layer that collect management data. This information from agents is collected by the element managers in EML. In a network with multiple network elements, there will be multiple element managers. To provide management data at a network level by aggregating information of multiple network elements in the network, a network manager is used that sits in NML. Over the network infrastructure, there can be multiple services that will be hosted. The management of the service happens in the SML. Finally the upper most layer of TMN, the BML is to handle business data in service provider space. The BML takes feed from the SML and it is the service reports that are used as input in the generation of billing reports and invoicing at BML.

The TMN from ITU-T provided a means to layer the management stack and also helped in defining the basics management blocks using FCAPS (fault, configuration, accounting, performance, and security). It still was not elaborate enough to map the various telecom processes into the management layers. So TMF expanded the TMN layers to form a new model called telecommunications operations map (TOM model). This model brought in more granularities in defining processes in service and the network layer and added modules related to customer interface and customer care process. TOM was later expanded from just an operations management framework to a complete service provider business process framework called eTOM (enhanced telecom operations map). Some of the new concepts brought in to encompass all service provider processes were to introduce supplier/partner management, enterprise management, and also adding lifecycle management processes. The eTOM is currently the most popular business process reference framework for telecom service providers. A separate chapter will handle telecom business processes and give a detailed explanation on TOM and eTOM. This section is intended to give the reader an understanding on how the basic TMN management stack evolved to the current eTOM.

## 19.7 Few OSS/BSS Processes

Some of the OSS/BSS processes are discussed on a high level in this chapter. Most processes are performed using a solution, which means that in most scenarios there is a management application mapped to the service provider process. The intent is to familiarize the reader with some solid examples to get a better understanding of the concepts that will be discussed in the chapters that follow.

- Provisioning: This process involves service and network management functions. The provisioning activity requires setting up the infrastructure to offer the service. So the provisioning module has to interact with an inventory module to check on what resources—switches, routers, connections—need to be utilized in setting up the infrastructure. Once the resources are allocated, next the resources need to be configured. This includes configuration of service parameters for the various customer instances and the configuration of elements making up the network. Next the service is activated. The element managers send activation commands to the configured resources. The agents feed the equipment status and other management data to the element managers. Service providers look for flow-through provisioning and activation solutions that require minimal or no manual intervention. These solutions will interact with inventory and monitoring modules, so that a single mouse click can result in selection, assignment, and provisioning of resources.
- Order management: The solution for order management will have a Web user interface (UI) or graphical user interface (GUI) that guides the user through the ordering process. The range of services that can be offered has increased, so has the value-added services offered with the basic service. There are also bundled services offered where a single order would have multiple services like phone, broadband internet, TV subscription, and so on. The order management is the main input feed for offering a service and there would be multiple inputs required in order management that may not be relevant for all customers. So the order management system needs to provide default values to ensure that order is completed in the minimum time possible and also be robust to avoid any errors in the order. Once the order is taken, the order

management solution also has to trigger modules like service provisioning to complete the order.

- Trouble ticket management: Centralized management of defects including ensuring defect closure and providing consolidated reports is required for infrastructure and product planning. The trouble ticket manager interfaces with fault management modules in the network and provides a centralized repository of troubles that needs to be fixed. This helps to improve the service quality and the response time in fixing a trouble condition. There will be different kinds of reports that can be generated by trouble ticket manager, which will give information on network equipments that get faulty most of the time, wrong configurations that degrade performance, efficiency of bug fixing team in fixing problems, map faults to modules or release products from specific vendors, customer quality of service, and issues reported by customer including outage of service. A trouble ticket can be raised by the customer, the service provider, or some predefined fault logs will automatically register as a trouble ticket. These tickets are then assigned to the resolution team. Based on how critical the trouble condition, different levels of supports will be offered and there is usually a dedicated team to handle each level of support. Support level can map to the services offered like level zero support might involve calling up the person who raised the ticket within an hour and providing support or it could mean that the trouble needs to be closed within a day. These items are explicitly slated in the SLA.

  When a trouble ticket is raised and needs to be assigned to a support person, the work force management module provides necessary inputs on the best person to solve the trouble condition. The work force management module has details on the skill sets and location of the technicians. It also maps technicians to specific trouble tickets, like whether the technician is free or working on a ticket, was there a change in the ticket status, or was the ticket reallocated to some other technician, and so on. Consider the trouble ticket deals with a faulty switch in Dallas, Texas. So the trouble ticket manager assigns the ticket to a technician in Dallas based on input from the work force module. After fixing the switch, the technician can change the status of a ticket, for reconfiguration of switch by an operations team in London. Now again the trouble ticket manager can use the work force management module to assign the ticket to a professional from an operations team in London. This way the trouble ticket manager can track the ticket from creation to closure.
- Inventory management: Cost saving in telecom space can only be achieved when the available inventory is properly utilized. The inventory needs to be properly allocated and deallocated for offering the service. Inventory data if not managed is really complex. For example, inventory will include a wide variety of network elements like switches, hubs, servers, trunks, cards, racks, shelves with different capacity, address range, access points, and so on. The inventory would be placed at different locations and the quantity of each of

these elements may be different at each of the locations. Again some of these elements may be faulty and some of the working elements might be already in use, booked for a specific provisioning, being fixed, and under maintenance leading to a different status at a given point of time.

When the provisioning module needs resources based on parameters like status, location, capacity, and so on, a decision is made on the most appropriate resources to be used. This information is provided by an inventory management module. The inventory data can also include details on serial number, warranty dates, item cost, date the element was assigned, and even the maintenance cost. All inventory information is logically grouped in the inventory management module. This makes it easy to view and generate reports on the inventory. Capacity planning can be performed based on inventory data and new inventory can be procured as per need before an outage occurs. Inventory affects the CAPEX and OPEX directly and hence inventory reports are of much importance to senior management. Procuring inventory from a new vendor usually goes through a rigorous vendor analysis followed by legal agreements for license including maintenance and support.

- Billing: Records are generated by the elements in the network that can be used for billing. In switches where calls are handled, call details records (CDR) are generated that have billing information. CDRs can have multiple call records. The CDR is parsed to for parameters like calling party number, destination number, duration of call, and so on, and the customer is billed. A mediation module performs the activity of parsing and converting the information in CDR to a format that can be used by the billing system. Billing modules have a rating engine that applies the tariff, discounts and adjustments agreed upon in the SLA as applicable, and create a rated record on how the bill was calculated. This record is stored in a database and aggregated over the billing cycle. At the end of a cycle (like a month for monthly billing of calls), the aggregated records are used by modules like an invoicing system to prepare an invoice for the customer. There can be different billing models that can be applied based on the SLA with the customer. It can be flat bill where a fixed amount is billed for usage of service, or a usage-based bill where the bill reflects the customer's utilization. Another example along the same lines is real-time billing where the debit from a prepaid account or amount to be paid for on a postpaid account can be viewed as soon as a service is used.

Now let us see an E2E flow of how the modules discussed in this section (see Figure 19.3) interact to offer a service like telephony. Every activity starts with a business requirement. So first the customer contacts the sales desk and uses a Web interface to request the service. Once the credit check is performed and the customer can place an order, the required customer details and service information is feed to the order management system. The order management system creates a work order and sends the details to the provisioning system. The provisioning system sends

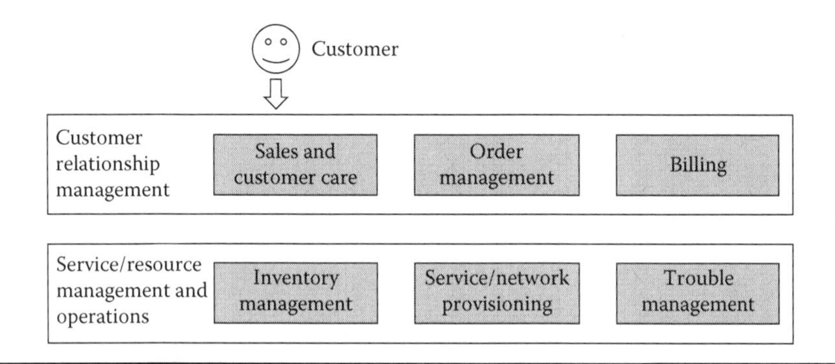


**Figure 19.3 Mapping of the support system process.**

data to the inventory system for allocation of appropriate resource. If some manual setup is required before starting the automated provisioning, then a work request or a trouble ticket is created. Using the workforce module, the request is assigned to a technician for resolution. Once the setup is ready, the provisioning module initiates provisioning of service and the network. Once provisioned, the service is activated and the billing module is triggered to collect and work on the billing records.

## 19.8 Conclusion

The overview presented in this chapter is just the foundation for the chapters that follow. The difference between OSS and BSS was discussed in the context of support systems. The ambiguity associated with OSS as a complete support system including network management, against a support system for service management alone was also handled in this chapter. The shift from the popular TMN to eTOM was also handled. Support systems are a vast subject and once the reader gets the fundamentals from this book there are many support processes where the reader can specialize as a professional.

## Additional Reading

1. Kornel Terplan. *OSS Essentials: Support System Solutions for Service Providers.* New York: John Wiley & Sons, 2001.
2. Hu Hanrahan. *Network Convergence: Services, Applications, Transport, and Operations Support.* New York: John Wiley & Sons, 2007.
3. Kornel Terplan. *Telecom Operations Management Solutions with NetExpert.* Boca Raton, FL: CRC Press, 1998.

# Chapter 20

# OSS/BSS Functions

This chapter is intended to give the reader an understanding of the fulfillment, assurance, and billing support systems. The building blocks in each of these support systems and the activities performed by the building blocks are explained in this chapter. There is fulfillment, assurance, and billing associated with both business and operation support systems. The reader will get a chance to understand where the different OSS/BSS solutions available in the market fit in with the holistic support system perspective.

## 20.1 Introduction

Usually operation and business support systems deal with fulfillment, assurance, and billing (FAB) for a service associated with a customer requirement. The backend planning to do the real-time support is more a process rather than the actual support system. The business map as in eTOM brings in an E2E perspective for all processes in service provider space having FAB as a major module. This chapter details the basic building blocks for fulfillment, assurance, and billing that make up the major portion of all support systems. For ease of understanding, the explanations are given using activities in each process without reference to any standard models like eTOM.

## 20.2 Fulfillment Process

The fulfillment process involves the set of operations used to fulfill a customer order (see Figure 20.1). These operations use a set of solutions like order management

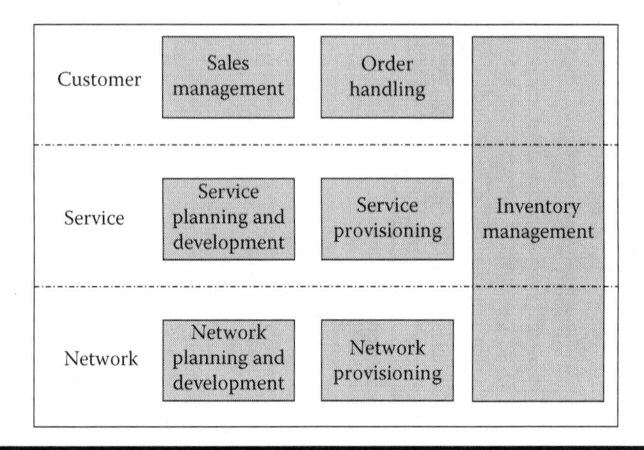


**Figure 20.1 Fulfillment process.**

module, inventory management module, provisioning module, service configuration module, and so on to complete the fulfillment process. Fulfillment spans business support systems with modules such as sales and order management and it spans the operation support systems with modules like provisioning and inventory management.

It is first the operations of sales to work with the customer to ensure that an order can be placed. The order management system (OMS) then captures all relevant information from the customer to offer the requested service. The OMS also has the capability to support changes in the order based on customer requests. Though the OMS is mainly for the service provider, the customer also gets a restricted view of the OMS. This customer view of OMS gives information on the status of the order and ensures that the customer can also track and manage the order. With this functionality, in case the order is not completed on time, the customer can take it up with the support desk and resolve any issues including a change or cancellation of order. There is usually a presales and sales activity before the order, though some services can have a direct order, such as service orders that can be placed online from a Web site. When placing an online order, the customer enters all the relevant details that the service provider can work on to offer the service requested.

Once the order is completed and a customer starts using the service, then the order number is used to collect data from the trouble management module, performance reporting module, and billing modules, so that the customer can view any information corresponding to the specific order that the service provider opens up to the customer. A customer can place more than one order and there would be a multiple order status that the customer will track even if the services are offered as a single bundle. While provisioning is the key process dealing with configuring the network to deliver the service desired by the customer, the order management system provides the relevant information to provisioning module to trigger the provisioning operation.

Some of the features of an order management system are:

- Preorder data collection: This involves collecting the customer details and validating credit (credit check) to ensure that the requested service can be placed.
- Accepting orders: This activity includes creating the order and making changes (editing) to the order based on a customer request.
- Integration with inventory database: The order can be successfully completed only when the required inventory to offer the service is available. To satisfy the order, the required resources are reserved or when the resources are not already available they are procured. In the usual scenario, the inventory management system ensures that the resources to offer a service is always available by issuing reports when a specific resource is low and needs to be stocked.
- Initiate provisioning: When the service provider has flow through provisioning modules, the order management system triggers the provisioning modules to complete the order. The input to provisioning with details of the service and the SLA parameters comes from the order management system.
- Customer view of order: The customer needs to be notified on the status of the order. The order management system centrally tracks and manages the order and notifies the customer on the order status.
- Initiate billing: There are multiple modules that integrate with the order management system. One of the modules is the billing system. To service an order, the OMS first triggers the provisioning module. When the provisioning module notifies the OMS that provisioning is complete, then the service is activated. Activation of service should also start billing the customer for the service. So the OMS initiates the billing system as part of completing the order placed by the customer.
- Plan development: The order management system helps in services and network planning. Based on orders that are placed, the key services that customers are interested in can be identified. Based on customer records these orders can give statistical inputs on what age group prefers what type of services and to identify new service areas that will generate business. The order information will also help in forecasting the services that will be in demand for the next business cycle and resources (both manpower as well as equipment) to offer the service in time can be made available. This planning will have a direct impact on reducing the CAPEX and OPEX by proper procurement and allocation of resources.
- Integration with workflow management system: When automated provisioning is not possible after placing the order and when a technician needs to be involved in setting up resources to offer a service, then the order management system sends data to the workflow manager so that an appropriate technician can be allocated to fix the issue before auto-provisioning can be triggered.

The same is the case when an auto-provisioning action fails and a technician needs to fix the issue, so that auto-provisioning can be retriggered.

- Interactive user interface: Web-based ordering is most popular in customer space and most service providers offer their customers the capability to place and track their order on the Web. The sales, operations, order management, and customer support team in the service provider space would mostly have an application or Web-based user interface to create and edit orders placed by methods other than World Wide Web (www) like a telephone order, order by mail, or order placed at a sales center.
- Ease of integration: The order manager has to integrate with multiple support solutions. The solutions OMS has to integrate with may be from different vendors. So it is important that OMS has its APIs exposed for other modules to integrate. Standard APIs and a scalable architecture is a usual trait of OMS, which make it easier for other modules to integrate with it. Some OMS solutions also support multiple platforms that makes it easier for integration in a multivendor environment.
- Interaction with trouble ticket manager: The OMS also has to interwork with the trouble management module. The service terms and QoS parameters captured in the order needs to be communicated to the problem handling modules. In case of any trouble condition as part of completing an order or during the service execution linked to an order, the trouble ticket manager will raise a ticket and assign it to a technician for resolution. Each order placed by the customer is uniquely tracked and any trouble conditions are mapped to the order.
- Integration with customer relationship manager (CRM): The customer support desk will get all types of queries with regard to an order like the date of completing the order, the quote, or the price the customer has to pay for the service in the order, the status of the order, changes to be made in the order, cancellations to the order, discounts or adjustments to the order based on promotions, confirmation on the order, troubles with regard to the order after service is activated, and so on. In all cases where the customer interacts with the support desk, the specific order has to be identified to respond to the query. So the CRM module interacts with the OMS module to get information on the order.
- Reports: The OMS issues different types of reports that are useful for customers as well as internally by the service providers. The reporting engines usually offer the capability to generate reports in a variety of formats like pdf, word doc, excel, image file, and so forth. The reports from OMS can be the order status, the orders that were not completed and the reasons, the trouble conditions that occurred during the completion of an order, the total revenue generate from orders raised on a specific service, and many more reports that help in proper planning and development of service along with an increase in customer satisfaction.

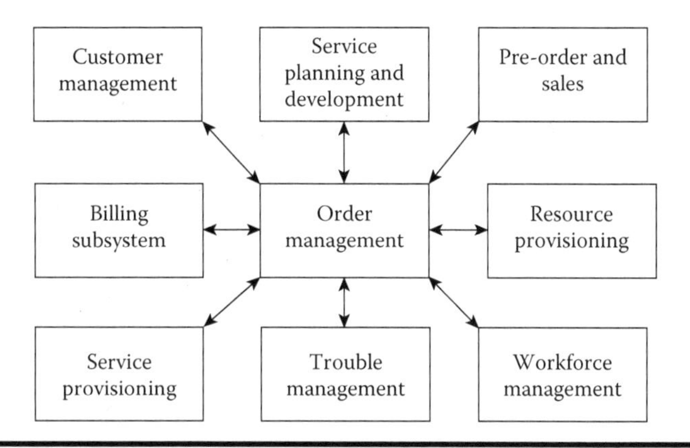


**Figure 20.2 Order management module interactions.**

An order management system offers a set of advantages that make it an important component for the service provider (see Figure 20.2). The advantages are:

- Reliable service to the customer: The service requested by the customer is captured as an order. The order management system has predefined formats for collecting orders for specific services. So when an order management system is used the completeness of the service details is guaranteed. In addition to clarity in collecting an order, the order management module also tracks the order to completion, thus ensuring that the service is rendered to the customer.
- Quick delivery of service: The use of an order management system automates the order collection and service activation cycle. As a result, the time to deliver services is minimized. The OMS does quick credit checks to ensure the customer has sufficient credit, it can check the inventory module for resources to complete a service request, and the OMS can also trigger flow through provisioning. The order management module communicates (integrates) with a variety of modules to reduce the net time to deliver a service to the customer.
- A lot of manual processes are eliminated: Manual processes are not just time consuming, they are also error prone. In the absence of an order management module, the customer requirements need to be captured manually. Also it should be noted that an order management module is like a central control module that triggers and coordinates activities with multiple modules. Absence of the order management system would mean that these triggers need to be manually activated based on requirement.
- Proper planning: This helps to reduce time-to-market service. Based on the orders, the customer interests can be identified. So the right set of resources could be made available and innovative services based on the customer order trends can be introduced.

- More customer satisfaction: Online self-ordering OMS make the customer feel they are more in control from service selection to activation. The OMS also gives the customer the capability to track their order status and keeps the customer updated on the progress of the order. Trouble conditions during service provisioning and issues raised by the customer as trouble tickets are tracked and resolved by the trouble ticketing module in reference to the order details provided by OMS. A quick delivery of service is also a value added to the customer, which increases customer satisfaction levels.

At a high level, the steps performed by the order manager can be listed as follows (see Figure 20.3):

1. Validate the order for completeness and correctness.
2. Identify the parameters and the steps for provisioning the services requested.
3. Check if sufficient inventory is available and allocate resources to fulfill the service request.
4. Do any manual installation and setup, if required to the identified infrastructure. This is performed by interacting with the workforce management module.
5. Trigger provisioning and activation of requested services.
6. Update the status of the resources with the inventory module.
7. Do end-to-end, real-time testing of the configuration and fix any issues.
8. Initiate billing for the activated services.
9. Communicate order status as completed to customer management modules.

The provisioning system plays a critical role in fulfillment of service. The provisioning process in itself is quite complex where the activities would be across different kinds of networks and can involve a variety of tasks related to the service. For example, when a service is provisioned it might involve configuring elements

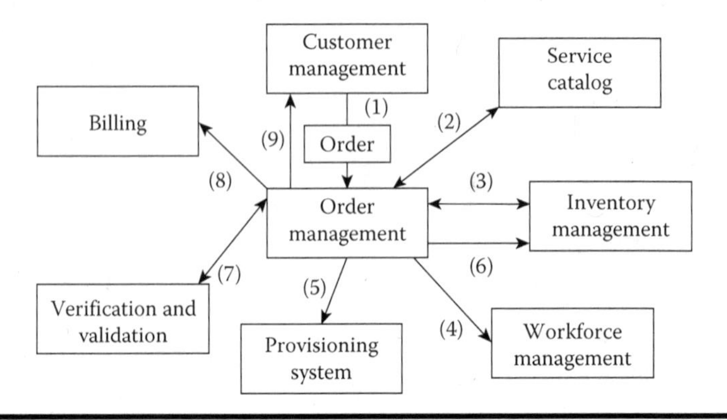


**Figure 20.3 High level steps in order management.**

in ATM or FR for transport, RAS for access and some elements in a wireless network. The actual task to be performed in the element could include, but not limited to, configuring an application, setting up parameters for security, connectivity setup between elements, associating storage with server or application, and so forth. The provisioning OSS (see Figure 20.4) can also handle configuring of network for dynamically oriented transactions like authentication, accounting, and authorization. To summarize, the provisioning system can do application provisioning, service provisioning, resource/element/server provisioning, network provisioning, storage provisioning, security provisioning, dynamic SLA-based provisioning, monitoring provisioning and any other setup activities that will be required to provision a product package consisting of multiple or at least one service, which is described as an order. An operation support system will usually consist of multiple provisioning modules specialized for a specific activity. Provisioning with minimal or no manual intervention is a goal that most service providers try to achieve in their OSS to bring down the OPEX.

The inventory management system supports and interacts with multiple modules relevant to fulfillment by keeping track of all the physical and logical assets and allocating the assets to customers based on the services requested. Order manager interacts with inventory manager (IM) to check if sufficient inventory exists to complete an order. The purchase and sales module updates the IM with new assets added, old assets sold, assets given to an outside party on rental, and assets that were borrowed from a third-party vendor. Provisioning module interacts with IM to work on allocated resources and also for updating the status of allocated resources. The next module that gets feed from IM is the network manager. The network manager contacts the IM to get static information on the resources being monitored and managed by the network manager.

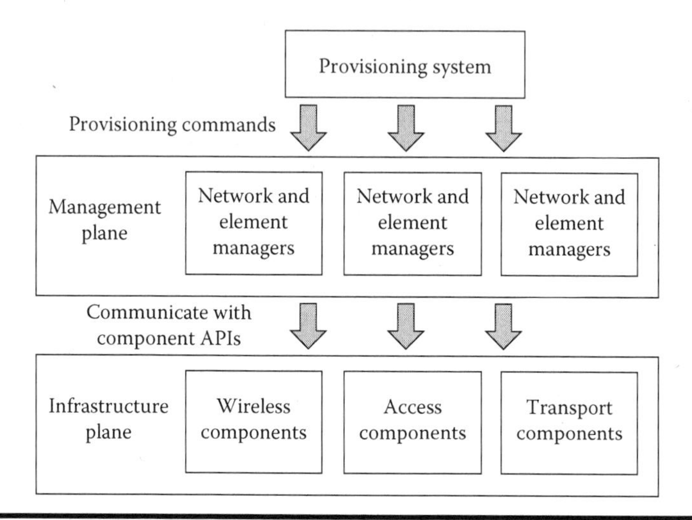


**Figure 20.4 Provisioning OSS.**

The network manager user interface will have the capability to display the attributes of the resources being managed. The value of some of these attributes can be queried from an IM module. Inventory management is mostly automated with minimal or no manual entries. This is because small errors while manually feeding inventory data can result in major trouble conditions and network failure. For example, two bridges intended for the same functions, with a similar look and feel, and with similar specifications and cost. The only difference is a single digit difference in the product code. This might correspond to a major impact if the bridges are intended for working in two different configurations, making one nonsuitable for working in the other's configuration. While there is a high probability that based on external features, the two bridges may look similar and the manual entry in IM might result in a bug. This problem does not arise when the product code and other parameters are read using a scanner even when the product code is an alpha numeric string or a barcode. Auto discovery and status update is available in some inventory management modules, where the IM module interacts with element managers or the agents in the network elements to get status updates and to automatically discover a new resource that is added to the network and not updated in IM module.

Automation of the inventory management system is a necessity considering the evolution in communication networks to offer complex services on a multivendor environment. This increased complexity demands simple and flexible ways to allocate resources and to manage, maintain, and plan how networks are configured. Planning requires operators to maintain configuration records, to search the records with simple queries, and refer consolidated reports when required. A proper planning process would be possible only by integration of records by a common system like the inventory manager (see Figure 20.5), where these records rapidly become manageable even when they are stored at multiple localized databases.

An automated inventory management system is a key module for having flow through provisioning. The IM interacts with an order module to allocate the resources. The provisioning system checks the inventory modules for allocated resources to do provisioning and finally updates the IM with the status of the resources after provisioning or failures during provisioning. Inventory hence is a key module and needs to be automated for quick provisioning without manual intervention. The quality of work is also improved when the administrative burden in managing inventory records are performed by an automated inventory manager.

On a single click, the IM can provide the operator information on whether proper equipment is in place to offer a specific service or new equipment needs to be installed. The links and capacity circuits that provide backbone transport can be checked and assigned. Commands to provision systems to do configuration can also be triggered from the inventory management system. There are many vendors who specialize in inventory management products. Telcordia is one of the market leaders in network inventory management. Inventory management can also be a bundled solution offered as part of a complete OSS solution suite.

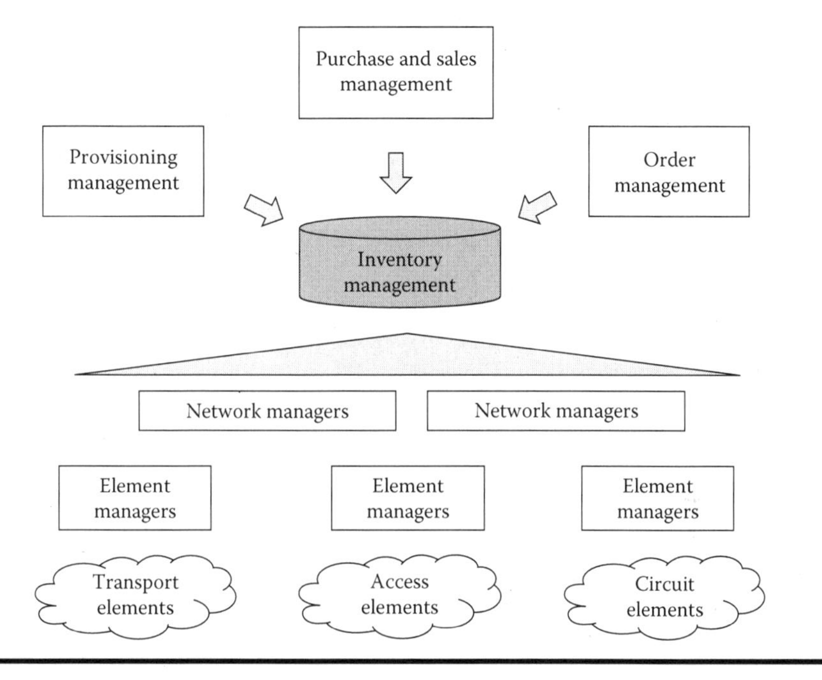


**Figure 20.5 Interactions with inventory management.**

An inventory management system is expected to have the following features:

- Standard interfaces: The inventory management system should have standard interfaces to easily integrate with modules like order management, customer care, and provisioning. MTNM/MTOSI is an example of interface standards defined by TMF (TeleManagement Forum) for inventory management.
- Dynamic discovery of network elements: When a new element is added to the network, the inventory management system should discover the new element and update the inventory database with attributes of that element. In addition to performing dynamic discovery that falls under configuration management, the inventory management system also can collect logs on inventory status, to update its database thus performing activities that fall under fault management.
- Web-based solutions: Most inventory management solutions currently available in the market are Web based or support Web launch along with the ability to be launched as a separate application. The reason for this is that most operators want to have the flexibility to work on inventory manager from anywhere, which is possible when the solution is Web based and can be launched with a Web browser to which data can be sent using the Internet infrastructure.

- Scalability: The huge volumes of inventory data and associated reports lead to an inventory database that can scale easily. The storage requirements should be planned properly when deploying an inventory management solution.
- Ease of use: The inventory information needs to be properly sorted based on device types. The data representation should be user friendly with the use of maps, drawing, and symbols wherever required. The inventory data represented must support different views, like a physical view showing where a particular device is placed (location) in the chassis, a hierarchical view that can be browsed down to get the attributes of individual elements or a service view where the elements are grouped based on services that are running/installed/offered on the elements. There should be built-in libraries of maps, locations, floor drawings, and network object symbols available in the inventory management system. This makes it easy to feed and represent data. In short, a proper use of text and graphics is expected in representing inventory data.
- Distributed architecture: Telecom inventory of the service/equipment provider may be distributed at different geographical locations. The inventory management system should adopt a distributed architecture in collection of data and updating of data. For data lookup, the inventory management system should provide a centralized view of all data available to the service/equipment provider.
- Assist network design: The network design is dependent to some extent on the available inventory. To realize a service, the supporting network design involves effective use of available resources. Pre-planning based on existing inventory data for an improved network design is also possible.
- Inventory creation and optimal allocation: The inventory management system needs to send alerts to add inventory that is scarce and reallocate inventory that is released after a service is terminated, thus creating inventory for allocation. Powerful and proven inventory algorithms need to be used to design and allocate network components. Proper allocation ensures rapid and accurate service delivery.

Network maintenance and restoration is performed by some inventory management systems to maintain the operational integrity of the network. This functionality cannot be specified as an essential feature of inventory management as separate solutions exist and are usually used to handle network maintenance and restoration.

We have discussed network provisioning, planning the network, and also mapping the network design to inventory. It would be good to have a short discussion on network design. There are several network design creation modules that can deliver network designs with different configurations to choose from, so that the service provider can make an optimal selection based on the service provider's requirements. Since the network design has to represent the configuration of all the elements in the network, the designs are mostly graphical with capabilities to drill down on specific equipment and get details on its suggested configuration. The planning performed prior to network design helps the service provider to make appropriate change in configuration for a smooth provisioning and good network performance after deployment.

Event or rule-based network design is usually used for network design. Event driven systems can be programmed to the dynamic of taking corrective actions to fault scenarios. This intelligence is specified as a rule. It is also common to do logical modeling whereby elements that can be grouped are considered one logical entity based on the services that are provided or executed on these elements. This makes it easier to define configurations, where attribute values are set on the logical collection rather than on individual elements that make up the collection. Network templates make design much faster and easier by offering the ability to define and maintain rules governing the ordering, design, and provisioning of various logical network systems. These templates have default values for configuration that can be reused without defining the value each time a new design has to be created, which is similar to a previous setup. The network design module can be used not just for setting the configuration of elements in the network; it also supports design of connections such as links between the elements.

Network design reference models currently available are mainly focused on the service provider and are intended to help them evaluate market penetration as well as other user-related capabilities of applications or services. The three parties—the service provider, service developer or software and equipment provider, and the customer—should be considered for introducing changes in the network. There should be an evaluation of end-to-end flow from services to network capabilities. When service providers want to offer a new and innovative service, in most cases the network to support the same will not be available and when the operational equipment provider comes up with an innovative technology in the network there might not a proper business case resulting in a suboptimal market fit. This affects profit margins of the service developers as well as impedes rapid service development.

For example, some of the widely used reference models for wireless network design include WWRF/WSI and I-Centric. One of the areas that these models do not address is the evaluation of the customer space and the network space as one full flow, which is required to bring out business-focused technology development. With the advent of the converged network, the service providers, content providers and network equipment providers have to rapidly conceptualize, evaluate, and integrate their service offerings. Also due to increasing competition, there is additional pressure to deliver the "right" product/service "on time, every time." It should be kept in mind that any focused development outside the realms of pure research needs to have a supporting business case and technical roadmap evaluating the value add it will offer to both technology and business. In coming up with a proper business case and technical roadmap, the customer space, service space, and network space need to be considered even when the actual

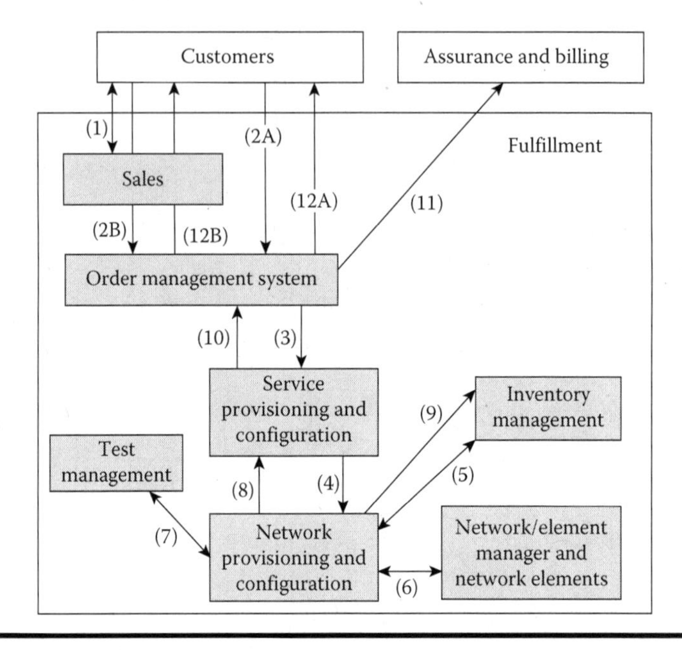


**Figure 20.6 Fulfillment process flow.**

development is restricted to one of these entities/spaces. For example when a new security router is envisioned, the services that can be offered on this router and the change in end customer experience should be evaluated to develop a product that can generate business.

The different steps in a fulfillment process can be summarized as follows (see Figure 20.6):

1. First the sales and the customer have an interaction on the various service offerings. In this interaction, sales will provide details on available services and answer the queries raised by the customer. The result of this step is that the customer will be ready to place an order if there is any service that meets customer requirements.
2. In this step the customer places an order, either directly with the order management system (2A) or places the order from sales (2B). After clarification of queries with sales on a phone, the customer places an order for a service package with sales on the same phone call, then this would be an example of ordering through sales. It is also possible that after the customer clarifies queries with sales, the customer logs onto the Internet and places an order online with an online order management system, which would be an example of direct order placed by the customer with order management system.
3. As part of placing the order, the customer and service related details are saved in the order management system. This information will be used for fulfillment,

assurance, and billing functions. This step involves the order management system requesting the service provisioning and configuration module to activate the service.

4. To configure the network and its elements, the service provisioning module sends commands to the network provisioning and configuration module. Network provisioning, as already discussed, takes care of elements in the network and the connectivity between the elements. The intent here is to configure the service at a network element level.
5. Before configuring elements, the network provisioning and configuration module first interacts with the inventory management module to check the availability of resources. The inventory management system does all the functions to make the resources requested for offering the service and informs the network provisioning module when it can go ahead and configure the selected resources.
6. Based on inputs from the inventory module, the network provisioning module now configures the network to offer the requested service. This includes configuration of elements, setting up the connectivity, and ensuring that the services are offered on a secure platform.
7. Next a test management module checks the configurations performed by network provisioning and the configuration module and ensures that the service is working as intended. The interaction between the test management module and network provisioning and configuration module ensures that all bugs are reported and fixed.
8. The network provisioning and configuration module sends a status completion message to service provisioning and the configuration module once a confirmation is received from the test management module that the network has been properly configured. The service provisioning module now performs any service level configuration required and activates the service.
9. The network provisioning module also updates the inventory management module with information on the status of the allocated resources.
10. Service provisioning and the configuration module now informs the order management system to update the status of the order. The order management system sets the order status as completed.
11. The order management system will then inform the assurance solution to start collecting reports on the services offered as part of the completed order and the billing solution to start billing the customer as per the SLA associated with the order.
12. Finally the order management system will directly update the customer (12A) or inform the customer through sales (12B) about the activation of the service package requested by the customer when placing the order.

Before we move from fulfillment to assurance, let us look into the key vendors providing solutions for fulfillment process. Telcordia would probably top the list with solutions for order management, inventory management, and provisioning. Telcordia is one of the major market leaders in network inventory management solution. Amdocs has fulfillment, assurance, and billing solutions. Cramer, which was the major competitor for Telcordia in network inventory management, was acquired by Amdocs. The huge market share of Telcordia can be attributed to its acquisition of Granite Systems. Metasolv and NetCracker are two other key vendors that have products on order management, provisioning, and inventory management. Some other popular vendors in network inventory management are Incatel, Viziqor, Axiom, and Syndesis. Sigma Systems and ADC are well known for their provisioning solutions. Amdocs is also a major player in order management. Eftia OSS Solutions is another well-known vendor for order management solutions. The vendors specified here are just a few of the key players in fulfillment space and this should not be considered a conclusive list.

## 20.3 Assurance Process

The assurance process involves the set of operations used to assure that the services offered to the customer are meeting expected performance levels (see Figure 20.7). The "assurance" process embodies a range of OSS solutions to ensure that a network is operating properly and that service quality thresholds are maintained. Two main areas covered under assurance are problem handling and data management across customer, service, and network. Once an order is completed, continuous monitoring is required to ensure that the SLA agreed upon with the customer is being satisfied, the service quality is optimal with the usage of hardware within permissible limits and the network performance guarantees maintaining the service as per SLA without any outages. Associated with monitoring is the identification

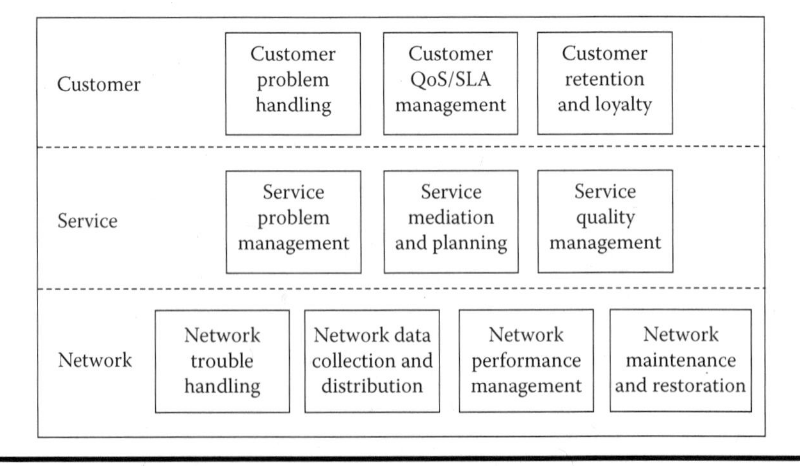


**Figure 20.7 Assurance process.**

of bugs and corrective actions to fix fault scenarios. The bugs can be identified and reported by the customer, or it could be identified as a service or network log as part of monitoring. Managing the lifecycle of the bugs identified is an essential part of assurance.

Network and service management is mostly covered under assurance. The collection of management records, distribution of records to other management applications, performing maintenance operations like installing patch and changing configuration, restoring an element or connectivity that failed, and monitoring performance records are all covered as components in the network management part of assurance. Service assurance is more on monitoring the service KPIs (key performance indicators) to ensure that the intended service quality is achieved. The numbers of outages, the maximum transfer rate, the number of fault scenarios are all indicators of the quality of service. At the customer level, the assurance process is more targeted to meeting an SLA. The parameters specified in SLA are monitored and any slippage in the SLA needs to be communicated to the customer and compensation. Customer retention and loyalty are also a part of assurance process. Sending reports, mail alerts on schedule maintenance, providing incentives and discounts, quick closure of troubles reported all help to improve customer satisfaction. A well-defined assurance process can be a key differentiator in increasing a service provider's competitiveness.

The network operations are monitored and maintained by administrators working from the network operations center (NOC; see Figure 20.8). To control and maintain the network activity there can be multiple NOCs for a single service provider. The device status data, alarms generated, performance records, and critical element failure information are all collected by the network management system at the NOC. The records collected by the NMS are used to generate reports that will be used for network planning and service management. The fault information from NMS can be aggregated to provide service level fault data. So a trouble condition in an offered service can be mapped to the associated network and further drilled down to the specific element that caused the issue. The NOC professionals have to continuously monitor the critical performance and fault information required to ensure smooth network operation.

The management solution in NOC deals with:

- Collection: The data from the network elements has to be collected. All fault, configuration, accounting, performance, and security related data are collected using management solution. Proprietary protocols and standard protocols like SNMP, CMIP, and T1 can be used for collecting the data. There can also be multiple networks involved in offering a service, in which case data from all participating networks needs to be collected by NOC.
- Consolidation: The data from different element managers need to be consolidated and fed to the network manager. CORBA IDL or XML based interfaces are used to make this communication possible where data needs to be

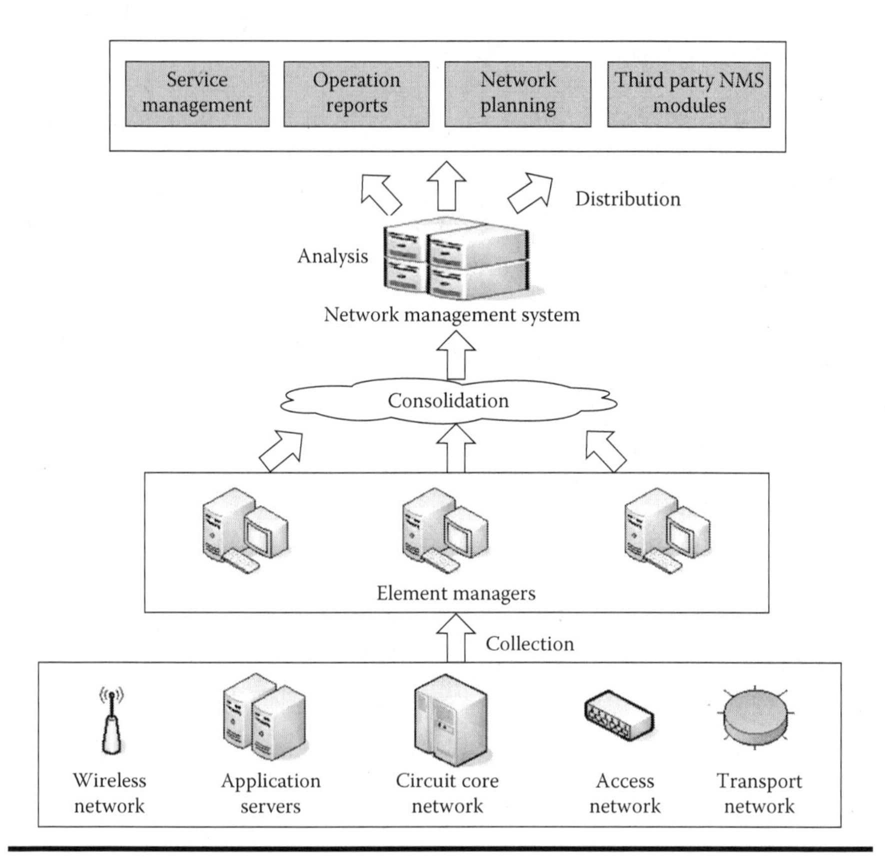


**Figure 20.8 Network operations center.**

consolidated from distributed systems. The requirement of NOC to have a holistic view of the network makes data consolidation a critical activity.

- Analysis: Filtering of records and correlation of records are components in data analysis. From the large amount of data that gets consolidated, relevant information has to be captured by filtering out unnecessary information. When a trunk shows a trouble state, the basic debugging approach involves a check on the status of its associated nodes like the related links or the link interface units. Fixing the link might bring the trunk back to service. So trouble scenarios need to be analyzed against different patterns that can cause the event. The data also need to be correlated to get relevant information. For example, when the connection between a media gateway and a MSC in the network goes down when a new module is loaded in the media gateway, there are various events that are generated during this scenario. The MSC will generate logs showing connection time out, link down, and so on and the media gateway in addition to connection logs will show the log with the module, which caused it to go down. So different logs needs to be filtered and correlated to identify the actual problem that needs to be fixed.

- Corrective action and maintenance: When analysis of data identifies a trouble condition that can be mapped to a predefined corrective action, then an event handling module will execute the steps to complete the corrective action. For the above example of media gateway failure during a new module load, the event or action handler will trigger the command on the media gateway to perform a module replace or fallback to the old module/load. An administrator can also execute commands to take corrective action for a trouble condition. Service providers aim in having minimal manual intervention for fixing trouble conditions. The actions can involve repair, restoration, and maintenance activity.
- Distribution: Data filtered at the network manager or service logs created by aggregating multiple network level logs gets distributed to several service management modules and other network management modules. the distribution channel can lead to business level modules like order management and customer trouble management.

Next let us look into quality assurance, which ensures that the service package after activation meets the guaranteed performance levels as per the customer specific SLA. Quality assurance is handled by customer QoS/SLA management, service quality management, and network performance management modules. A set of attributes that are classified under network, service, and customer levels are defined for measuring quality. These attributes are monitored by the quality assurance management modules.

Some of the attributes monitored by network performance management module are:

- Latency: It is the time delay between the initiation of an event and the moment the effect begins. There are different types of latency.
  - Propagation latency corresponding to time delay for travelling from source to destination. Again propagation latency can be calculated as one way or round trip.
  - Transmission or material latency is the time delay cased in transmission due to the medium used for transmission.
  - The response from a system may not be instantaneous upon receiving the request. The delay in response caused due to processing of request leads to processing latency.
  - There are other forms of latency like synchronization latency, caused when communication occurs between two systems that have different processing capability due to factors like difference in buffer size or processor speed. The latency caused due to the mismatch can lead to flooding of the system with lower capacity.

- Throughput: It is a measure of successful message delivery over a communication channel. The most popular unit of throughput is bits/sec. There are different types of throughput.
  - Network throughput: It is the average rate of successful messages delivered in a communication channel. The channel is usually a network link or node.
  - System throughput: It is the aggregate of data rates delivered to all terminals in a network. System throughput is also referred to as aggregate throughput.
  - Normalized throughput: Throughput expressed as a percentage is called normalized throughput or channel utilization.

  Several attributes linked to throughput like maximum achievable throughput and peak measured throughput help the operator in network planning and quality measurement.
- Error rate: This attribute is a measure of the data blocks with error. The data blocks can be bits in which case the term used is bit error rate, it can be words corresponding to work error rate or error in data packets measured as packet error rate. Error is mostly expressed as a ratio of error data blocks against the total number of data blocks. Total and average value of error rate is also used as an attribute for measuring errors.
- Network failure: Different flavors of network failure is monitored for performance measurement purpose. The most common ones include dropped calls to identify the total number of calls dropped, link down to identify disruption of service in connection links, packets lost or packets dropped corresponding to the number of packets that required resending, and node down to identify the time duration when a network element was down causing network failure in offering a specific service.

Some of the attributes monitored by service quality management module are (see Figure 20.9):

- MTTR: This attribute is a measure of the availability of the system. For systems that have a backup facility where a secondary machine takes up control without delay when the primary goes down, the MTTR is zero. Here the system has zero MTTR while the individual devices (primary and secondary) have nonzero MTTR.
  - Mean time to recover: It is the average time a system or network device/element takes to recover from a trouble condition or failure.
  - Mean time to respond: The average time the support or maintenance team takes to respond to a request to fix a network issue is calculated in mean time to respond. The MTTR value is usually mentioned as part of the support agreement the service provider has with a network or equipment support vendor.

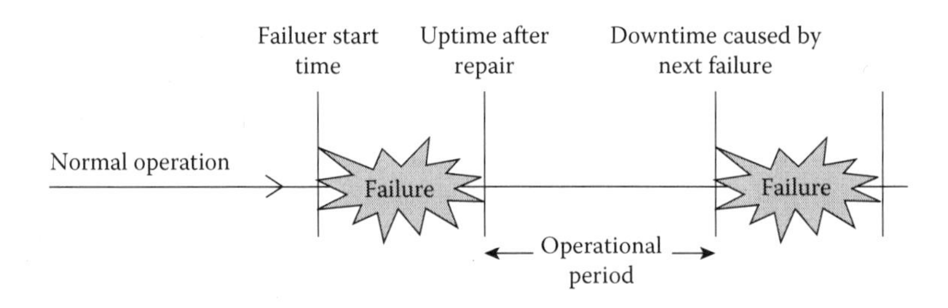


**Figure 20.9 Attributes monitored for service quality.**

- Mean time to replace/repair/resolve: A system or equipment trouble might sometime require replacement of the system/equipment or repairing the same. Some other methods like installing a patch or upgrading a component can also resolve the trouble condition. This leads to different acronyms like mean time to replace/repair/resolve corresponding to whatever may be the case the service provider wants in the support agreement.
- Maximum time to recover/resolve/repair: Mean time is mostly an unreliable parameter to enforce. For example, mean time to resolve in 12 hours does not necessarily mean that when a trouble condition occurs the issue will be fixed during 12 hours. The support personal might solve one minor bug in one hour and later take 20 hours to solve the next issue, which might have been a critical issue that required a quick resolution. Even when the support team takes 20 hours, mean time to resolve within 12 hours is not violated as the average time taken to resolve is less than 12. To avoid this kind of issue, maximum time to resolve an issue is specified in the agreement. Maximum time to resolve/repair/recover is a more reliable parameter than can be easily monitored and enforced on all trouble or failure scenarios.

■ MTBF: Mean time between failures is the average time between system failures. MTBF is usually measured at the system level and is a part of the contractual agreement. MTBF is a measure of the reliability of the system. For computing the value of MTBF, the sum of operational periods is divided by number of failures. Different flavors of MTBF based on the failure scenario include mean time between system aborts (MTBSA) and mean time between critical failures (MTBCF).

$$\text{MTBF} = \sum (\text{downtime} - \text{uptime})/\text{number of failures}$$

■ Service downtime: This is the time interval during which a service is unavailable. High availability of systems is now a mandatory requirement while offering most services and a specific value of maximum service downtime is specified as part of the contractual agreement. This attribute is similar to MTTR, except that repair and recovery usually deals with a system and its elements, while service downtime looks at the entire service offering without any reference to the systems that make the service possible.

- Service order completion time: There is a time gap between placing an order and completing of the order. This time gap is required to do the provisioning activity for activating the service. The customer placing the order is provided with an order completion time. This helps the customer to know when the services will be ready for use or when to check back on the status. There could also be price packages associated with order completion time. Some customers might be in urgent need of a service and can request premium processing of the order. Rather than waiting for all orders that were placed by other customers before the current order to be processed, the order with premium processing is given a higher priority and will be processed as soon as possible leading to shorter service order completion time.

Some of the attributes monitored by customer QoS/SLA management module are:

- QoE: Quality of experience is also referred to as quality of user experience and perceived quality of service. It is a measure of the end user's experience with the vendor. QoE is usually identified using a survey and by getting feedback from the customer. It is different from QoS, which is contractual and can be directly measured and tracked. It can happen that the vendor is satisfying the QoS but the customer may not be satisfied with the service leading to low QoE or the vendor may be violating some of the QoS attributes but the customer might still be satisfied with the vendor leasing to high QoE. Customer loyalty and continued business from the customer is only possible when QoE is high or in other words the vendor meets the customer expectations.
- Customer service satisfaction: This is another important quality parameter monitored by the customer quality management modules. For customer service through a hot line, the call is recorded or monitored to check the way the customer or technical support person attends the call and solves the customer's problem or answers the queries raised. Customers may also be requested to fill out a hard copy or online feedback form on the satisfaction levels on the service offered by a customer service team. Most service providers also have automated customer service so that the customer does not have to wait for getting a free operator to get a response to the query. In most cases the customer just has to make a few selections from an automated menu to get a response to the specific query. Only when the customer is unable to get a proper response from the automated system will the customer service personal need to get involved.
- Proactive reporting: Keeping the customer informed about the subscribed service is a very important part of quality at customer level. This would involve proactive reporting to the customer on possible outages to services

due to reasons like maintenance or system failure. Providing online reports on the service usage and capability for the customer to manage or customize the service based on needs.

- Billing articulation and accuracy: The bill presented to the customer is the source of revenue to the service provider for the services consumed by the customer. the customer may not always properly understand the break downs involved in the bill in coming up with the final amount to be paid. This is mostly the case with the first bill from the service provider, where a customer has to understand the information articulated in the bill. Accuracy in calculating the bill amount and the way information is articulated makes an impression of the service provider on the customer. Regular discounts and promotions, including incentives and compensations for SLA breaches go a long way in improving customer satisfaction.

There are different types of reports created as part of an assurance process, some of which are used by the service provider and others useful for the customer. Let us briefly touch upon some of these reports:

- Network usage reports: The network utilization is evaluated using this report. The network planning can be done with this report. Any additional usage can be identified and billed. Under utilization can be monitored and resources can be reallocated for better utilization without compromising on SLA.
- Performance dashboards: The performance of the network elements are represented in dashboards. The dashboards are prepared using performance reports collected from the network elements at regular intervals of time.
- SLA report: This report shows the performance of a service against the SLA. It helps to identify SLA breach and SLA jeopardy scenarios. The service provider can take corrective actions to ensure that service levels remain within a predefined boundary values. SLA reports are sometime also shared with the customer to show transparency of information and improve customer satisfaction levels.
- Customer usage patterns: This report is published to the customer providing details of the usage. The reports are mapped to the billing cycle. For example, when the customer is billed monthly, the customer usage report will show the variations in usage based on preceding months. The change in a billed amount is also provided as a pattern for customer viewing by some service providers.
- Service providers also publish reports on general statistics on most popular services, feedbacks/satisfaction levels from customer, annual or quarter wise financial reports of the company, and so on.

The next important activity in assurance process is trouble handling. This involves customer problem handling, service problem management and network

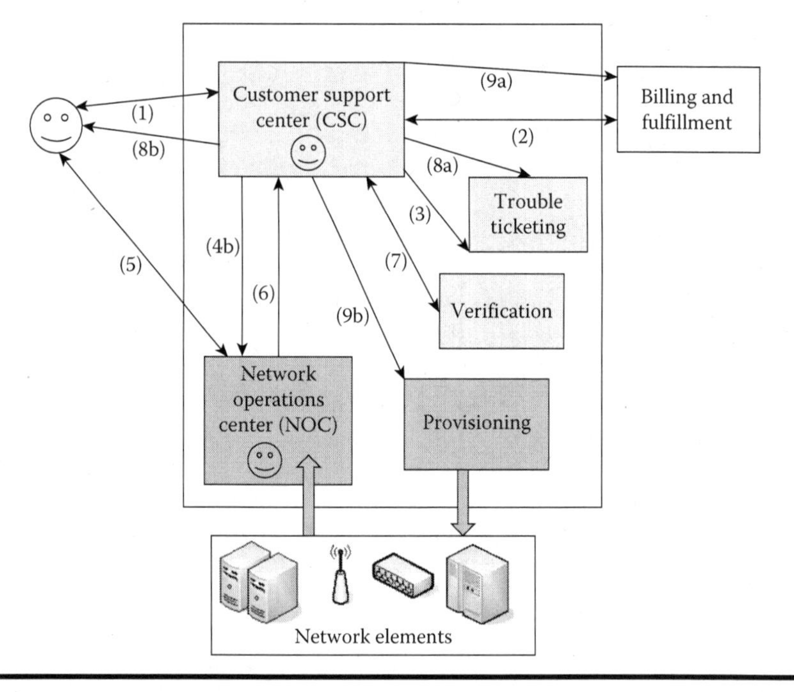


**Figure 20.10 Trouble handling.**

trouble handling processes corresponding to a customer, service and network layers, respectively (see Figure 20.10). Let us consider a typical workflow of how an issue raised by a customer gets handled in the assurance process.

The steps are:

1. Customer calls up the customer support center (CSC) and reports an issue about the service offered along with details on the customer account.
2. The CSC will verify the account details of the customer with the billing and fulfillment modules.
3. Next the specific problem and the history of the problem is documented with the trouble ticketing module.
4. CSC will then try to fix the issue (4a) or hand off the call to technical support (4b) in the network operations center (NOC).
5. Assuming CSC hands off the call, the NOC will interact with the customer and try to do the following checks based on professional support expertise and diagnostic tools available:
   a. Identify if this needs to be fixed at service level.
   b. Identify if this is caused by a problem with the network.
   c. Determine if this problem is caused due to a problem already identified and tracked by an existing trouble ticket.

   d. Check if a similar trouble was previously reported and an already existing fix can be applied.

   If an immediate fix cannot be applied the ticket is assigned to an appropriate team and the customer is given a ticket number that can be used to track the status of the fix.

6. After the issue is isolated and fixed, NOC informs the CSC to close the trouble ticket.
7. CSC informs the verification module that checks the fix and ensures that the problem no longer exists. If verification responds stating that the issue is not fixed then the ticket is assigned back to NOC. This activity continues until NOC can identify a fix that passes verification.
8. CSC will then close the ticket (8a) and inform the customer (8b) that the issue has been fixed.
9. Based on SLA, the CSC might update billing records (9a) to credit the customer for the reported trouble and the provisioning modules (9b) to take corrective action so that the same bug is not reported by another customer using a similar service.

There are several issues that come up as part of trouble handling. Some of them are:

- Customer support team has limited access to data: In most CSCs the support team only has access to the customer relationship management and billing data. So the customer might have to elaborate the issue first to the customer support and then to the billing representative. Even normal queries that involve the signal strength may require intervention of technical support.
- Customer Interaction: The customer will always want an immediate solution to the problem and in some cases may not be able to correctly communicate the issue. The support team should be well trained to handle even the most frustrated customers.
- Most calls are queries rather than reporting an issue: The calls from customer to support desk can be for more information on a product, details on using a service, balance query, interpreting a report, and so on. To avoid time loss in dealing with queries and to devote more time of professional support on actual trouble scenarios, the support system is mostly automated. So when the customer dials a customer support number, the customer is requested to make a selection to describe the reason for the call. Wherever possible the automated support module will try to satisfy the customer.
- Call handoff not seamless: There are multiple sections in the customer support team to handle issues on specific topics like billing, new product information, service activation status, and so on. It can happen that the multiple call handoff happens first at the customer support in CSC before the call goes to a technical support professional in NOC. A typical example is the customer

calling up support to check why the new service is not working. This would first go with the billing team to check credit, then to the service activation team to check the status, and if the service is already activated then to the technical team to isolate and fix the problem.

- Proactive identification: The problems that are reported by a client and need a fix are definitely cases that could have been identified by the service provider professional itself. Proper validation and verification can reduce the number of bugs reported. When a bug is fixed with regard to a specific service, suitable actions need to be taken that ensures the same issue in another service or for another customer is also fixed. Proactive fixing of issues by the service provider and service developer can reduce the number of bugs reported by the customer.
- Isolating the issue: This involves mapping network issues to a service and more specifically to identify the node in the element that has caused the service failure. When there are multiple issues to be fixed then a decision has to be made on what issue needs to be fixed first. A discussion might be required between teams to identify the best way to solve a set of network level issues to bring back a service. Some equipments and solutions used in service provider space may be from third-party vendors. To fix some bugs and to isolate the issue itself, interaction with third-party vendors may be required. Most service providers outsource their support activities to IT companies to have a single point of contact for inquiring on the status of a ticket and the IT company has to manage the different external vendors.

Customer centric assurance is currently a necessity. Most service providers have a dedicated team to identify strategies for service differentiation based on the assurance process. Standardizing forums are also giving due importance to assurance with dedicated groups like ICcM (integrated customer centric management) in TeleManagement Forum (TMF) working on customer related issues in assurance. The goal of a customer centric assurance is to:

- Fix customer issues in minimum possible time: The customer is worried about his service and will be calling the support desk for a quick and concise solution to the problem. The quality of customer service depends on how much the customer is satisfied at the end of the call. For this reason, extensive training is offered to call center professionals with the intent to improve customer experience.
- Real time monitoring of customer experience: Customer service can be a key differentiator in making a service provider more lucrative to a customer compared to its competitor. Similar to monitoring the service and the network, the customer experience also needs to be monitored for customer retention and satisfaction. For this reason the customer calls may be recorded for analysis of customer service quality.

- Identify the service and network performance: After deployment, the issues raised by the customer is a valuable source to identify the quality of service, the bugs in the service and the underlying network and the issues that needs to be fixed in similar service offerings that are active or will be provisioned in the future. Based on bugs reported by customers, if the bug density of a product is relatively high and leading to law suits, then those services can be revoked for releasing an improved version and for planning toward new products.
- Identify reliable third-party vendors: There can be multiple third-party vendors handling support and assurance functions. These could be suppliers of the network equipment, external call center professionals, or developers of the services. The equipment failures gives an indication of which vendor can be approached for future opportunities. Time to solve a reported issue by replacing or fixing the equipment becomes a critical factor. The number of satisfied customers after a call with a support agent is an indication of how well the call center professionals are handling support tasks. Finally the number of bugs reported on the service is an indication of the development and test effort performed before the service was launched. Identifying reliable third-party vendors for future opportunities can be done as part of monitoring the assurance reports.

Most customers have high expectations on a subscribed service, leading to low loyalty and much dissatisfaction when a service does not perform to the expected levels. The assurance helps in clarifying the customer queries before they subscribe to a service and later helps in bringing their expectations in par with what is actually offered and determined by SLA. Quality of customer service is always a key differentiator for service providers, which affect their revenue and profitability.

The increase in complexity of network technology is also affecting the assurance with an increased time to fix issues. Management of the complex network in a multivendor environment also adds to the effort in assurance. New and compelling services are being introduced with advanced equipment making up the underlying infrastructure. The monitoring and trouble management system should be such that the underlying complexity is abstracted to a level that makes the management of services and network easy. Standards on next generation network management are toward this end for smooth operation of next generation services on IP based next generation networks.

Service developers can also play a role in ensuring integrated customer centric experience. The high degree of complexity accompanying telecom industry with converged services, heterogeneous networks, and multiple vendors necessitates effective service developer strategies to ensure that the service provider can offer integrated customer centric experience.

Some of the issues that generate dissatisfaction to customers are:

- Many interactions with support team without a positive outcome.
- Inconsistent information provided on interaction with different channels.
- Increased time for service fulfillment and customer service.
- Multiple services offered on divergent platforms like mobile, PSTN, Cable, and so on, which have to be handled separately by the client.
- Competitors offering better customer experience.

The number of support calls from customers has increased recently and the main cause of these issues can be attributed to the following changes in service provider space affecting customer experience:

- New complex convergent services are being introduced that the customer is not familiar with. Support calls can be with regard to ordering, service usage, billing, and so forth.
- Different types of networks are coming up and the service provider professionals may not be able to fix all bugs before deployment.
- Many suppliers and partners with regard to services being offered and the content associated with the services.
- Acquisitions and mergers in service provider space, which require aligning business process and integrating solutions.
- Nonscalable legacy applications for management impedes proper monitoring on next generation network and services.

Service developers can contribute to better assurance in current telecom scenarios by offering seamless integration of services and better management at the service provider space. Toward this goal, the following strategies are becoming increasingly popular on the service developer space that will impact the service provider:

- SOA (service oriented architecture) in development can lead to smoother integration of services. SOA promotes reuse of services, which can bring down the CAPEX and OPEX. Service development and lifecycle needs to be well managed and network data collection and control should be separated from the service implementation.
- Centralized service management interface that can make the monitoring process much easier for professionals at the NOC. This can reduce the fulfillment and service time and provide more consistency in information. Cost reduction is also achieved by saving on the expense in training personals with various management interfaces.
- Management solutions that can hide the network complexity and provide abstract information that would be easier for the technical and customer support professionals to work with. Complex services would impose more

limitations on the technical desk in identifying the network issue. This would affect the problem handling and service fulfillment time. Filtering the most relevant events for the issue raised by the customer would be a challenge if sufficient event abstraction is not already in place.

- Auto recovering and self-managing network elements can reduce the work of the support professionals and fix issues of the customer without having to call up the support desk. With an increase in network complexity, the cost of maintenance and time to fix issues will increase. Auto recovery mechanism and presence of self-managing network are key functionalities in a well-managed network
- Standard based development and well-defined contracts for interaction should be followed by the service developer to ensure interoperability and ease of integration. The issues that originate in a multisupplier/partner environment with acquisitions/mergers can only be addressed using standardization.
- Transformation of legacy application is also a strategy followed by many service developers. Since legacy solutions are functionality rich, many service developers don't opt for new SOA-based solutions, rather they transform their legacy solution in stages to an SOA-based framework.
- COTS software and hardware developed with holistic telecom vision should be used by service developers in the solutions they offer to service provider. Off the shelf implementations will reduce cost and integration time. The challenge is to ensure that a COTS product will interoperate as expected with an existing product. NGOSS based applications should allow solutions to be rapidly implemented through integration of off -the-shelf software components. ATCA (Advanced Telecommunications Computing Architecture) is a modular computing architecture from PICMG for COTS telecom hardware development.

Let us conclude the discussion by mentioning some of the key players that offer assurance solutions. After acquiring Tivoli Systems Inc. and Micromuse, IBM has emerged as the market leader in offering a variety of assurance solutions. HP Openview products of Hewlett-Packard (HP) can manage and monitor a wide variety of network elements used in data, access, and transport network. Computer Associates have a suite of solutions for management of application servers. TTI Telecom offers OSS service assurance solutions suited for fixed, mobile, and cable networks. Telcordia is another OSS solutions provider worth mentioning in the market leaders for assurance solutions.

## 20.4 Billing Process

Billing corresponds to the processes in service provider space for getting revenue from the customer for the services consumed. The service used by the customer is metered for generating the bill. There is a rate associated with the service plan

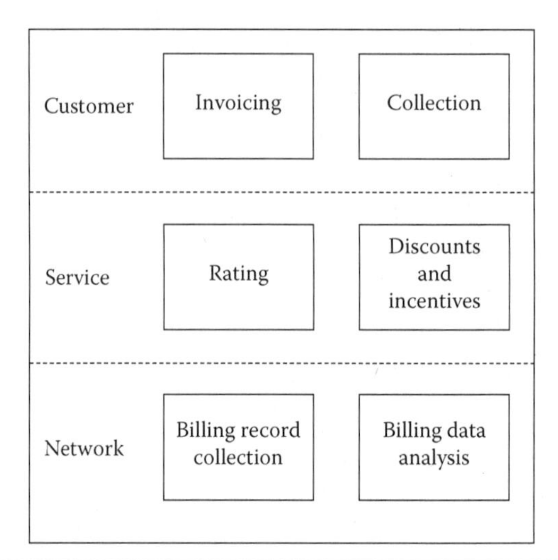


**Figure 20.11 Billing process.**

selected by the customer. This rate is applied on the service consumed to compute the customer amount due to the service provider at the end of a billing cycle. The billing system should allow easy management of customer accounts. A centralized repository that maintains a record of customer payments will be queried and updated by the billing system.

The three main processes in billing are (see Figure 20.11):

1. Invoicing and collection: This is a customer level process. An invoice is generated for the customer with information on the bill amount for the subscribed service. The invoice is then sent to the customer. The bill amount is then collected from the customer using the deferent payment options supported by the service provider and the customer account is updated.
2. Rating and discounting: This is a service level process. A predetermined rate is applied on the service consumed to come up with the amount due for a specific service usage. On the computed amount, discounts or other rebates are applied to come up with the billable amount for which an invoice can be generated.
3. Billing data collection: This is a network level process. Service usage is calculated using records generated at the network elements on which the services are offered. For simple voice service, call detail records (CDRs) make up the billing data (records) that is collected. Information in the billing data generated by the network elements are processed to get usage information on which rates can be applied.

Billing is no longer limited to a method of collecting money from the customer for services used. It has evolved into a process that has its own independent standards and can be a key differentiator for a service provider. Billing is often used as a strategic tool to get and retain business. The results of the renewed approach to billing has led to different types and models.

Some of the key terminologies used in billing that the reader has to be familiar with are:

1. Account: Each customer is handled as an account in the billing system. While multiple services can be associated to an account, usually one account will have only one customer. For competitive pricing, some service providers are offering family level accounts, where different users are linked to a consolidated account for generating a single bill.
2. Service: Any tangible or nontangible offering from a service provider for which the customer is billed makes up the service. The package that is offered as a service varies based on the service provider.
3. Plan: This is a bundle of services mapped to a pricing schema. The customer is offered various plans by the service provider from which the customer selects a plan to be mapped to the account. The account is billed based on the selected plan.
4. Call detail record (CDR): This record is generated by the exchange switch with details of the call between source and destination. CDRs have valuable information about the call that can be used for billing.
5. Recurring charge: The charge the customer has to pay periodically for a product or service offered by the service provider.
6. Billing cycle: It is the periodic cycle during which the customer is billed for the subscribed service. That is, a billing cycle can be calculated as the time interval between two bill generations.
7. Accounting cycle: In order to generate a bill, a set of accounting activities needs to be performed. The cycle for which the activities to identify charges against an account are calculated. Accounting does not involve bill generation and accounting is only about tallying the customer account.

Some of the billing types are:

- Wireline billing: The billing of utilization in a wireline network. The call records are generated at the switch or elements handling billing in the core network.
- Wireless billing: The billing of utilization in a wireless network. Similar to wireline networks, billing records are collected to compute the utilization.
- Convergent billing: In convergent billing all service charges associated with a user or user group is integrated onto a single invoice. This will give a unified view of all the services provided to that customer. Services in a convergent bill can account for voice, data, IPTV (Internet Protocol Television), internet,

Video-On-Demand (VOD), and many more services that a customer is using from the service provider.
- IP billing: The billing for IP-based services like VoIP (Voice over Internet Protocol). In IP billing it is usually the content that determines the bill amount rather than the distance between the origin and destination as in traditional billing methods. This is because IP distance will not be a key factor and it will be parameters like quality of service and security level that determines the tariff to be imposed on the service subscribed by the customer.
- Interconnect billing: Services can span across multiple service providers. For example, a simple phone call between two people can involve two service providers, when each of them is serviced by a different service provider. Outgoing call charges are imposed by one service provider at the call originating end and incoming call charges are imposed by another service provider at the call terminating end. This setup connecting the network from different service providers/operators requires some kind of billing mechanism to monitor the actual amount of data (sent and received). It is known as interconnect.

Some of the billing models are:

- Volume-based billing: Here the bill is calculated based on the volume of data (total size). An internet service provider charging the customer based on the volume of data upload and download is an example of using volume-based billing.
- Duration-based billing: In this billing, time is the determining factor. An Internet service provider charging the customer based on duration of sessions the user is connected to the internet is an example of using duration-based billing.
- Content-based billing: This model is most common in IP billing. In this billing, the type of content offered to the customer is the determining factor. It can be applied on services like video broadcast, multimedia services, gaming, and music download.

Charging process can be:

- Off-line charging: In off-line charging the service rendered cannot be influenced by the charging information in real-time.
- Online charging: In online charging the service rendered can be influenced by the charging information in real-time. Online charging is becoming increasingly important with convergent services where a third-party content provider will be paid by the service provider based on revenue sharing agreement for third-party content offered to a subscriber by the service provider.

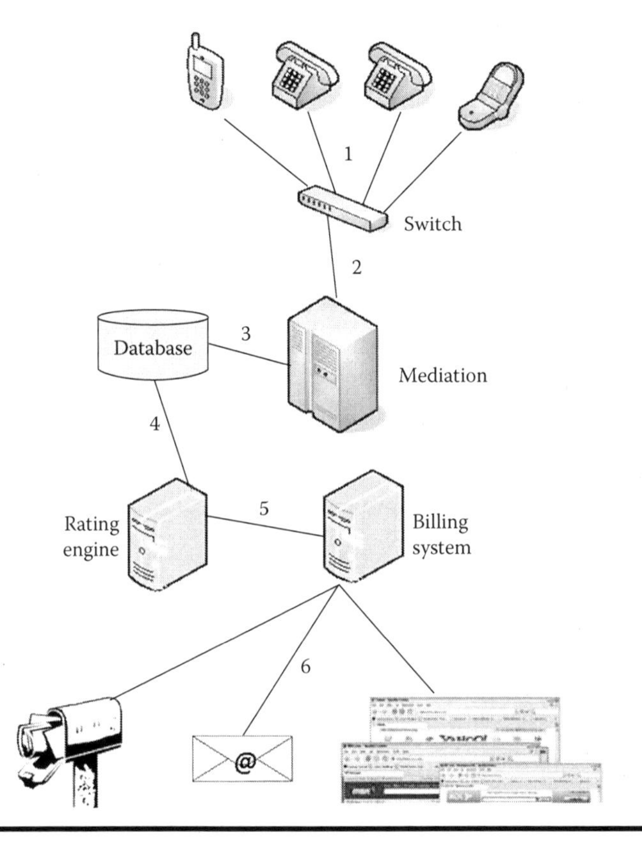


**Figure 20.12 Steps in billing.**

Bill payment methods can be:

- Prepaid: In this payment system the customer has to pay an amount in advance for service usage. The service usage is tracked against the amount deposited and continuity of service is ensured only when customer credit is above a predetermined value.
- Postpaid: In this payment system the customer receives a bill for service already consumed. The payment is only after service usage and continuity of service is ensured when the previous usage bill is settled before a predetermined date.

Let us next look into the steps (see Figure 20.12) involved in a typical billing process. The steps are:

1. Calls are made between mobile equipment through a switching center. The switch generates CDRs on the calls.
2. Mediation device collects call records from the switch. The mediation system can collect data from multiple switches handling call routing in the service provider network.

3. After processing of the call records, the mediation device stores relevant information to a database.
4. Information in the database is accessed by the rating engine to rate the calls. The rating engine will apply the payment rules on the transactions. Discounts and promoting can also be applied.
5. The information from a rating engine is sent to the billing system for sending the invoice. Any bill related settlements can be done by the billing system.
6. Invoices prepared by the billing system are sent to the customer as post, e-mail, or uploaded to a secure Web page. Most service providers allow the customer to register in the service provider Web site and give them the capability to securely access and pay their bills through the Internet.

Some of the essential requirements in billing are:

- Multiple rating techniques: In a competitive market it is essential to provide the customer with different rating plans. Customers can be segmented based on:
  - Geography: The parameters used for this segment include region, population size, density, and climate.
  - Demography: The parameters used for this segment include age, family size, income, gender, occupation, education, and social class.
  - Psychograph: The parameters used for this segment include lifestyle and personality.
  - Behavior: The parameters used for this segment include evaluation of benefits, analysis of usage rate, and attitude toward a specific product.

  Different customer segments have different needs and a single rating for all customers will not satisfy customer needs. Service providers are coming up with innovative plans that best satisfy the customer segments.
- Convergent billing: There would be multiple services from a service provider that a customer would be using. The customer might opt to use broadband internet, IPTV, Call Service, and On-Demand services from a single service provider. These services will be offered as a package by the service provider for a rate that would be less compared to the amount the customer has to pay when these services are subscribed on an individual basis. Rather than separate bills for individual services, the customer wants to get a single bill that details usage for all the subscribed services. For the customer this single bill is easier to track, maintain, and pay. Convergent billing for a package of services is currently offered by most service providers.
- Ease of understanding: There would be multiple items under a single service for which the customer gets charged. For example, in a simple call service a different rate is applied for interstate call, intercountry calls and international calls. Also special charges are applied for calling specific service numbers. In the first bill from a service provider, in addition to the rate wise breakdown

there would be charges against initial provisioning, equipment supplied, service charges, and so on. Now, how complex would the bill be for the case when the segregation of charges for a package of services is provided in a single bill. The billing information should be presented to the customer in a format that is easy to understand and helpdesk/online/call support should be offered to customers for clarifying bill related queries.

- Multimode, multipayment, and multiparty options: In multimode billing, the billing system must be capable of handling "real-time or nonreal-time" billing of "single or batch" billing records. Multipayment option, also known as unified accounting gives the customer the ability to do payment using postpay, prepay, nowpay, or combination of these payment options. The multiparty settlement option gives a customer the flexibility to settle multiple bills related to the customer or multiple bills incurred by multiple parties. For example, all bills associated to different members in a family can be linked to a single payee. This leads to ease in payment of bills. It should be kept in mind that the billing system in service provider space should have the required capabilities to offer multipayment and multiparty options to the customer.
- Billing system should have carrier-grade performance: Current billing systems have to support complex billing models. Even when advanced functionalities are offered, the billing system should still be scalable. An increased number of billing records should not affect the performance of the system. The billing system should provide enough abstraction of information to make it easy for the operator to work on the system. Availability is the key requirement of a billing system, as a few seconds of discontinuity in billing can result in a major financial loss to the service provider.

The billing module interacts with many other support system modules (see Figure 20.13). The main interactions at the business level happen with customer care and marketing and sales modules. At the service level, the interaction is with

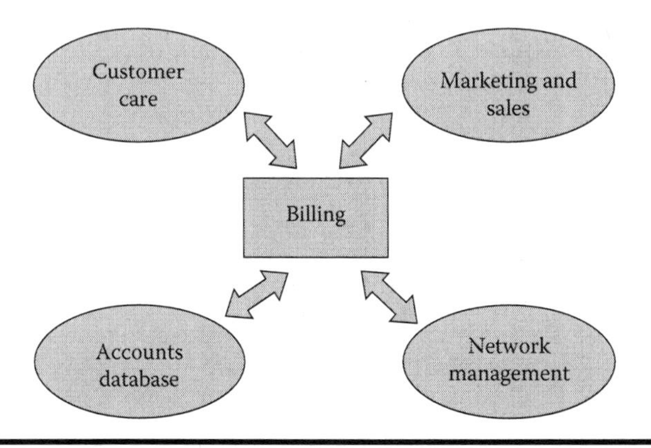


**Figure 20.13 Interactions with billing module.**

an accounts database and at the network level the interaction is with the network management module.

The billing system can generate trend reports of services that are generating more revenue that can be used by the marketing and sales team for designing strategies to increase business. The billing reports help in planning for investments based on market trends. The billing data is to be used by a sales team in providing information to customers on what could be the best service plan for a customer based on trends in the specific customer segment that the user falls in. Impact of promotions and discounts on sales of a specific product can be computed from the billing summary report for a specific product.

The customer care modules will interact with billing to apply service order adjustments and any specific discounts that need to be offered to the customer on a need basis. For example, a customer might want to cancel a specific service before the end of the billing cycle. In this case the number of days for which the customer has to be billed has to be updated in the billing system based on information provided to the customer support executive. The customer can also call the support desk for clarification on billing related queries. The customer support desk will have access to specific (subset) billing information to service the queries. However, most support professionals ask sufficient authentication queries to ensure that private information is disclosed only to the intended owner of an account. Some support professionals also request permission from customer, if the support personal can access the account owners billing data and if required, modify it.

The billing system will interact with the accounting modules in NMS to get the billing records. The billing records generated by the elements in the network are handled by the element managers and network management systems to provided relevant billing data to the billing system. There is an account database that the billing system will query to get information on customer account balance, credit limit, billing rules to be applied as per SLA, tax to be added as per regulations, additional charges for usage of specific service, and so on that are associated with generating a final bill to the customer.

Two important modules in the billing OSS are the mediation system and the rating system. All service providers have a mediation system that consists of several subsystems to capture usage information from the network and service infrastructure and distribute it to upstream BSS modules like billing, settlement, and marketing (see Figure 20.14). The subsystems collect billing records from the various network infrastructures supported by the service provider and provide feed to the central mediation system that aggregates the records, performs correlation between records, repairs records with errors, filters the records based on predefined rules, and maps the data to a target format required by the business support systems. Some of the popular formats handled by the mediation systems include CDR, IPDR, and SDR (see Figure 20.14).

Mediation functions can be classified into usage data collection and validation functions.

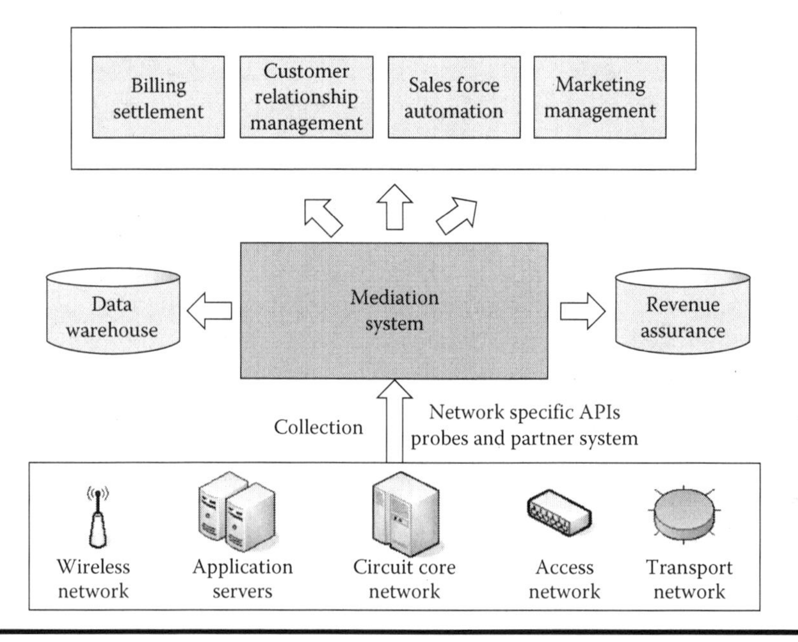


**Figure 20.14 Mediation system.**

1. Usage data collection: The usage records may be based on number of calls, messages sent, traffic utilization, or events based. A set of usage types can be defined based on originating/terminating point and the service being used. In fixed to mobile type the originating point is a fixed network while the terminating point is a mobile network. Similarly there can be other types like fixed to fixed, mobile to fixed, and mobile to mobile. When the user keeps on switching networks for a single usage record itself, the type is Roaming. SMS and call forward are grouped under value added service type.

   A part of the usage data collection the mediation system performs is a set of key activities:

   a. Polling for data: The mediation system continuously looks for data on usage records from network infrastructure. While polling is a pull based mechanism where the mediation system checks for new record availability and collects data, a push based mechanism can also be implemented, where the billing agents in the network infrastructure actively sends data to the mediation system when a new record is generated with details on the usage. To make the billing agents independent of the mediation system implementation, it is a usual practice to implement a polling mechanism for data collection where responsibility to collect and perform mediation to map to a format lies with the mediation system.
   b. Consolidation of data: Once data is collected by a "push or pull" mechanism, next the data across network elements needs to be consolidated. The

information sent from the mediation system to the upper business layer modules needs to be formatted data that can be directly interpreted. So the aggregation, correlation, and filtering between billing records takes place in the mediation system to send consolidation information to other business modules.

c. Standard interface for interaction: Multiple business modules make use of the information from the mediation system. Hence it is important that the mediation system is not designed for communication for a specific module. The interfaces for data request and response need to be standardized. These standard interfaces are published for business modules to easily interoperate with the mediation system. The function here is a seamless flow of information between the mediation system and business support systems.
d. Drop nonbillable usage: All information in a usage record will not be useful for billing. Only specific fields in the raw records collected from the network elements can be utilized for billing. As part of formatting the records, the mediation system will drop nonbillable usage information and the unused fields are filtered. It should be understood that this activity is quite different from rating. In rating some usage may not be billed based on terms in the SLA, but the mediation systems needs to collect and send all billable usage information for rating.
e. Support for real-time applications: Mediation systems have to support real-time application billing. This requires immediate processing of usage records with minimum processing time. The usage records are computed while the service is being used and information of the usage bill has to be generated at the end of each usage session.
f. Abstraction: One of the most important applications of a mediation system with regard to usage data is to insulate the biller from the network elements generating the raw data required for billing. In the absence of a mediation module, the biller will have to work with individual data coming from each network element to compute the usage data. This abstraction of generating useful information for the biller makes it easier for business support systems to work on the billing information.

2. Validation functions: The mediation system performs multiple validation functions on the usage records. Some of them are:
   a. Duplicate checks: This involves identifying duplicate records and eliminating them to avoid discrepancies in billing the customer.
   b. Eliminate unwanted record: Call failure and dropped calls should be dropped as nonbillable records based on validation checks on the collected records.
   c. Edit error records: Some records might contain errors that make it noncompliant for mediation processing tasks like correlation and aggregation.

In these scenarios the mediation system will perform edits and translations to make the content in the record suitable for interpretation.

d. Unique tracking: All data after mediation validation and processing are assigned a unique tag number to make it easier for business modules to identify if the relevant records have been received from the mediation system. This helps to re-transmit lost records or synchronize records with other management modules. A table look up can be implemented for proper tracking of records.

Design of the mediation systems tend to be complex due to different factors that needs to be considered when implementing this critical module that has high impact on the final bill presented to the customer (see Figure 20.15). Some of the factors that influence the complexity of mediation system are:

- Granularity of pricing: The granularity of pricing creates a variety of information that needs to be collected and formatted before feeding the data to business support systems.
- Real-time constraints: Data from the network elements has to be processed at real-time when the mediation system has to support a real-time billing

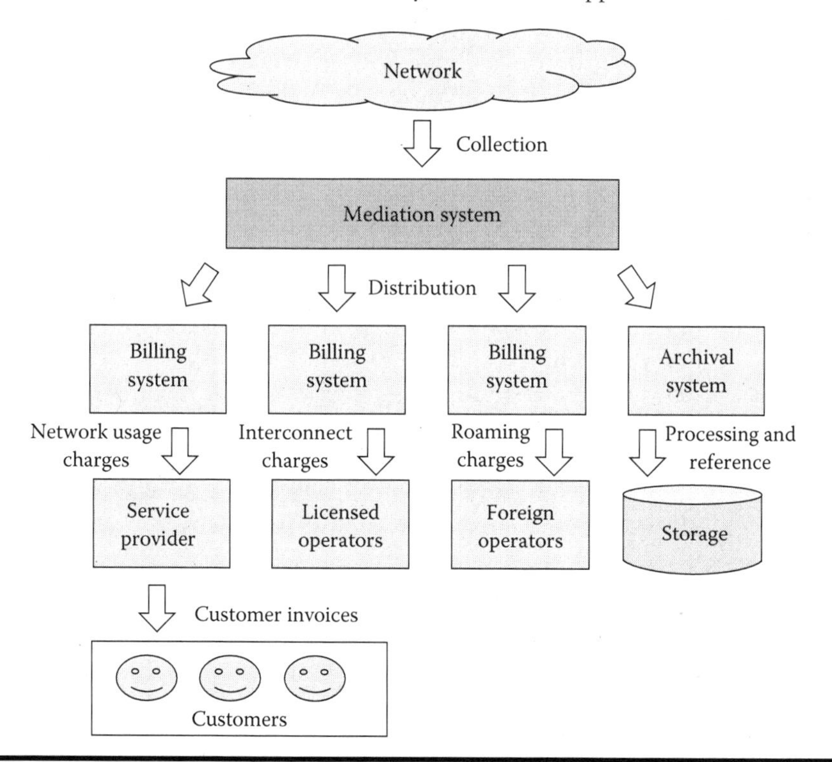


**Figure 20.15 Mediation data usage.**

model. This requires the mediation system to scale up in performance based on requirements. The processing should ensure that all relevant billing information is mapped at real-time without any loss of data.

- Regulatory requirements: The billing systems have strict regulatory requirements to comply with in terms of satisfying the SLA and security of content. The mediation systems that supplies data to the billing modules must ensure that the standards of compliance are satisfied. Any security or compliance failure at the mediation will have a huge impact on billing.
- Multiple connection points: On the network side the mediation system has to collect data from multiple network elements and on the business modules side the mediation system has to interact with multiple billing and business support solutions. The mediation system should be designed to provide easy integration with its interacting modules. Use of standard and published interfaces can reduce the implementation overhead, however in most cases there will be multiple interfaces defined by standardizing bodies and the mediation system, and hence may have to support multiple interface standards or have adapters to work with different solutions.
- Heterogeneous infrastructure: The scale and diversity of network elements is continuously increasing with new technical innovations. The mediations systems have to support aggregation and correlation of billing records on the new elements in the network. Most next generation networks have a new billing defined by standardizing bodies, which the mediation system has to adapt while formatting the records.
- Complex validation checks: There can be scenarios where correction is required for inconsistent formats, access, and semantics. Identification of these issues and fixing them are not straight forward and will require intelligent algorithms. The mediation system should be designed to handle multiple records in parallel to ensure that maximum processing is performed to support real-time billing.
- Rich information records: The information content in the records is quite huge compared to the actual billable usage information. The additional information is mainly for debugging and validation purposes. The mediation systems should filter out relevant data based on predefined rules. When new record formats are introduced, the filtering logic needs to be updated to ensure that these records are not omitted as error records. The validation, filtering, and formatting logic is usually implemented to work together to ensure that there are minimum iterations performed on the records there by increasing performance and decreasing processing time.
- Enormous CDR flow: One of the biggest challenges in designing mediation systems is to make it scalable enough to handle enormous CDR flows. There are a large number of call or usage records that are pumped to the mediation systems at any given point of time. The mediation systems should

have a mechanism to collect all the records and process them in the shortest possible time.

- Data correlation: The correlation between records to send consolidation information is also a challenging activity considering the wide variety of services being offered and the network infrastructure deployed by the service providers in current telecom industry.

There are several billing systems that take feed from the mediation system as depicted in the figure on mediation data usage. Data from the mediation systems is used to compute the network usage charges, interconnect charges, roaming charges, and other value added service charges. The data is also achieved in a storage system for future reference or for processing by business support solutions like planning, which try to identify usage statistics based on aggregated information in historical records.

Next we will discuss the rating engine whose role it is to apply pricing rules to a given transaction, and route to the rated transaction to the appropriate billing or settlement system. Rating takes input from a predefined rules table and applies it on usage information provided by a mediation system. The charge computed by the rating engine is not just based on usage. The final charge has various rating components taken from modules like the order management, marketing management, and SLA management system.

The main activities performed in rating are:

- Applying rating rules on the usage information for each of the customers.
- Applying any credits for outage.
- Applying discounts that are agreed upon as part of the customer order.
- Applying any promotions or offers on the charge.
- Applying credits for breach of terms in the SLA.
- Resolving unidentified and zero billed usage scenarios.
- Adding/reducing the charge based on special plans selected by the customer.

In rating, first the value of the attributes that are used for rating are identified from the formatted records forwarded by the mediation system (see Figure 20.16). Some of the attributes that effect rating are:

- Connection date and time: There could be special rates imposed on specific dates. For example, the service provider might option to give a promotional discount to all subscribers on Thanksgiving Day. The discount could also be associated to time (like 6 AM–6 PM) on the specific date.
- Time of day (TOD): Service usage has peak hours and off peak hours in a day. For example the number of users using the Internet would be more during office hours. To make effective utilization of resources, service providers offer a low rate for service usage during off peak hours. It is quite

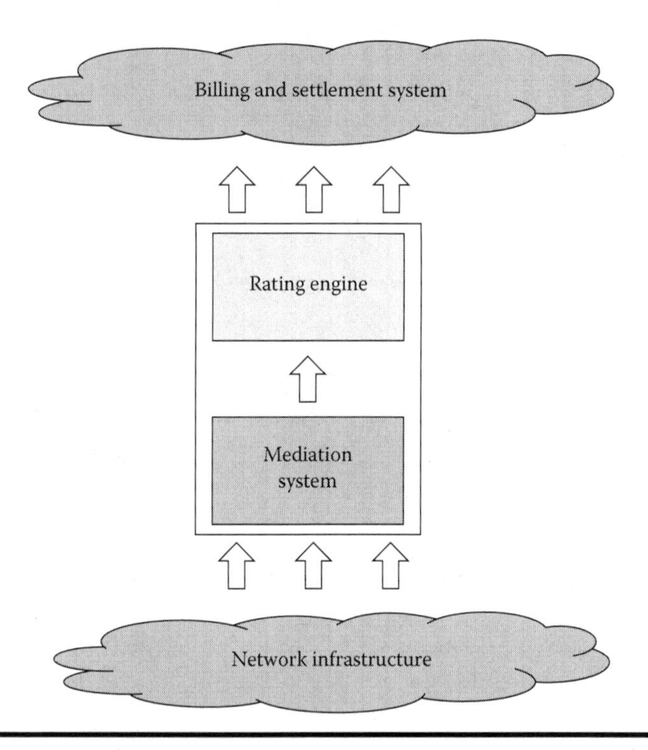


**Figure 20.16 Message flow with rating engine.**

common to see some content providers specifying a download limit during peak hours while giving unlimited limits during off peak hours, another common scenario is reduced call rate for calls made between 10 PM and 6 AM.

- Day of week (DOW): The same scenario of peak usage during office hours has an implication on special rates on specific days in a week. The usage of some services like Internet will be less on holidays compared to regular working days. Service providers account for these usage patterns by preparing rate plans to suit user needs.
- Duration of usage: Rating can be based on duration of usage. The total time is a direct indicator of usage interval. When a flat rate is applied for a predefined interval of time, charge to customer would be a simple multiple of rate against the number of intervals in the duration of usage. It can also happen that the rate plan follows a set of rating increments. That is, rates keep on incrementing every 30 or 60 seconds. This could be the case for high demand services.
- Type of content: With an IP-based telecom framework a wide variety of contents are processed in the network. Different types of content involve different levels of bandwidth utilization. It could also happen that some content

has a higher demand for which the users are willing to pay a premium. This leads to different rates based on the type of content.

- Jurisdiction: Rates can vary based on geography. The regional rate plan will be different from the national rate plan, which would be different from the international rate plan. The state level taxes, rules, and regulations can also impact rating.
- Volume of data: Rating could also be based on volume of data. That is the rate would be applied on per packet or per byte basis. This could be extended to specifying usage limits and incremental rates based on usage volume.

After the value of attributes to be rated are identified, the next step in rating is to look up the rate table for applying the rate against the value of the attribute. The rate tables corresponding to different rate plans are identified using the unique rate table ID. Once the value of attribute value and rate is identified, next the event charge is computed. From the event charge the final charge is then calculated by applying discounts, taxes, promotions, and other parameters that impact the final bill to customer but may not be directly linked to usage.

Let us now take a specific example of how a call is rated. The following are the high level steps:

1. Determine originating charge point: The originating point would include the country code, city code, and exchange number. The originating cell site address is identified in this step.
2. Terminating charge point: The terminating cell site address is computed where the terminating number will also be a combination of country code, city code, and exchange number.
3. Determine the duration of the call.
4. Determine the rate plan to be applied. The band could be based on networks (fixed or mobile), TOD, DOW, terminating network, and so on.
5. Based on the rate plan, the rate table lookup is done. This could be based on the rate table ID assuming the table id can identify specific plans.
6. Apply the rate plan on the usage to compute the rate.
7. Apply any other charges like discounts on the rate to determine the customer payable for the call.

In short, rating mainly loads the transactions, does a lookup on external data, then applies the business rules and routes the rated transactions to accounting for bill settlement. Figure 20.17 shows that the interaction of other OSS modules with rating engine. Some of the key players developing stand-alone rating solutions include RateIntegration, OpenNet, Am-Beo, Boldworks, Highdeal, and Redknee. There is more competition in mediation space, where some of the key players are Ventraq (combined companies of ACE*COMM, TeleSciences, and 10e Solutions), Comptel, HP, Ericsson, and Intec.

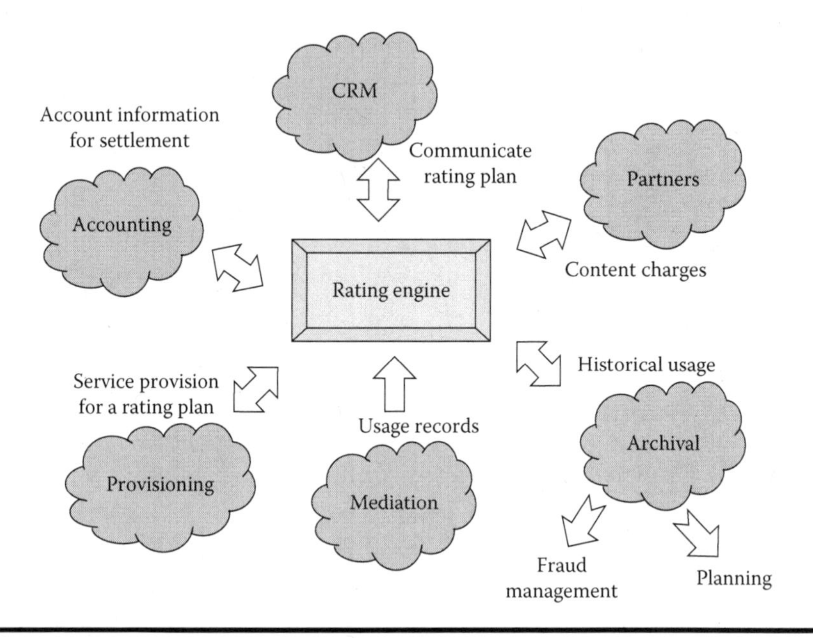


**Figure 20.17 Interactions with rating engine.**

To summarize, the billing system is used to manage the receivables invoicing, payment, collection, and crediting functions and the billing support systems also includes billing data collection, mediation, and rating (see Figure 20.18). The key players in billing can be segregated under the following heads:

- Billing products: Companies like Amdocs, Telcordia, and Convergys are among the top players developing billing products for the telecom industry.
- Billing services: Service bureaus for billing offer both tier 1 and tier 2 services. Amdocs, CSG, and Convergys are among the top tier 1 service companies. There are many players offering tier 2 service—a few notable companies being Alltel, VeriSign, Intec, and Info Directions.
- Solution integrators: Billing integration is another hot area and Oracle is one of the leaders in telecom billing integration.
- Legacy/custom solution development and support: Some of the service providers opt for custom solution development or relay on their legacy application to meet the billing needs. This is mainly attributed to the high level of security in content associated with billing and the company specific billing process that may not easily be satisfied by an off the shelf product. In house developers or developers from service and consulting companies are employed by the service provider initially to develop custom solutions and later to support the solution.

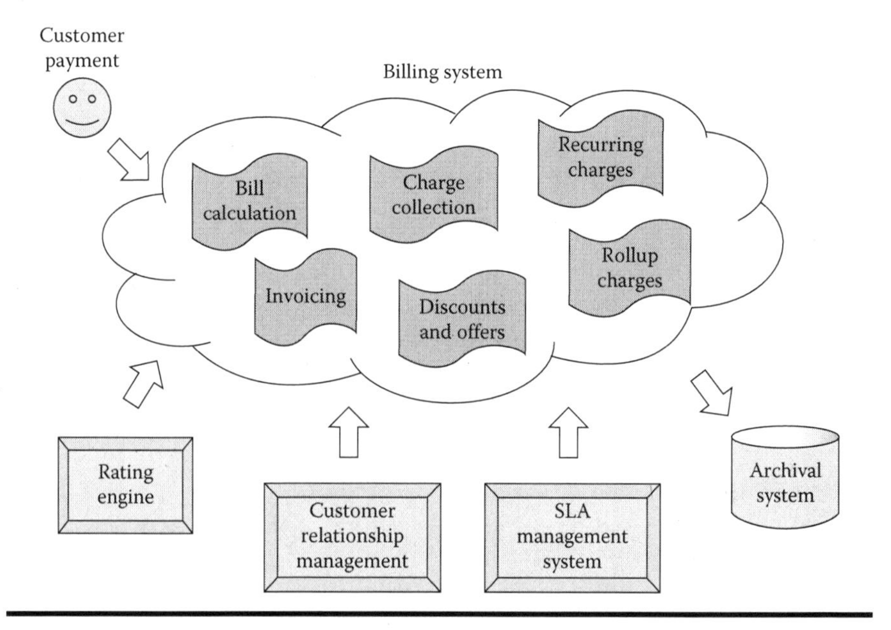


**Figure 20.18 Billing system.**

## 20.5 Conclusion

OSS deals with the range of operations in a service provider space that is critical to the operator's success. Fulfillment, assurance, and billing (FAB) systems make up the major modules in operations and business support systems that are used by the service provider on a regular basis for seamless business continuity. A closer look into FAB reveals several core modules like order management, network management, and mediation system that interoperate to fulfill a customer order, assure service continuity, and bill the customer for the ordered service.

The order management systems allow operators to perform the process of configuring the network to deliver the services ordered by a customer. Activating services on the allocated resource infrastructure is performed by the operator using provisioning systems. The allocation of resource and tracking the assets itself is performed by the operator using an inventory management system. After activation, the task of monitoring the network to detect and isolate network failures and quickly restore services is performed using a network management system. A service management module works above the network manager to correlate network performance with service quality requirements, and appropriately communicate and compensate customers for any disruption. Finally, mediation systems allow operators to determine service usage. This information is taken utilized by a rating system that allows operators to determine how much to bill based on the service usage for a particular customer.

Interaction between support system modules can be made simpler when service oriented architecture (SOA) is used, where the modules are developed as services with defined interfaces for interaction. Integration activity can be made easier by deploying an enterprise application integration (EAI) layer. The EAI allow operators to quickly integrate diverse support systems and network elements. Electronic integration of support systems is required for an automated back-office in large-scale operations. This integration layer creates an open data exchange platform optimized for heterogeneous interprocess communication. The EAI is not limited to OSS and can be used to support communication between software systems in other domains also. Some of the key players in EAI development are Tibco, BEA Systems (acquired by Oracle), and Vitria.

## Additional Reading


1. Kundan Misra. *OSS for Telecom Networks: An Introduction to Network Management.* New York: Springer, 2004.
2. Dominik Kuropka, Peter Troger, Steffen Staab, and Mathias Weske, eds. *Semantic Service Provisioning.* New York: Springer, 2008.
3. Christian Gronroos. Service Management and Marketing: *Customer Management in Service Competition.* 3rd ed. New York: John Wiley & Sons., 2007.
4. Benoit Claise and Ralf Wolter. *Network Management: Accounting and Performance Strategies (Networking Technology).* Indianapolis, IN: Cisco Press, 2007.
5. Kornel Terplan. *OSS Essentials: Support System Solutions for Service Providers.* New York: John Wiley & Sons, 2001.

# Chapter 21

# NGOSS

This chapter is about TeleManagement Forum's (TMF) New Generation Operations Systems and Software (NGOSS). The chapter is intended to give the reader a basic understanding of NGOSS concepts and life cycle. This author suggests reading TeleManagemet Forum publications for a deep understanding of NGOSS concepts and components.

## 21.1 Introduction

Let us start by discussing why TMF introduced NGOSS. The telecom industry was changing rapidly both from technical and business perspectives. On the technical front, the focus shifted from simple voice service to data services, which was required on an IP-based platform as against the previous circuit platform. Also the focus shifted from a fixed to wireless network and then to convergence of both fixed and mobile termed FMC (fixed mobile convergence). On the business front, prices were falling due to competition and there was a demand for higher return on investments. This new landscape brought in different needs for each of the telecom players. The suppliers wanted to have standard interfaces for interaction to reduce the integration costs and ensure that there is a higher chance for operators to buy a product that will easily integrate and interoperate with other products. The operators wanted to reduce service life cycle and achieve minimal time to market by reducing integration time. Another important need of the operator was to reduce the operating cost and spending on OSS.

To meet the needs of the key players, the OSS expected to offer rapid service deployment, customer self-care strategies, flexible real-time billing models and support for the multiservice, multitechnology, multivendor environment. The business models and systems available were not able to scale up to these new requirements. The suggested solution was to change the approach toward management solution development in a way that will ensure that OSS solutions can:

- Adopt changing business models and technologies.
- Reduce CAPEX and OPEX on OSS/BSS.
- Integrate well with existing legacy systems.
- Offer integrated billing.
- Can function as off the shelf components.
- Offer flexibility, interoperability, and reusability.

NGOSS, driven and managed by TeleManagement Forum, is the new approach that meets these challenges using a set of well-defined methodologies and practices. Key elements of this approach include a business model, system integration architecture, information model, and an application map. This chapter first outlines NGOSS concepts, the life cycle showing a holistic telecom view, NGOSS adoption methodology, and the NGOSS key focus areas. The TMF offers a collaborative environment where service provider's work closely to solve their business and technical issues using NGOSS principles. Wide spread adoption of NGOSS has led to significant reduction in operation expenditure and reduced integration cost. NGOSS offers an integrated framework that looks into capturing needs, designing, implementing and deploying operational and business support systems.

## 21.2 NGOSS Overview

The service providers following legacy OSS process and solutions are having considerably high operational costs. The lack of automated flow-through process has resulted in higher manpower costs. Time to market is significantly high in these industries due to rigid and inflexible business processes. Legacy systems do not scale with change in requirements leading to higher costs for enhancements and integration to support changes in technology. The impact of having legacy process and rigid systems becomes evident in a multiservice, multitechnology, and multivendor environment where adaptability and interoperability are key concerns, and mergers and acquisitions are quite common. These can also result in poor customer service because of poorly integrated systems with inconsistent data.

The NGOSS approach to this problem is to adopt the lean operator model in telecom industry, which has already proved its effectiveness in automobile, banking, and retail industries. The main attributes of a lean operator are reduced operational costs and a flexible infrastructure. The reduced cost of operations in a lean operator model is achieved with high levels of automation. Seamless flow of information across modules is a major requirement of the lean model, which can be achieved with standard information model and well-defined interfaces for

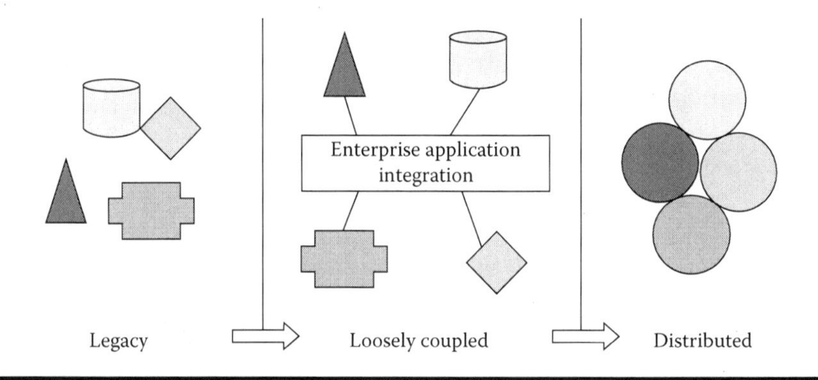


**Figure 21.1 Changes in architecture.**

interaction. The seamless flow of data also ensures that data is consistent leading to better customer service. Customer self-care is also suggested to bring down operational costs, reducing the need for more customer desk executives.

Commercial off the shelf products that has already been a valuable tool for ease of integration is another factor to be looked into for reducing operational costs. When the business and technical requirements change, the infrastructure should be flexible enough to adopt the changes. The faster the resources can adopt a change in requirements the more rapid will be the time to market. This means significant reduction in the service development and delivery time. The adoption of lean practices should be done without compromise in service quality.

The OSS architecture has migrated over the years from legacy to loosely coupled and finally to a distributed model (see Figure 21.1). Legacy systems were inflexible to change and were poorly integrated. There was minimal or no distribution of data. This changed in a loosely coupled system where service did interoperate but there existed multiple databases and user interfaces. The NGOSS approach is for a distributed environment as advocated in SOA, where interfaces are published and there is a seamless flow of information between modules, which are policy enabled.

## 21.3 NGOSS Lifecycle

There are a number of techniques for developing and managing software projects. Most of them deal with either the management of software development projects or methods for developing specific software implementations including testing and maintenance. Some of the most popular frameworks currently used are the Zachman Framework, Reference Model for Open Distributed Programming (RM-ODP), Model Driven Architecture (MDA), and Unified Software Development Process (USDP). Most of these techniques look into a complete life cycle of software development from specification of business requirements to deployment of the software. NGOSS tries to incorporate best practices from these techniques and tries to streamline the life cycle to telecom industry.

Some of the important concepts from the existing techniques that were incorporated in NGOSS lifecycle and methodology are:

- Model based development as in MDA. A set of meta models are used for developing business and technology artifacts. This way changes can be tracked and easily updated.
- Separate emphasis to enterprise as business, such as in Zachman framework.
- Support for distributed framework as in RM-ODP.
- Use case based iterative approach used in USDP.

Some of the goals of having an NGOSS lifecycle were to:

- Identify different levels of process and information decomposition.
- Define specifications for business model and solution.
- Provide traceability between the different stages in the life cycle.
- Have a formalized way for NGOSS solution development.

NGOSS lifecycle is built on two sets of views. In the first set there is a physical or technology specific and a logical or technology neutral view. The technology specific view handles implementation and deployment and the technology neutral view handles requirements and design. When a problem is identified, the requirements that are captured are not specific to a technology. NGOSS suggests the design definition also to be independent of the technology to facilitate multiple equivalent instantiations from a single solution definition. In the implementation and deployment phase a specific technology is adopted and used.

The second set consists of a service developer view and a service provider view. While the service provider identifies the business issues and finally deploys the solution, the design and implementation of the solution is handled by the service developer. It can be seen that both service provider and service developer has interactions with the logical and physical views of the NGOSS lifecycle.

The overlap of service provider view with the logical view happens in the business block where the business problems are identified. All the activities of the service provider, for business definition including process and preparation of artifacts for service developer to start work on the design, happens in the business block. The overlap of service developer with the logical view happens in the system block where the technology neutral design is defined. All the activities of the service developer for system design including preparation of artifacts to start work on the implementation happens in the system block. The overlap of service developer with the physical view happens in the implementation block where the technology specific implementation occurs. All the activities of the service developer for implementing

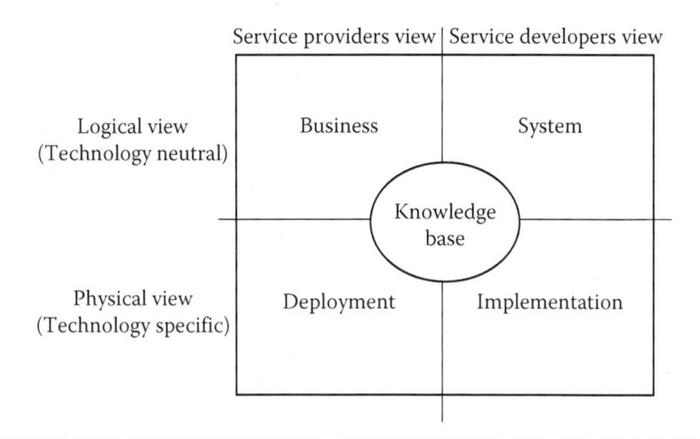


**Figure 21.2 NGOSS lifecycle.**

the solution as per the definitions in system design happen in the implementation block.

The overlap of service provider with the physical view happens in the deployment or run time block where the technology specific deployment is performed. This block is expected to capture all the activities of the service provider for deploying the solution as per the definitions in business, system, and implementation block. Though the model appears to be hierarchical in passing through the stages of defining, designing, implementing, and deploying one after the other, it is possible to move into any one of the intermediate stages and then work on the other stages. This means that there is traceability between blocks and at any stage a telecom player can start aligning with NGOSS. The overlap of views then creates four quadrants each corresponding to a block in the telecom life cycle (see Figure 21.2).

There are many artifacts that are generated as part of the activities in the various blocks (the blocks are also referred to as views). The business view will have artifacts on business requirements, process definition, and policies. In the system view as part of modeling, the solution there will be artifacts on the system design, system contracts, system capabilities, and process flow represented as information models. In the implementation view the system design is validated and implemented on specific technologies that will generate artifacts on implementation contracts, class instance diagrams, and implementation data models. The deployment view has artifacts on contract instances, run-time specifications, and technology specific guidelines. These make up the NGOSS knowledge base.

NGOSS lifecycle knowledge base, which is a combination of information from various views, offers end-to-end traceability. It can be considered as a central repository of information. It contains three categories of information:

- Existing corporate knowledge: Knowledge from corporate experience.
- NGOSS knowledge: Collection of artifacts identified by NGOSS.

- Community or shared knowledge: Knowledge common to both NGOSS and corporation. This is represented as an intersection of existing corporate knowledge and NGOSS knowledge.

## 21.4 SANRR Methodology

The NGOSS defines a road map that can be used by current systems to align with the reference models defined by NGOSS (see Figure 21.3). This roadmap is based on SANRR (scope, analyze, normalize, rationalize, and rectify) methodology. The methodology can be used for iteration at life cycle level or view level. The outcome of the iteration is to identify the activities and steps required to align with NGOSS components. Let us look into this methodology in detail:

1. Scope: In this phase the area that needs to get aligned with NGOSS is identified. There are multiple processes spread across customer, service, and network management. A modular approach is required to make changes to

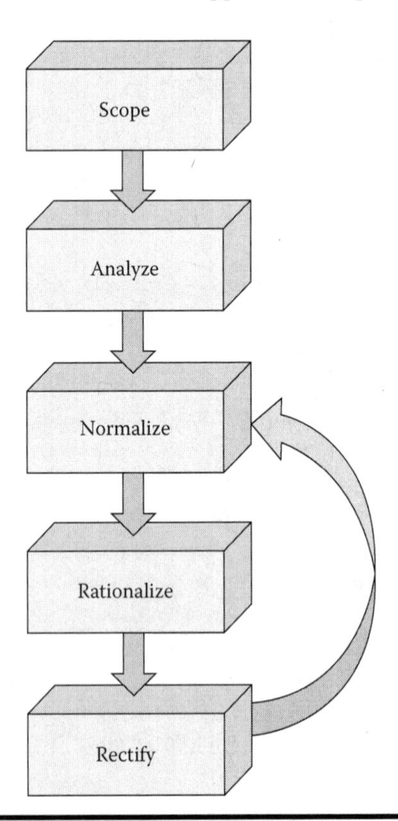


**Figure 21.3 SANRR methodology flow diagram.**

the process, architectures, and contracts. In scope, the organization identifies what module needs to be worked on. Although multiple modules would need to be worked on, each module is considered to define an element of scope.

2. Analyze: In this phase the scoped area is analyzed using cases and maps to identify the current system. This activity is intended to get the current model as it is in place, so that the model based on NGOSS can be created. This would involve preparation of process maps, information models, and all artifacts that best describe the current system. A clear analysis needs to be performed by a dedicated team who needs to work on the technical and business aspects of the scoped area.
3. Normalize: The current description of the system in place based on available artifacts and those prepared as part of analysis will not be in a vocabulary as used by NGOSS. This makes it difficult to identify what is missing in the system or where an overhead is occurring. In the process of normalization, the current view of the area that is scoped and analyzed is converted to NGOSS vocabulary.
4. Rationalize: The first step in rationalizing is to examine the normalized view of the current area for changes needed. In rationalization, activities like gap analysis, replication analysis, and conflict analysis is performed. The result of the rationalization is to identify what changes needs to be brought out for aligning with NGOSS. The first level of rationalization may not result in an alignment with NGOSS. The intent is to do a staged transformation to NGOSS.
5. Rectify: In rectification, modifications, deletions, and additions are performed on the scoped area based on changes identified during rationalization. The rectification will bring in changes to the business and technical aspects of the company for the scoped area.

This is an iterative methodology in that once an activity from scoping to rectification is performed, the system is not immediately aligned with NGOSS. As already discussed, the intent is a staged migration where one round of rectification generated a new view of the current system. This is the new current view that is put back in the normalization phase and again normalized, rationalized, and rectified. In addition to NGOSS lifecycle, the other important elements in NGOSS are eTOM (enhanced telecom operations map), SID (shared information and data), TNA (technology neutral architecture) and TAM (telecom applications map).

## 21.5 eTOM Model

The eTOM is a business process model or framework that defines the telecom and enterprise processes in the service provider space. It is an enhancement of the TOM model. eTOM serves as an industry standard framework for service providers to procure resources and to interface with other service providers. The complex network of business relationships between service providers, partners, and other third-party vendors are also accounted in the model.

eTOM provides a holistic view of the telecom business process framework. The normal operations process consisting of fulfillment, assurance, and billing is differentiated from the strategy and life cycle processes. There are three main process areas in eTOM comprising of strategy infrastructure and product management, operations, and enterprise management. Strategy infrastructure and product (SIP) management covers planning and life cycle management of the product. Operation covers the core of operational management and enterprise management covers corporate or business support management.

The real-time activities in operations are classified under fulfillment, assurance, and billing (FAB). The planning to ensure smooth flow of operational activities is captured as part of the operation support and readiness. This is separated from FAB in the eTOM model. In order to support operations for a product, there needs to be a planning team working on product strategy and processes for life cycle management. This is the strategy component in SIP. The processes that take care of infrastructure required for the product or directly relate to the product under consideration makes up the infrastructure part of SIP. Finally, processes for planning the phases in the development of the product make the product group of SIP.

Telecom companies would have a set of processes common to an IT company like human resource management, knowledge management, and so on. To represent this and to keep the definition of a telecom process separate from an enterprise process, a separate enterprise management process group is created in eTOM. This grouping encompasses all business management processes necessary to support the rest of the enterprise. This area sets corporate strategies and directions and provides guidelines and targets for the rest of the business.

While OFAB (operation support and readiness, fulfillment, assurance and billing) and SIP make up the verticals in eTOM, there is a set of functional process structures that form the horizontals in eTOM. The top layer of the horizontal process group is market, product, and customer processes dealing with sales and channel management, marketing management, product and offer management, and operational processes such as managing the customer interface, ordering, problem handling, SLA management, and billing.

Below the customer layer is the process layer for service related processes. The process grouping of service processes deals with service development, delivery of service capability, service configuration, service problem management, quality analysis, and rating. Below the service layer is the layer for resource related processes. The process grouping of resource processes deals with development and delivery of a resource, resource provisioning, trouble management, and performance management. The next layer is to handle interaction with supplier and partners. The process grouping of supplier/partner processes deals with the enterprise's interaction with its suppliers and partners. It involves processes that develop and manage the supply chain, as well as those that support the operational interface with suppliers and partners.

The entities with which the processes interact are also represented in the TOM model. The external entities of interaction are:

- Customers: The products and services are sold by the enterprise to customers.
- Suppliers: The services and resources used by the enterprise directly or indirectly to support its business are brought from a supplier.
- Partners: The enterprises with which a telecom enterprise cooperates in a shared area of business are partners of the telecom enterprise.
- Employees: These are human resources that work in the enterprise to meet the organizational goals of the enterprise.
- Shareholders: The people who have invested in the enterprise and thus own stock are the share holders of the enterprise.
- Stakeholders: When there is a commitment to the enterprise other than through stock ownership, the individuals are called stakeholders.

## 21.6 SID

The NGOSS SID model provides a set of information/data definitions and relationships for telecom OSS. SID uses UML to define entities and the relationships between them. The benefit of using the NGOSS SID and its common information language is that it offers business benefits by reducing cost, improving quality, reduced timeliness for delivery, and adaptability of enterprise operations. Use of SID allows an enterprise to focus on value creation for its customers. Even the attributes and processes that make up the entity or object are modeled with UML. SID provides a knowledge base that is used to describe the behavior and structure of business entities as well as their collaborations and interactions.

SID acts as an information/data reference model for both business as well as systems. SID provides business-oriented UML class models, as well as design-oriented UML class models and sequence diagrams to provide a common view of the information and data. Mapping of information/data for business concepts and their characteristics and relationships can be achieved when SID is used in combination with the eTOM business process. SID can be used to create cases and contracts to the four views: business, system, implementation, and deployment in the NGOSS lifecycle.

SID has both a business view and a system view. The SID framework is split into a set of domains. The business view domains include market/sales, product, customer, service, resource, supplier/partner, enterprise, and common business. Within each domain, partitioning of information is achieved using aggregate business entities. Use of SID enables the business processes to be further refined making the contractual interfaces that represent the various business process boundaries to be clearly identified and modeled.

## 21.7 TNA

TNA offers three types of capabilities as shown in the high level view (see Figure 21.4):

- Framework services: These services support distribution and transparency.
- Mandatory services: These services support decision, sequencing, and security.
- Business services: These are general services provided by the NGOSS solution.

The high level view is supported by a number of artifacts that make up technology-neutral architecture. TNA artifacts are either core artifacts or container artifacts. Core artifacts are the basic units of composition in the architecture and container artifacts are the artifacts that encapsulate or package core artifacts.

The TNA is built on a set of concepts. Some of them are:

- NGOSS component: An NGOSS component is defined as a software entity that has the following characteristics:
  - It is independently deployable.
  - It is built conforming to the specifications of component software model.
  - It uses contracts to expose its functionality.
- NGOSS contract: The interoperability of the NGOSS architecture is defined using NGOSS contracts. TNA provides the specification of NGOSS contract operations and behavior. It exposes functionality contained in an NGOSS component.

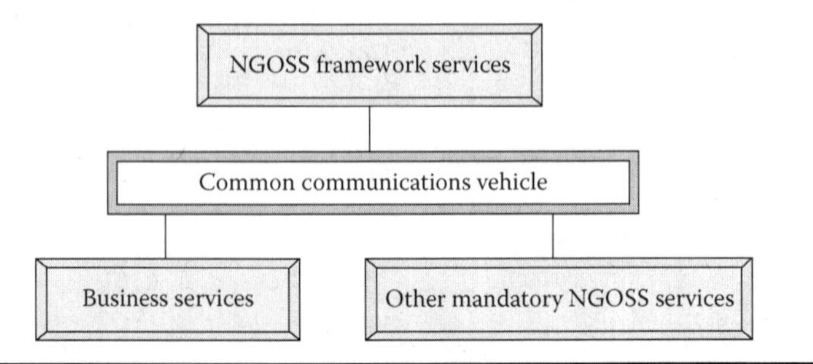


**Figure 21.4 High level view of TNA.**

- NGOSS contract implementation: This is a technology-specific realization of the NGOSS contract. The implementation is based on requirements and specifications provided by the NGOSS business and system view.
- NGOSS contract instance: This is a runtime manifestation of a contract implementation that is manageable. It provides one or more functions to other runtime entities and represents the unit of binding.
- NGOSS announcement: This is an invocation initiated by a client component instance. The innovation will result in the conveyance of information from the client component instance to the server component instance. It can be a request to the server component instance to perform a function on its instance.
- NGOSS announcement signature: This defines the name of the invocation and the parameter types.
- NGOSS interface: The interfaces represent functionality provided by managed software elements that reside in a component.
- NGOSS interface signature: The interface signature is a set of announcement and interrogation signatures for each operation type in the interface.
- NGOSS operation: Operation involves interaction through a contract invocation. Operation happens between a client component and a server component that is either an interrogation or an announcement.
- NGOSS interrogation: Interrogation is similar to an operation, except that it consists of two interactions, one initiated by a client component instance and the second initiated by the server component instance.
- NGOSS application: This is a container artifact that provides an encapsulation of the requirements, specifications, and implementations. The artifacts define desired functionality from the perspective of service providers that is needed to support a specific business goal within their operating environments.
- NGOSS service: An NGOSS service can be either a customer facing service or a resource facing service. An NGOSS service is created, deployed, managed, and torn down by one or more NGOSS contract operations.

## 21.8 TAM

The telecom applications map is a reference framework on the set of applications that enable and automate operational processes within a telecom operator. TAM is mainly intended to address the need for a common language for information exchange between operators and system suppliers in developing OSS applications. Having common map component standardization for the procurement process is expected to drive license prices down. Having a standard application model will ensure that investment risks and costs are reduced by building or procuring systems technology that aligns with the rest of the industry using a well-defined, long-term architecture. Some of the advantages in having a common reference application model are:

- Reduced development cost with more reuse of components developed on a common standard.
- Volume economies of scale to suppliers with high volume of similar applications.
- Reduction in operational costs with more automation.
- It is easier to adapt to change when an application is developed on a standard application model. The change can be organization changes or changes brought in by acquisition or mergers. This means that the defined frameworks are independent of the organizational structure.
- Standard interfaces for development and well-defined business to business interfaces will reduce development costs.
- Flexibility is the key, where applications act as building blocks that can be easily added, replaced, or modified without impact on other blocks.
- When workflows become repeatable and easy to modify there is increased efficiency in operations.

The different layers used for grouping applications in TAM are similar to those used in the eTOM model. This ensures close mapping between business process (eTOM), information model (SID), and applications (TAM).

## 21.9 Conclusion

A layered approach is required to ensure focus on various aspects of the telecom domain. This is achieved in NGOSS, where eTOM, SID, and TAM have defined telecom under a set of domains and functional areas to ensure sufficient depth of focus. In a converged environment multiple services have to interoperate in a multitechnology, multivendor environment. NGOSS offers a common framework of methodologies and models to support a converged service environment. An important requirement on next generation OSS is the presence of customer self-care strategies and integrated customer-centric management. Customer experience will be a key differentiating factor between service providers and will determine profitability. This is one of the key focus items in NGOSS and TMF has dedicated teams working on this area to improve customer experience with NGOSS.

Fast creation and deployment of services would include service activation on demand and rapid personalization of service. NGOSS is working on service life cycle management with SOA, which would make service introduction and management easy, resulting in rapid roll in/out of services. The next generation OSS/BSS solution should be able to monitor quality of service (QoS) and quality of experience (QoE) and ensure that customer service-level agreements are met.

NGOSS addresses the end to end SLA and QoS management of next generation support systems.

NGOSS addresses the issue of standard information model and interfaces for interaction. Data used by different services need to follow a common format as defined by SID for proper interoperability between services. The UML in SID helps to build shared information and data paradigm required for the telecom industry. NGOSS interfaces ensure ease of integration and better interoperability. A service provider might use OSS solutions from different vendors. For example, the provisioning system may be from one vendor while the inventory management solution may be from some other vendor. To enable flow through provisioning in service provider space, these solutions from different vendors need to integrate seamlessly. This seamless integration can be achieved if both the vendors adhere to NGOSS methodologies and models. In short, NGOSS provides the necessary framework to develop next generation operation support systems.

## Additional Reading

1. www.tmforum.org
   Latest release of these documents can be downloaded from the Web site:
   a) NGOSS Lifecycle Methodology Suite Release
   b) GB921: eTOM Solution Suite
   c) GB929: Telecom Applications Map Guidebook
   d) TMF053: Technology Neutral Architecture
   e) GB922 & GB926: SID Solution Suite.
2. Martin J. Creaner and John P. Reilly. *NGOSS Distilled: The Essential Guide to Next Generation Telecoms Management.* Cambridge: The Lean Corporation, 2005.

# Chapter 22

# Telecom Processes

This chapter is about telecom process models. The influence of IT models in telecom process models is discussed in detail. The main models that have been used in this chapter for reference are eTOM and ITIL. A sample business flow and how it can be aligned with telecom standard model is also taken up in this chapter. The chapter is intended to provide the reader with an overview of how different models can be applied together for effective business process management.

## 22.1 Introduction

There are multiple telecom and enterprise specific business process models that are used in a telecom company for effective business management. Some of the most popular ones include eTOM (enhanced Telecom Operations Map) for telecom process; ITIL (Information Technology Infrastructure Library) for IT process, and quality certifications like ISO/IEC20000 and CMMI. The usage of eTOM and ITIL used to be in two different domains, where telecom companies continued using eTOM and IT companies used ITIL. The change in telecom space brought out by convergence and flat world concepts has resulted in compelling IT enabled services. Most telecom service providers then started outsourcing IT business activities in the telecom space to external IT service providers. Many IT companies play a significant role in current telecom industry. This is because the IT infrastructure is the heart of every business.

The increasing participation of IT companies in telecom industry has brought confusion on what process models need to be used to manage the telecom business. IT Infrastructure Library (ITIL) developed by the Central Computer and Telecommunications Agency has been adopted by many companies to manage their

IT business. eTOM the popular process framework of telecom service providers is the most widely applied and deployed for end-to-end service delivery and support framework for ICT (Information and Communication Technology) service providers in the telecom industry. So identifying the purpose of ITIL and where eTOM can be complemented with ITIL is an area of concern in service provider space.

Another concern for service providers is the migration strategy for moving from legacy to next generation OSS standards. Service providers have to align to standard business models to meet the needs of managing next generation services. The main issue most service providers face is the migration of existing assets and services to a next generation process and information platform. There will be multiple broken data dictionaries, lack of a common definition of data across systems, nonstandard interfaces for data exchange between the various systems, duplication of records across the systems, lack of transparency in business processes and systems, and improper mapping of data with business process.

To address the main issues faced by service providers, this chapter will first discuss the overlaps between ITIL and eTOM. The issue of aligning a legacy process to standards like eTOM is then handled. Finally the methodology to adopt eTOM with ITIL is discussed. Similar to the other chapters in this book, a basic overview of concepts is presented before a discussion of the telecom process issues in service provider space and the solution adopted by companies to solve these issues.

## 22.2 eTOM

The eTOM is a business process framework for development and management of processes in a telecom service provider space. It provides this guidance by defining the key elements in business process and how they interact. The important aspect of eTOM is a layered approach that gives enough focus to the different functional areas in a telecom environment. As discussed in the previous chapter, eTOM is part of the NGOSS program developed by TeleManagement Forum (TMF). It is a successor of the TOM model. While TOM is limited to operational processes, eTOM added strategy planning, product life cycle, infrastructure, and enterprise processes in the telecom business process definition.

eTOM has a set of process categories across various functional areas like customer, service, and resource management. This makes it a complete enterprise process framework for the ICT industry. eTOM is used by all players in the telecom space including service providers, suppliers, and system integrators. Based on the holistic telecom vision in eTOM an end-to-end automation of information and communications services can be achieved. eTOM has been adopted as ITU-T (Telecommunication Standardization Sector of International Telecommunication Union) Recommendation M.3050 making it an international standard.

Though eTOM provides a business process framework for service providers to streamline their end to end processes, it should be kept in mind that eTOM is a framework and its implementation will be different from company to company. It facilitates the use of common vocabularies for effective communication between business and operations. eTOM is limited to business process modeling and the aspects of information and interface modeling is handled as part of other elements in NGOSS. However, the implementation of eTOM is supported by other NGOSS specifications like the shared information/data model (SID), telecom application map (TAM), NGOSS Lifecycle & Methodology, and other NGOSS specifications. A direct mapping between these specifications is also provided by TeleManagement Forum.

Being a generic framework, eTOM can be applied on different telecom operators offering a wide variety of services. This implies that compliance cannot be certified at the process level. eTOM compliance is achieved through the NGOSS Compliance Program offered by TMF. The certification is based on tools and is independent of the organizations and the process flow. However, the conformance tests check the business objects used and the operations framework to ensure that process implementations are based on eTOM.

The successful development of the telecom process documentation gives the ability to accurately decide on a new IT service provider to support the entire range of business processes. Customers are outsourcing the delivery and support of their IT services to telecom companies where the telecom company offers IT service management as a sellable service. eTOM helps to eliminate process gaps through visualization of interfaces for interaction.

Most service providers using eTOM were able to reduce IT costs, improve the product quality, and reduce time to market the product. Use of eTOM as a foundation for process identification and classification ensures that the process repository will be able to evolve with industry best practices. Though eTOM has a functional area on enterprise process, this part is not defined to the level of granularity offered for telecom processes. So though eTOM can be considered a complete process framework, there are enterprise processes that will require more levels of definition. TeleManagement Forum definitions have eTOM level 0, level 1, level 2, and level 3 process maps.

## 22.3 ITIL

The ITIL is a public framework describing best practices for IT service management. ITIL facilitates continuous measurement and improvement of the business process thereby improving the quality of IT service. ITIL was introduced due to a number of compelling factors in IT industry:

- Increase in cost of maintenance of IT services.
- Increase in complexity due to changes in IT infrastructure technology.
- The frequent change management required in IT services.

- Emphasis from customer on quality of IT services.
- Flexibility to align the IT services based on business and customer needs.
- Need to reduce the overall cost of service provisioning and delivery.

The ITIL processes represent flows in a number of key operational areas. The main emphasis is on the stages in service life cycle, namely:

- Service strategy: IT planning is the main activity in service strategy. The type of service to be rolled out, the target customer segment, and the resource to be allocated is part of the strategy. Market research becomes a key ingredient in service strategy where the market competition for the service and plans to create visibility and value for the service is formulated.
- Service design: This stage in the service life cycle is to design services that meet business goals and design processes that support service life cycle. Design measurement metrics and methods are also identified in this stage with the overall intent to improve the quality of service.
- Service transition: This stage of the service life cycle is to handle changes. It could be change in policies, modification of design or any other changes in IT enterprise. The reuse of an existing process and knowledge transfer associated to the change is also covered as part of service transition.
- Service operation: This stage is associated with management of applications, technology, and infrastructure to support delivery of services. Operational processes, fulfillment of service, and problem management are part of the service operation stage. The key IT operations management is covered under service operation.
- Continual service improvement (CSI): This stage deals with providing value add to the customer offering good service quality and continued improvement achieved by maturity in the service life cycle and IT process. The problem with most IT service companies is that improvement is looked into only when there is considerable impact on performance or when a failure occurs. The presence of CSI ensures that improvement is a part of the service life cycle and not a mitigation plan when an issue arises. Improvement will require continuous measurement of data, analysis of measured data, and implementation of corrective actions.

The service delivery management is defined in a complete manner giving the processes, roles, and activities performed. The interactions between processes are also defined in ITIL. Similar to eTOM, ITIL is also a framework and hence its implementation differs from company to company. However, ITIL is mainly focused on internal IT customers without giving a complete consideration to all the players in the industry as in eTOM. ITIL is not a standard, but standards like ISO 15000 is based on ITIL. So organizations are usually accessed against ISO 15000 to check on its capability in IT service management and direct compliancy or certification from ITIL cannot be achieved. Other process models/standards include ISO/IEC 20000, Six Sigma, CMMI, COBIT, and ISO/IEC 27001. ISO/IEC 20000 is related to IT service management and most of these standards support certification.

## 22.4 Integrated Approach

From the discussion so far it would now be clear that eTOM and ITIL were developed for different domains and the framework for these models was developed separately. However integrating the two models into a unified view offers added value to a telecom enterprise (see Figure 22.1). Standardizing bodies like TMF are actively pursuing this goal of an integrated model. Compelling companies to select one of the models and adhere is not feasible as both the frameworks provide insight and guidance to the enterprise to improve the service quality. Another reason for integration is that the ICT industry has a community of users working with both techniques. These community users work on both to offer a common reference/standard in defining a business process.

The focus of standardizing bodies is to identify and remove obstacles in eTOM and ITIL integration. The GB921 document of TeleManagement Forum provides a workable method for implementing eTOM process flows with ITIL. This document demonstrates overlaid on ITIL process steps as a background that enterprises can directly build support for ITIL best practice services on a process modeled with eTOM.

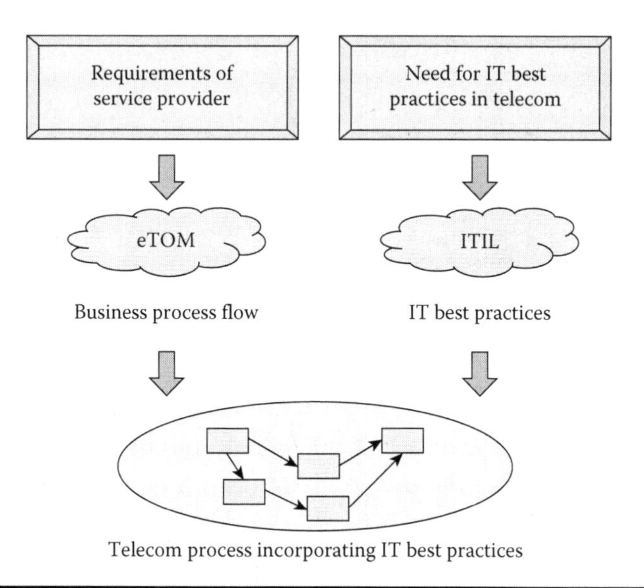


**Figure 22.1 Integration of eTOM with ITIL.**

It should be kept in mind that merging or direct end-to-end integration of these models is not possible due to structural issues and domain areas where these reference frameworks are used. The initiatives performed by standardizing bodies try to identify how these reference frameworks can complement each other in offering telecom processes that utilize IT best practices. Detailed recommendations and guidelines on using eTOM with ITIL are available in GB921 that can be downloaded from the TeleManagement Forum Web site. The sections that follow will give more details on the process segments where eTOM and ITIL overlap.

## 22.5 eTOM and ITIL in ICT

There are multiple ways in which the activities in an ICT industry can be segregated. Let us take one such approach by dividing the ICT industry into a set of layers (see Figure 22.2). The layers are:

- Customer: This is the top most layer of the ICT industry. This corresponds to the end consumer that uses the products and services offered by the ICT industry. This is shown as a separate layer due to its importance in ICT industry where multiple standards and certification techniques are available to measure the effectiveness of interaction with customers. The customer is

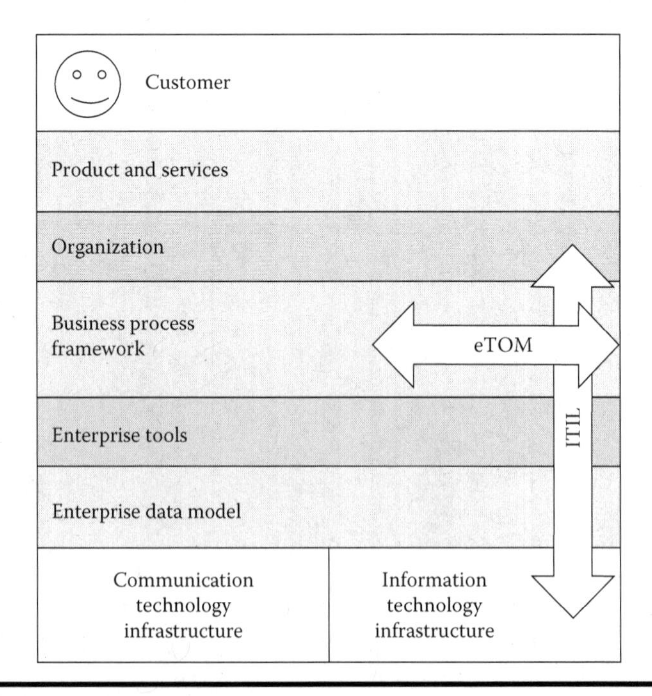


**Figure 22.2 Use of eTOM and ITIL in ICT.**

an entity that interacts with the ICT and the customer layer incorporates the activities involving the customer and the ICT industry.

- Product and services: The end offering of an ICT industry can be a product or a service. While product is the common terminology used for tangible offering of the provider, service is the terminology used for nontangible offerings. To site an example, online/mobile news is a content-based service while a software game to be played on a mobile is a product offering.
- Organization: This is the glue layer between the service provider offering and the backend business process that makes the smooth delivery of product or service possible. The employees and their capabilities make up the organization of the ICT industry. The organization is effective in setting up business goals and the business process is intended to provide a framework to achieve these goals. Customer interactions take place with the organization layer and the interaction is on topics in the product/service layer.
- Business process framework: This is the most important layer in context with the discussion in this chapter. It covers the business processes used in the ICT industry. These include both telecom and IT business processes used in the ICT industry. The eTOM model comes in this layer and defines the E2E telecom business process framework.
- Enterprise tools: This layer has the management applications and tools that are internal to the enterprise. It uses the data coming from telecom and IT to effectively manage the enterprise. This is a glue layer that brings in automation and links the infrastructure with the business process.
- Enterprise data model: Data from the resources is expected to comply with the enterprise data model. Similar to multiple process models like eTOM and ITIL, there are multiple data models. For telecom companies the shared information and data (SID) is the most popular data model and for IT companies the common information model (CIM) is most popular. So enterprise data model can encompass multiple data models.
- Infrastructure: Enterprise infrastructure consists of communication technology infrastructure and information technology infrastructure. This is the lowest layer consisting of the infrastructure used for product and service development.

The infrastructure is expected to support the execution of business processes. The infrastructure related information used by the business process moves through different layers. While eTOM work is limited to the business process framework layer, ITIL can influence multiple layers from the IT infrastructure to the organization. There is an overlap of eTOM and ITIL in the business process framework layer. However the expanded scope of ITIL can be beneficial for eTOM.

An alignment between eTOM and ITIL at the business process framework layer will enable multisourcing and partnerships between companies following these models. This will remove the existing confusion on which model to adopt and align with, and to comply with requirements of next generation operational standards.

One of the main goals of harmonization between the business models is to meet customer driven requirements. Though customer is shown as the top most layer consuming the products and services, the customer also interacts with the enterprise through the organization layer and at times the customer can directly interact with the business process framework layer. Interaction directly with process is achieved using customer self-service where customer interactions become a part of the business process.

Both models increase operational efficiency and foster effective productivity. They can result in considerable cost reduction without negatively impacting the quality of service. Implementing eTOM telecom process that incorporates ITIL best practices will result in considerable improvement of service quality. The improvements in quality are mainly attributed to the presence of a specific continuous service improvement stage in the service life cycle defined by ITIL.

To summarize this section, both ITIL and eTOM can be used in conjunction in an enterprise for effective management of services. The effective method of using both is to keep eTOM to define the telecom process and ITIL for incorporating best practices. So the steps in preparing a business process that uses both eTOM and ITIL will start with preparation of a process flow. Make it compliant with eTOM. Then filter and merge the process with ITIL best practices. Verify the new model and ensure it still complies with eTOM.

TR143: "ITIL and eTOM—Building Bridges" document from TeleManagement Forum has many recommendations and guidelines on using eTOM with ITIL. The document proposes extensions to eTOM framework to incorporate ITIL process and best practices. Convergence is bringing in a lot of changes in process and information models to ensure that a single model can be used across multiple domains.

## 22.6 ITIL Processes Implemented with eTOM

Some of the ITIL processes that can be implemented with eTOM are discussed in this section. Service design, service transition, and service operation are three high level ITIL service life cycle processes that can be implemented in eTOM. The subprocesses under these high level ITIL processes can fit into different levels of eTOM model. Let us look into these processes in much more detail.

1. Service design: This process can be further subdivided into the following processes.
   a. Service catalogue management: This process involves managing the details of all operational services and services that are planned for deployment. Some information from the catalogue is published to customers and used to support sales and delivery of services.

b. Service level management: This process mainly deals with agreements and contracts. It manages service level agreements, operational level agreements, and service contracts. Service and operational level agreements are made with the customer who consumes the product and services, and contacts are usually established with suppliers and vendor partners.
c. Capacity management: This process is expected to ensure that the expected service levels are achieved using the available services and infrastructure. It involves managing the capacity of all resources to meet business requirements.
d. Availability management: This process is expected to plan and improve all aspects of the availability of IT services. For improving availability, first the resource availability is measured. Based on information from measuring, an analysis is performed and an appropriate plan of action is defined.
e. IT service continuity management: This process is intended to ensure business continuity. It looks into risks that can effect delivery of service and develop plans for mitigating the identified risks. Disaster recovery planning is a key activity performed as part of the IT service continuity management.
f. Information security management: This process is expected to ensure that a secure frame is available for delivery of services. The security concerns are not just restricted to the services offered. Ensuring security is more an organizationwide activity.

2. Service transition: This process can be further subdivided into the following processes.
   a. Service asset and configuration management: This process deals with managing the assets and their configuration to offer a service. The relationship between the assets is also maintained for effective management.
   b. Change management: This process deals with managing changes in an enterprise in a constructive manner that ensures continuity of services. Change can be associated with organization changes, process changes, or change in the system caused changes in technology.
   c. Release and deployment management: The product or service after development is released to either test or to live environment. This release activity needs to be planned and scheduled in advance for proper deliver of product or service in time to meet market demands.
3. Service operation: This process can be further subdivided into the following processes.
   a. Event management: In this process, events are monitored, filtered, grouped, and handled. The action handling will be set based on the type of event. This is one of the most important processes in services operations. It effects quick identification of and resolution of issues. Events can be a key input to check if there is any service level agreement (SLA) jeopardy to avoid a possible SLA failure.

b. Incident management: It is critical that any incident that might disrupt the service is properly handled. Incident management is the dedicated process that handles this activity in ITIL.
c. Problem management: This process is intended to analyze incident records. Incidents that have occurred can be analyzed to identify trends. This will help to avoid problems that can cause incidents effecting business or service continuity.
d. Request fulfillment: Change request and information request management need to have a process in place to ensure timely request fulfillment. The request fulfillment process takes care of this activity.

There are many more processes in ITIL. However, in this section the processes covered are the ones that can be incorporated to work with eTOM.

## 22.7 Business Process Flow Modeling

A typical trouble management process flow in a company that has not aligned with eTOM is shown in Figure 22.3. Let us consider that the service used by the customer is experiencing failure in operation due to a bug in the vendor software. The usual process flow in resolving this issue will start with the customer calling up the customer support desk to report the problem. The support desk will verify

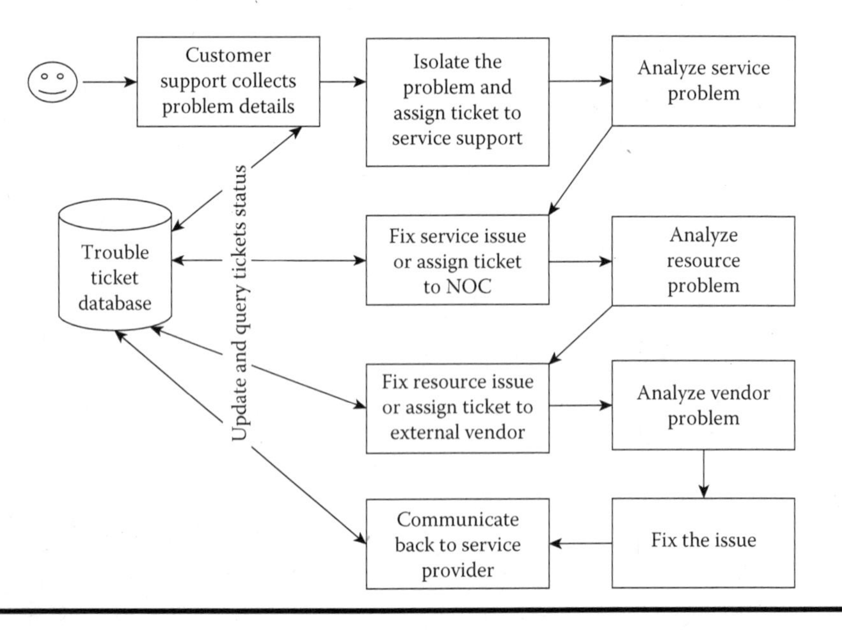


**Figure 22.3 Business process flow.**

details of the customer account and collect all possible information to describe the trouble condition to a sufficient level of detail and create a trouble ticket for the issue. This ticket number is given to the customer, so that the customer can call up and check the status of the issue resolution easily by quoting the ticket number to the customer support desk.

From the account details and the problem description on service having issue, the ticket is then assigned to the service support desk. If the service problem is caused by some provisioning or service configuration fault, the service support desk will reprovision or reconfigure the service to bring it back to service. In the specific problem scenario being discussed, the problem is with vendor software running on one of the resources. Upon analysis conducted by the service support desk, the resource that is causing the issue is isolated and the trouble ticket is assigned to the resource support desk sitting at the network operations center (NOC) owning the resource that has been identified as the cause of the problem.

The identification of the resource that caused problems in a service is done by analyzing event records generated by the resources. It is expected that the developer creates enough logs to troubleshoot and fix an issue. The resource team will then analyze the issue and isolate the cause of the problem. If the problem is caused by a module developed by the service provider, then the ticket is assigned to the team that developed the module to fix the issue. On the other hand (as in our example) if the problem is caused by a module developed by an external vendor, then the trouble ticket is assigned to the vendor trouble submission queue.

When the vendor is notified about trouble in their product, based on the support level offered, the vendor will work with the service provider to resolve the issue. Once the issue is fixed, the information is communicated to the service provider. The service provider will update the status of the trouble ticket as closed in the ticket database. This activity will generate an event to the customer support desk, so that they call up and inform the customer that the issue has been resolved. The customer can call the customer support desk not just to enquire the status of the trouble ticket but also to close the ticket if the issue was temporary or was caused due to some issue that the customer was able to fix as part of self-service. It is therefore important that all the processes that update the ticket status with the database, first queries the database and checks the status before making any updates. In addition to ticket status and problem description, the trouble ticket will also contain a history of activities performed to resolve the trouble ticket, dependencies on other trouble tickets that need to be resolved first, the person who is currently working on the ticket, and so on.

Now let us look into the flaws in this business process for trouble management. Though the explanation gives a clear idea of the flow, the process by itself does not give details of the entity responsible for a specific block. For example, it is not clear if analysis of the service problem is performed by the customer support desk as an initial activity or by the service support team. Another flaw is that several processes that could have improved productivity of trouble management like centralized tracking and managing that involves searching for similar trouble conditions is missing. Also the customer support desk does not always try and solve technical issues. The customer support desk isolates the issue and assigns it to the service support desk. A problem resolution at customer support could have save more than 70% of the cost the service provider will incur when the problem resolution happens only at vendor support level.

It is in these scenarios where standard models like eTOM will help. The current process needs to be aligned with eTOM using SANRR methodology. Here scope (the S in SANRR) is to align trouble management with eTOM. First of all, eTOM has a detailed process map for each area that helps to clearly identify what entity is responsible for what. For example there are four layers for trouble management in eTOM. At customer level (customer relationship management layer) there is problem handling, at service level (service management and operations layer) there is service problem management, at resource level (resource management and operations layer) there is resource trouble management and at external vendor level (supplier/partner relationship management) there is supplier/partner problem reporting and management. There are multiple levels of process definition from Level 0 to level 3 with eTOM. The process effects can be improved by removing invalid processes and adding new processes as per eTOM (see Figure 22.4).

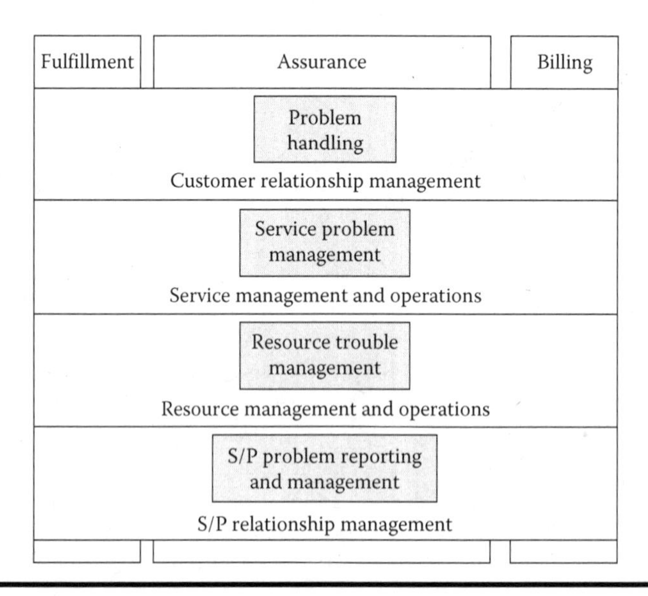


**Figure 22.4 Trouble ticketing in eTOM.**

## 22.8 Conclusion

Let us conclude the discussion with the three step methodology from TeleManagement Forum to apply eTOM in conjunction with ITIL. The first step is to ensure that the best practice specification using ITIL exists with regard to the processes in eTOM model. The second step is to identify the enterprise area in an application where eTOM needs to be applied in conjunction with ITL. The third and final part involves refining the identified area using eTOM process elements on the ITIL process steps.

While there are several books and white papers on eTOM and ITIL, the intent of this chapter is to introduce the reader to the most popular models. A simple process diagram is also explained along with guidelines on how these models can be used together to achieve business goals.

## Additional Reading

1. www.tmforum.org
   Latest release of these documents can be downloaded from the Web site:
   a) GB921: eTOM Solution Suite
   b) TR143: Building Bridges: ITIL and eTOM.
2. Randy A. Steinberg. *Servicing ITIL: A Handbook of IT Services for ITIL Managers and Practitioners*. Indiana: Trafford Publishing, 2007.
3. John P. Reilly and Mike Kelly. *The eTOM: A Business Process Framework Implementer's Guide*. New Jersey: Casewise and TM Forum, 2009.

# Chapter 23

# Management Applications

This chapter is about telecom management applications. The intent is to randomly list some of the common management solutions that are available on the market and to create awareness for the reader on what the functionalities performed by these applications are now. Though standard application models like the telecom applications map (TAM) of TeleManagement Forum exists, this chapter handles the application landscape in a more general manner without reference to any standard model.

## 23.1 Introduction

There are a wide variety of telecom management applications currently available on the market. Most telecom application developers concentrate on building a few management solutions. It can be seen that the top player in inventory management may not be the top player in the development of billing management applications. In most scenarios a vendor specializing in development of inventory management may not even have billing or customer relationship management applications in its portfolio of products.

In general, management applications can be classified under the following heads:

- Customer management applications: These relate to the customer and deal with applications to manage the customer. Some of the applications in this category are customer information management, customer SLA management, customer QoS management, and so on.

- Business management applications: These are the applications that deal with business for the service offered to the customer. Some of the applications in this category are billing management, fraud management, order management, sales management, and so on.
- Service management applications: These relate to the services offered to the customer and deal with applications to manage the service. Some of the applications that fall in this category are service quality management, service provisioning, service performance management, and so on.
- Resource management applications: This category has applications to manage the resources in the telecom space. Some of the applications that fall in this category are resource provisioning, resource specification management, resource testing, resource monitoring, and so on.
- Enterprise management applications: Every Telco has a set of enterprise applications to manage the enterprise. Some of the applications that fall in this category are knowledge management, human resource management, finance management, and so on.
- Others: While most applications fall under one of the categories listed above, there are some applications like integration tools, process management, supplier management, transportation management, and retailer management that does not strictly fall in a specific category. In most cases, some of the features in these applications will either be part of business management or enterprise management.

Applications are discussed in this chapter with regard to the functionality supported by the application rather than category to which the application belongs.

## 23.2 Overview of Applications

1. Inventory management: This application is intended to maintain the active list of inventory or assets in the service provider space. It is also referred to as asset management application. Before provisioning a service, the provisioning module queries the inventory module to identify the set of resources that can be allocated for provisioning the service. The inventory module also performs the function of allocating resources and keeping track of the allocated resources. This application is also linked with taking key decisions of when new inventory is to be procured and provides insight on possible asset capacity to meet the requirements of a new service. Inventory management applications are no longer linked with keeping the data of static assets that can be put to use. Current inventory management applications perform auto discovery and dynamically updates asset status information.
2. Bill formatting: There are many vendors currently in the telecom industry that provide a solution for bill formatting. With new value added services

and practice of converged billing, the information content in a bill is quite complex. There is a need for proper bill formatting applications that can abstract relevant information based on the services subscribed by the customer. The actual bill format and the way information is arranged may vary from customer to customer based on their individual preference. The bill formatting application in most cases act as a COTS (commercial off the shelf) product that easily integrates with the billing system. This way the records and customer information can be properly formatted as per requirements of the service provider and the customer.

3. Fraud management: There are several types of fraud that the telecom service provider may want to monitor. Postpaid customers who don't pay the bill and key switching SIM cards, blocking of a stolen mobile and its SIM card, and call trace/call interception for assisting law agencies are just a few scenarios to list the cases for fraud management. A fraud management application has the capability to handle a set of fraud scenarios based on built-in algorithms. For the specific scenario of catching postpaid customers who don't pay their bill and keep switching SIMs, the fraud management application will try to obtain usage patterns like the common numbers that are called by a person who repeats this fraud activity. Once a usage pattern is matched, the calling party is marked as potential fraud and the SIM is disabled. This way the fraud management application is able to tag fraud users and disable them before significant loss is incurred by the service provider. While IMSI (International Mobile Subscriber Identity) can uniquely identify the subscriber, IMEI (International Mobile Equipment Identity) can uniquely identify the mobile equipment.
4. Route optimization: Using an optimal traversal path is critical in ensuring proper utilization of resources in service provider space. An issue with route planning at an equipment level like in a router is that the equipment is only aware of the status of its next hop and maybe the next to next hop. Let us consider an example where the data has to be send from A to F and the two traversal paths are A-C-D-F and A-B-E-F. B is having more traffic that C. So from the perspective of A, the best path would be A-C-D-F even when D might be faulty, as A is only aware of its next hop. The solution to this problem is to have route optimization applications that keep track of the status of all the routes. Based on real-time values of the status of all possible routes these optimization applications take decisions to route traffic in an optimal manner. Use of route optimization algorithms and applications implementing the algorithms traffic congestion can be avoided. Another important use is to avoid resending of information that is caused by faulty nodes and improper route selection.
5. Service provision and activation: Provisioning and activating the service is a key activity of the service provider for order fulfillment. Dedicated service provisioning and activation applications are used by the service provider to

enable flow through provisioning. Once the order is captured in the order management system, information on the services requested by the customer is passed to the service provisioning module. The provisioning module identifies the resources that can be provisioned by interacting with inventory management. The allocated resources are then provisioned, which also includes network provisioning for setting up the infrastructure to support the services. After network level provisioning is successfully completed, the service provisioning performs the required configuration to set up the parameters to be used when activating the service. Finally the service is activated. A validation module checks if the service activation is successful and the subscribed service is ready for customer use. The service provisioning module also informs the order management module that the order is successfully completed and triggers the billing subsystem to start billing the customer.

6. Service configuration management: Service configuration management involves configuring the resource components of a generic service built for a customer or a class of customers. After basic installation of components for operating a service, the specific parameters to be used while operating the service needs to be provided. Configuration applications put in the relevant values for offering the performance guaranteed by the customer SLA. Usually a centralized configuration application will do the configuration across application modules that deliver the service. Having distributed configuration managers is also common in service provider space. A generic service build will have a set of customizable parameters that are configured based on the customer subscription.
7. Service design: At a process level, service design involves planning and organizing service resources to improve the quality of service. The service resources can be manpower, infrastructure, communication, or other material components associated with the service. There are several methods in place to perform effective service design. The increasing importance and size of the service sector based on current economy requires services to be accurately designed in order for service providers to remain competitive and to continue to attract customers. Applications that aid the service design process is called a service design application. These applications can redesign a service that could change the way the service is configured, provisioned, or activated. The resources, interfaces, and interactions with other modules can also be modeled during service design.
8. Order management: This application is used for entering the customer order. It first captures customer information and account level information. Before completing the order, credit verification or payment processing is done to check for validity/availability of funds from the customer. After the order is placed, the order information is transferred to the provisioning system to activate the service(s) specified in the order. The order management needs to check if sufficient resources are available to fulfill the order. On successfully

activating the service, the order management system needs to close the order and maintain the records of the order for billing and customer support. Billing is related to order management as the customer is billed only for subscriptions specified in the order placed by the customer. Data from the order is used by customer support to validate the customer against the order number. Referring to the order placed by the customers, this helps the support desk professional to better understand the problem specification from the customer. Order management usually has a workflow that it follows. The workflow would involve order capture, customer validation, fraud check, provisioning, activation response, order closure, and notification of customer. It has multiple interactions with a variety of OSS modules to manage order life cycle.

9. Credit risk management: Applications handling credit risk management offer communications service providers the ability to continuously assess and mitigate subscriber credit risk throughout the customer life cycle. This helps the service providers to achieve greater operational efficiency. These applications tracks credit risk in near real-time, prior to subscriber acquisition, ongoing usage, and recovery. The real-time check is made possible by building an extensive customer profile consisting of demographics, usage patterns, payment information, and other relevant customer information.

   This level of detail gives the accurate information of the subscriber and enables operators to automatically track credit risk by automating the process with the value of service utilization. Most applications can also create credit risk assessment schemes. Risk variations can be tracked by configuring alert conditions.

10. Interconnect billing: The interconnect are settlement systems that handle a variety of traffic, from simple voice to the most advanced content and data services. They support multiple billing formats and have powerful preprocessing capabilities. Volume-based simple to complex rating, revenue-sharing, discounting, and multiparty settlement can be performed on many types of services.

    Interconnect billing applications help to track and manage all vital information about partners of the service provider, partner agreements, products, services, and marketing strategies. This helps to accurately track and timely bill the customer. Interconnect billing also performs cost management and can be used to easily identify rampant costs and route overflows in billing systems.

11. Invoicing: The invoice management solution speeds processing time by automating the bill payment process from receipt to accounts payable. Telecom bills are mostly complex and error prone. This leads to high invoice processing costs. Use of an invoice management solution reduces cost per invoicing. Other advantages include a faster invoice approval and collection process. This leads to an increase in efficiency caused by improved financial reporting.

Some of the key features of an invoice management solution are invoice consolidation, invoice auditing, invoice content validation, tracking of invoice, getting approvals, generating intuitive reports, and smooth integration with accounting system.

12. Billing mediation: A billing mediation application collects billing data from multiple sources, filters them, and does formatting and normalization before sending it as a single feed to the billing system. Converged billing mediation solution can collect billing data from IP, fixed, and mobile networks. The activities performed by billing mediation application can be summarized as gather, process, and distribute of usage data.

    The tasks start with collection and validation of usage records. This is followed by filtering out irrelevant records that are not relevant for billing. Records are then aggregated and correlated to obtain useful information. Finally the usage data is formatted and normalized. Then it is distributed to northbound OSS applications.

    It should be noted that all data mediated by the billing mediation system is not used by the billing system. For example, some traffic information can also be obtained from the records that could be used in capacity planning rather than billing. Also mediation systems can be used to execute commands like provisioning tasks.

13. Network optimization: The main aim of network optimization solution is to improve the network performance and ensure effective utilization of resources. Optimization would involve changes in operations modeling. This is done to ensure that the service and network performance achieves the service level parameters shared with the customer. Optimization of energy usage is another key factor looked into by many telecom companies. The green initiatives in companies have added stress to optimize energy usage. Energy-efficient and sustainable communication can be achieved through optimal network design, site optimization, and the right choice of energy sources.

    Maximizing resource utilization and implementing scalable solutions to meet the capacity needs of Telco is also an activity under optimization. By optimizing the network, operators can get the most out of their existing investment. Optimizing network performance ensures capacity being used in the most profitable way. Efficient use of energy will help in securing the supply.

14. Data integrity management: When multiple OSS solution interoperates there can be inconsistencies in data that can lead to inaccurate computations and reports. Data integrity deals with insuring that data spread across OSS is consistent and accurate. With real-time service and network provisioning, integrity of data is a key concern. There are applications that handle end-to-end data integrity across solutions and data correction in specific modules like inventory reconciliation for inventory management based on resource discovery.

Data integrity management application does integrity checks to fix revenue leakage and reduce access cost. In some cases, a database cleanup is also performed as part of data integrity management. These applications can improve tracking of assets and reduce the number of provisioning failures. The basic operation of a data integrity management solution is to identify discrepancy and do synchronization of data.

15. Sales force automation (SFA): This is a solution that automates the business tasks of sales. The SFA gives the sales team the ability to access and manage critical sales information. As sales involves frequent traveling, it is quite common to have sales force applications accessible through a PDA or laptop. In such scenarios, synchronization for sharing data and merging new and updated information between off-line users and the sales force application server needs to be implemented.

    SFA helps to track sales inquiries and seamlessly route sales personal to the right location by interaction with workforce management module. With sales force automation, sales representatives get instant access to the latest prospects that improve the chances of getting business. SFA give organizations clear visibility into their sales pipelines. Accurate, timely forecasts of revenue and demand using SFA help sales to achieve higher profits. Most importantly, workflow automation helps standardize sales processes for greater operational efficiency.
16. Collection management: This application is intended to manage the bill amount collection from customers. Regular audits using the collection system helps to identify the collecting debt. The presence of up-to-date records enables seamless recovery of amounts from the clients. The collection management solution helps to define the daily collection amounts and strategy. These activities are performed using real-time data and are expected to be accurate. Various distribution models are supported in most collection management solutions. It is a transaction management system that supports billing in a service provider space.
17. Billing management: This application is intended to perform and optimize the tasks involved to produce a bill for the customer. Billing solutions can produce invoices in a variety of formats similar to reporting engines. Changes in the customer account can be updated using a billing management system as it directly affects the delivery location of the bill. One of the major requirements of a billing system is to manage the services subscribed by the customer for which a bill needs to be generated.

    In supporting converged billing for the service provider, the billing management solution can prepare consolidated bills for a variety of networks and services that the customer has subscribed. With some of the current rate plans, bills may even be consolidated before sending to customer. Additional data to support hierarchy-based billing is available in a billing manager. Most solutions also support real-time billing and the ability to provide data for settlements with other operators and content providers.

18. Churn management: An important activity in most telecom companies is to manage churn. It can be any form of change that can impact the customer to switch to another service provider. It is also noted that the best practices used in managing churn can be reused effectively to solve similar scenarios. Applications that manage churn keep track of the activities involved in fixing the churn. Identifying similar scenarios enables solutions reuse.

    Telco(s) are using churn management systems to almost accurately predict the behavior of customers. Churn is a widely accepted problem in telecom service provider space. In simple terms, churn refers to customers cancelling their existing contract to embark on a subscription with a competing mobile service provider.
19. Sales management: This is more an application used across the enterprise and not just limited to telecom. A sales management application does sales planning, sales development, and sales reporting. Based on sales data, sales planning mainly involves predicting the sales figures and demand for a product. Sales planning will improve business efficiency and bring in focused and coordinated activity within the sales process.

    As part of sales development, the sales activities performed are tracked. All incomplete activities are worked to completion. History information is analyzed to bring in continuous sales development. Sales reports are generated to check operational efficiency of the sales process and to ensure that the results set forth in sales planning are achieved.
20. Customer contact management: These applications ensure effective communication with the customer and are usually slated as customer relationship management solutions. A contact management solution usually has an integrated call-tracking system that provided up-to-the-minute information on the status of customer calls. This helps to check the quality in handling customer calls based on records generated and feedback collected from the customer.

    To maintain customer communication data, a centralized database is implemented with contact management solutions with the ability to support multiple communication channels. Customer contact management solutions are usually used at the service provider help desk, which can be dedicated call/contact centers. Some contact center solutions can do intelligent call routing. For example, when a customer calls with regard to a specific service, the contact management solution tries to find the exact customer query. This is done by requesting the customer to make a selection from a set of options. After multiple selections are made by the customer, the solution either routes the call to the appropriate help desk or automatically checks customer account details to provide a response.
21. Interparty management: These solutions are used by the service provider to settle charges with their partners in addition to billing the customer. The demand for interparty billing solutions has increased with the need to bill

content based services along with voice settlements. Bundling of various services from multiple partners leads to multiple charges per transaction. This makes it increasingly difficult to reconcile partner payments.

An ideal interparty billing management solution is expected to calculate multiple charges for each transaction and perform correlation of retail revenues with interconnect cost. The interparty solution should be implemented in a way that the addition of support for other content like wholesale, intercompany, and IP should not impact ongoing operations. The solution can implement complex revenue sharing arrangements to simplify the service provider operations.

22. Customer information management: Managing information about the customer is a critical activity of the telecom company both for increasing the customer satisfaction levels and to plan services based on customer insight. Customer information management solutions help to efficiently leverage customer information to improve the operational efficiencies of the service provider. This results in enhanced customer satisfaction, improved workforce productivity, and revenue.

    Informed decision making, the key to continuous improvement, is done with managed customer information. These solutions offer better visibility and a unified view of customer information. Management solutions also analyze the customer information and perform correlation to generate reports that can leverage customer information in marketing and sales activities.

23. Revenue assurance: Solutions for revenue assurance are intended to maximize revenue opportunities and minimize revenue leakage. While revenue leakage can occur as part of operations planning and support, the most common cause is fault bill records. Leakage also occurs as part of the process and management practices including quality. The actual causes of leakage and the percentage loss from a cause, varies from telecom company to company.

    The usual process employed by a revenue assurance solution is to check for the most common revenue leakage scenarios, identify the actual cause that are affecting the company, make a measure of the scale of leakage due to each cause, correct the leakage cause, minimize the leakage, and reevaluate the causes. This leads to continuous improvement.

24. Transaction document management: This is used for applications that require invoices generated for customer billing. These applications convert text and image content to print streams. When the report document is a Web format, the content is converted to Web-ready stream. These streams can then be displayed on a variety of media including paper bills.

    The document production has three main steps. They are selection, conversion, and generation. In selection the format template, the data input source, and the rules for associating the data input with the format template are identified. In conversion the billing files are formatted based on the template.

These formatted documents are then stored for future reference. Using these documents in the final generation stage the desired file streams (print, web, xml) are generated.

25. SLA management: The service level agreement the service provider has with the customer needs to be constantly monitored for compliance. This is achieved by measuring attributes that can be monitored for business contracts made with the customer. The SLA management system identifies potential breach conditions and fixes issues before an actual SLA breach occurs. In cases were an actual breach occurs, the operator is notified so that the issue can be communicated to the customer and suitable compensations are provided as per breach terms defined in SLA.

    The QoS guaranteed to the customer is a form of the SLA. Degradation of QoS is a direct indication on the possibility of breaching SLA. Any service quality degradations that are expected should be identified and the customer informed. This means that the customer is to be informed of any planned maintenance or other scheduled events likely to impact delivery of service or violation of SLA. Preset thresholds are set for the SLA attributes to identify SLA jeopardy and SLA breach. SLA management also reports to the customer the QoS performance as per queries or customer reports generated for the subscribed service.

26. Customer self-management: Customer self-care applications are now used by almost all service providers. These applications provide a set of interfaces to the customer to directly undertake a variety of business functions. This way the customer does not always have to contact a support desk and some of the support operations can be performed by the customer itself. Access to the customer interface application needs to be secure and should have a solid authentication and authorization mechanism.

    These applications provide fully assisted automated service over various customers touch points. The customer self-management application interacts with the customer relationship management modules and performs the operations that was previously done by a support desk professional.

27. Customer problem management: This application is intended to handle customer problems. It is closely linked with trouble ticketing. When a customer calls with a service or billing problem, the customer information and problem details are collected using problem management application. When the reported problem cannot be fixed immediately, a trouble ticket is raised so that customer can refer to the ticket number for tracking the status of the problem resolution.

    The trouble ticket is assigned to the appropriate team in service or network support department and the problem management module continuously tracks the status of the ticket.

    Customer problem resolution applications are mainly concerned with how the problem would affect the relationship between the customer and service provider. It is quite common to integrate customer problem management with service problem management and resource/network problem management applications.

28. Partner management: The service provider offering maximum value for the money and has the maximum number of services are the best choice for most customers. To be competitive in the telecom market, most service providers offer a lot of services from partners to add to their service portfolio. This gives the customers the flexibility to choose from a wide variety of services. Service providers can also form channel partners to offer their products to other markets where they are not direct players. Revenue from the customer for service packages needs to be shared among these value chain entities based on predefined agreements. Partner management applications handle all interactions including revenue settlement and scenarios where the transactions with partners have to happen in real time.
29. Work force management: This application is intended to manage the tasks performed by field service executives. When a new work order or trouble ticket is raised that needs support from a field support executive or technician, an appropriate resource is identified for completing the task using a work force management application. This is achieved by keeping an active record of the resource mapping the field skill sets, work load, current location, and so on, which helps to map the task to most appropriate resource. The application is mainly intended for service provisioning for a new order or change in settings and for service assurance to handle a trouble condition.

    There are different attributes of a work force professional that needs to be considered in allocating tasks. These attributes include shift time, job functions, skill sets, and availability. Along with work force attributes the service provider attributes like optimal staffing, cost to resolve, and expected daily work load also needs to be considered. The results of the task are recorded for future planning and an analysis of resource effectiveness in completing an allocated task.
30. Financial management: These applications are used to manage the cost of telecommunication services. Cost of service is managed with evolution or change in service with telecom financial management application. Financial management deals with service inventory, contract inventory, and cost control. Service and contract inventory needs to be reconciled against the invoice to ensure that the services subscribed by the customer are invoiced correctly.

    It is essential that financial management generate savings from billing error identification and correction. Validating the accuracy of service charges includes investigating service charge amounts for contractual accuracy and confirming service charges are covered in financial management. Finance management is an enterprise level activity. Hence it is common to integrate telecom finance management application with the enterprise finance management application.

31. HR management: Human resource management applications are used at the enterprise level and not restricted to telecom. Human resources departments that are not automated lack the flexibility to adjust quickly to changing business or regulatory requirements and usually have higher operational costs compared to HR departments that used software solutions for administrating salaries, benefits, recruiting, and performance management.

    HR solutions usually streamline payroll processing. Some of the solutions have modules that handle hiring, can do employee performance analysis, and facilitate effective communication. Automation and the smart use of software tools can go a long way toward simplifying administrative tasks and controlling costs.
32. Knowledge management: There are solutions currently available for automating this enterprise level process. The knowledge management applications help telecom organizations to systematically collect, manage, and execute the numerous ideas and suggestions of its employees. The innovative solutions lead to an increased intellectual capital.

    Use of knowledge management solutions result in higher productivity, faster response, and delivery times. These solutions have a central repository for storing data and a management block captures and evaluates employee ideas for faster product introductions.
33. Customer qualification: These applications validate the customer credit eligibility and offering availability at the customer location. This application is expected to ensure that orders contain only products that are feasible at the customer location and the customer can pay for at a risk level the provider can accept.

    Customer credit eligibility checks the credit of the customer, considering the payment history with the service provider as well as the credit scores with external credit agencies. Offer availability checks the services offered by the service provider that are available in the customer area. Finally the customer is qualified against the order, which is validated based on contract and billing rules. The billing dates and the due dates for payment of billing are identified along with the service level agreement to be executed with the customer.
34. Capacity planning: When an order is placed by the customer, the order has to be completed before a date communicated to the customer as the order completion data. Improper capacity planning can result in the work order not getting completed on time. Capacity planning solutions consider the projected time of capacity restraints and helps service providers to determine the best way to run their work force management application to overcome the problem, either by running more shifts, adding capacity, or taking some other action.

    Capacity planning solutions are intended to accommodate current and projected demands while delivering consistently high performance. The planning applications help to forecast resource needs in order to avoid emergency provisioning of resources. Good capacity planning solutions increases production throughput, identifies and eliminates production bottlenecks, and increases visibility into production plan and work orders.

35. Product/service catalog management: The service provider offering can be a product or service. The management of catalogs associated with the service or product of the service provider can be done using catalogue management solutions. Each product/service will comprise of multiple components. Each of these components may be designed and developed by different departments in an organization.

    This makes it important to model the product/service and manage the catalogs associated to the components of the service/product and for the product/service as a whole. Some of the components may also be shared for reuse. The individual departments working on the components are responsible for the individual catalogues and store them centrally, while the product/service level catalogues are maintained centrally for reference and update of products utilizing available components.

36. Product life cycle management: These applications manage the entire life cycle of the product and its components. This means applications that support designing, building, deployment, and maintenance of the product. The applications have the capability to handle new products and updates to existing products. It can be seen that there will be significant use of project and program management activities and tools used as part of product life cycle management.

    While project management deals with the activities related to satisfying a set of requirements, the program management will encompass all activities in a product program. A program management can involve management of multiple projects. Requirements are also captured as part of product life cycle management. These requirements map to detailed specifications on the product/service that are catalogued. Some life cycle management solutions also support closure of obsolete products.

37. Product performance management: These applications as used to analyze the product performance in the telecom market. Information about the product is gathered and analyzed using tools to compute the performance. The three main factors that contribute to product performance are campaign, revenue, and cost. The product campaign will identify how the product was communicated to the customer and directly contribute to the performance in the market place.

    Product cost performance can be significantly improved by an active supply chain management. This starts with ensuring that the cost benefit is achieved with proper sourcing of components and services used in product and optimization of inventory. The performance tools help to identify potential points where the product development is not meeting the estimated capacity.

These performance attributes can be corrected to improve the investment and revenue generated from the product.

38. Product strategy management: The development efforts in a telecom company should be in line with the business strategy. This means that sufficient data needs to be analyzed to ensure that investment is allocated to the right technology areas and projects. Prioritization is also a key criterion in determining where investment needs to be done at a specific point of time and what decisions need to be delayed. Product strategy management solutions maintain a repository of data that will help to take effective decisions on product strategy.

    These solutions also ensure that strategic decisions are communicated properly across the enterprise after taking input from key stake holders and getting approval. Previous case studies can also be referred with these solutions to tailor an approach to the unique situation and requirements. Product innovation management is also a part of strategy management. There are specific solutions that handle product innovation management.
39. Network activation: As part of order fulfillment, the details of the order is communicated to the service provisioning module. The provisioning module sends a request to the network activation application for activating the service. The network activation system generates specific control commands targeted toward individual network elements for enabling the elements to activate the service.

    The activation module activates the elements, followed by modules to start billing data collection, the activated elements are updated with status in inventory management module, and finally the validation module checks if the activation was completed successfully. Most network activation solutions support multitechnology, multivendor, multielement activation.
40. Subscriber provisioning: These solutions mainly handle provisioning and migration of subscriber data. Provisioning would involve updating subscriber details in an HLR (home location register) or HSS (home subscriber server) and an example of migration is to move data from a legacy HLR to an IP HLR. The provisioning task requires considerable planning and coordination that is managed by the provisioning module.

    The subscriber solutions are usually generic provisioning modules that facilitate subscriber provisioning across multiple network elements. The migration source and destination can be set or defined by the user. The rich user interface makes it easy for the operator without much prior provisioning experience to effectively do subscriber provisioning and migration without taking inputs from an engineer. In subscriber provisioning, the management module directly interacts with the module that manages and maintains the subscriber data.
41. Network data mediation: These applications take data from multiple network elements and reformat the data to a format that can be used by multiple upper

layer service and business support modules. Network data mediations mainly provide feed to the monitoring system. The data from the different network elements will be in different formats and the mediation component does the job of converting to a common readable format.

Mediation applications are usually used with most OSS system that take data from multiple elements or element management systems. When standard interfaces are not used, change in network element brings in significant development to change the mediation system to support the new or changed element. Control commands and messages can be exchanged through the network data mediation solution. Most of the current mediation solutions are standards based to support the multiple northbound and southbound interfaces. In addition to formatting, some of the other functions performed by mediation system include parsing, correlation, and rule-based message handling.

42. Event management and correlation: Fault data is a critical source of information in automating management functionality. Predefined actions can be mapped to specific events to handle scenarios the same way it would be done by a network professional. Event management solutions display event records and take corrective actions for specific events. A network operator uses a sequence of events to identify a particular problem. Similarly, the event management system needs to collect and correlate events to identify the event pattern that maps to a predefined problem.

    Event management solutions collect data from multiple network elements. These events are filtered and the event is checked against specific event patterns for match. If the events can lead to a pattern, then such events are stored in a repository for reference. For example, if three SLA jeopardy events on the same service raises an e-mail to the operator, then each SLA jeopardy event or the event count needs to be stored.

43. Network design: These solutions help to develop a design model of new elements to be included in a network and to design network configurations that are needed to support new services. First the network and its elements are modeled and simulated before assigning the design to appropriate systems that support the implementation.

    Network design will include physical design, logical design, and software design. For example, multiple call servers need to be placed in a rack. The physical design will detail what shelf and slot each of the servers will be placed. The logical design would segregate the servers into call processing servers and management servers to detail the interactions. Software design will drill down into the software packages and features that should be available on the servers.

44 Network domain management: These applications provide interfaces to hide vendor specific implementations. Using domain management application, legacy functionalities are exposed as standard interfaces compliant to

standards like MTNM, OSS/J, and MTOSI, following standard data models like SID and CIM.

The net result is an in-domain activation, alarm collection, filtering, and QoS features. A key advantage with domain management solution is that it can be replicated across policies for different resource domains. An interesting feature is that domains can also be replicated to cover multiple vendors and equipment types. This makes it highly reusable and a versatile application used by most service providers.

45. Network logistics: These applications are used to manage the availability and deployment of network resources at site locations where the service is offered. These applications will coordinate the identification and distribution of resources to meet business demands in the minimum possible time. They are employed in stock balancing, resource or kit distribution, warehouse stock level projections, and capacity planning.

    It can be seen that network logistic applications have interactions with supply chain management applications to get the details for stock distribution. Network logistics also help in workforce management encompassing all the responsibilities in maintaining a productive workforce. An important module in network logistics management is automating the engineering work order. This involves automating many engineering processes and eliminating data reentry to increase engineering productivity.

46. Network performance monitoring: These applications collect performance data from the network elements. The performance records facilitate effective network management. Performance records give information about historical behavior and utilization trends. This can be used for capacity management and planning of resources. Problem identification and testing can also be performed using the records collected by performance monitoring systems.

    Network monitoring systems provide inputs to resource problem management, capacity planning, and forecasting applications. A set of key performance indicators and key quality indicators are monitored to identify a shift in network performance from a predefined threshold. As historical records need to be maintained for performance analysis, it is common to have a dedicated repository for archival of performance records. The network performance monitoring application will access the repository for storage and analysis of data.

47. Network planning: These applications deal with all types of planning associated with different technology layers in the network. Engineering works and meeting unpredicted demands cannot be performed without proper planning. There would be considerable network rearrangement and relocation of network capability that needs to be planned to meet operational scenarios. Use of network planning applications brings in considerable cost benefit to the telecom company.

Network planning is required in fulfillment and assurance functions. As part of fulfillment, network planning ensures that sufficient resources are available to meet the capacity requirements for individual services. Network planning applications support assurance process by doing fault planning for network repairs. It also does performance planning for traffic engineering.

48. Network problem management: These applications has the capability to record problems and provide necessary escalation of the problem based on predefined criteria. Problem management systems are able to associate the effects of problems to root causes based on root cause analysis systems or human deduction.

    The trouble ticket application records a trouble condition. If the trouble condition relates to a network level problem, then the ticket details are send to the network problem management module. The service or customer level description of the problem is mapped to network specific events and conditions at the network problem manager. The network problem manager internally tracks the problem resolution at network level between various departments and notifies the service problem management module on the problem status.

49. Network provisioning and configuration: Order fulfillment involves service provisioning/configuration and network or resource provisioning/configuration. In resource provisioning, the order details are converted to resource level requirements that need to be configured and activated. Network provisioning and configuration module works on the resource level requirements to configure the resource identified by the inventory management module to offer a specific service.

    The provisioning and configuration module has multiple interactions with inventory management module. Hence it is important to make these two applications interoperable. After completing provisioning and configuration of network resources, the provisioning application updates information on the status of the resources with inventory application. Some applications also support creation of resource configuration plan. This plan has details of the required resource actions that can be sent to an orchestration to trigger auto provisioning and configuration of resources.

50. Specification management: These applications are used for creating, editing, storing, and retrieving specifications. Specifications can be at the service level, capability level, and resource level. Service specifications deal with the service attributes and its values that are common across a specific set of services. Resource specifications will have technical details on the layout, attributes, and connection parameters of the resource. General characteristics across the resource that can be realized in multiple resource instances make up capability specifications.

    The specification management application can be used to create new service instances from the service specifications. Capability specifications can be accessed from specification management applications to perform strategic planning and implementation planning. The resource specifications are retrieved from specification management application for the creation of resource instance that comply to the specification.

51. Discovery and status monitoring: The discovery application is expected to automatically discover the resources and their attributes through a management channel. These applications will either communicate directly with the network resources or collect information using element managers.

    Most discovery modules will first get a status of the components or elements in the network and then drill down to get further details on the subcomponents of the discovered components. After discovery, the status of the discovered components and subcomponents are monitored using these applications. The status information collected is used by the inventory management solution to reconcile its data.

52. Network testing management: These applications ensure that the network elements are working properly. Testing management is associated with both the fulfillment and the assurance process. In fulfillment, after provisioning and configuration of the network, the testing management module is invoked to verify if the configuration is as expected and to confirm sanity of the network for activating the service.

    In the assurance part, the network testing management application does fault isolation and correction. An automatic test can be triggered as part of troubleshooting to fix a bug. Most test management applications support automatic and manual test initiation. The testing application should be able to execute test cases, interpret the test results, store the results obtained and notify other applications on the test results.

53. Element management: These are applications that connect to the network elements to collect management data. It can be used to perform FCAPS (fault, configuration, accounting, performance, and security) management functions on the network element. Network management system will connect to multiple element managers. That is on the southbound interface the element manger interacts with the devices being managed and on the northbound interface element managers interact with network and service management solutions.

    These applications fall in the EML layer of the TMN model. The element manager can manage one or more network elements. It discovers the managed elements and routing audits the status of the elements. The EMS is used to execute commands on the network elements as part of network provisioning, where the individual elements in the network need to be configured to interoperate and offer a service.

54. Customer query knowledge base: These applications are intended to help the customer support professional to efficiently handle the customer queries. Customer query solutions have a centralized knowledge base with details of

common customer queries that can be referred using links and search utilities to identify how to troubleshoot a problem. These applications support workflows to assign customer complaints to the appropriate help desk when the user makes a call.

Tracking of the status of a customer complain can be done with these applications. To achieve this functionality the support solution interacts with trouble ticket management solution. It is quite common to have the solution offered from a Web portal through which customer issues can be submitted and the central knowledge base can be accessed. Some of the functionalities in this solution will be available in an integrated customer relationship management solution.

55. Traffic monitoring: These applications are used to identify potential congestion in traffic over the network. Identifying a possible congestion helps to route traffic appropriately to avoid an actual congestion. The traffic monitoring solutions usually interact with the configuration modules doing traffic control and with inventory modules to allocate resources to fix traffic problems.

    Most traffic monitoring solutions are not limited to identification and analysis. The corrective action to fix a problem is either suggested by this solution to the operator or a predefined set of actions is executed to fix a potential congestion. The corrective actions are applied based on resource utilization values of the routes participating in the network. The services affected by a traffic problem can be identified and highlighted using the traffic monitoring solutions.
56. Retail outlet solutions: These solutions are used by wireless service providers to perform point of sales and point of service in their retail stores. The functions performed using these solutions are similar to what is performed at the customer support desk. Instead of placing the order by phone to a support desk, the customer visits the retail outlet to place the order.

    These retail solutions will interact with the order management module to place and manage the customer order. It can also access the customer information. In short, most of the interfaces offered in customer relationship management and customer query knowledge base solutions are provided in these retail outlet solutions. In addition to these capabilities local cash management, retail store inventory management, local promotions, and local pricing that make up retail integration are available in the retail outlet solutions.
57. Security management: The rapid spread of IP-based access technologies as well as the move toward core network convergence has led to a much wider and richer service experience. The security vulnerabilities such as denial-of-service attacks, data-flood attacks (SPIT), and so on that can destabilize the system and allow an attacker to gain control over it during in internet, is true for wireless communication over IP networks also. Hackers currently are not limited to looking at private data, they can also see their victims.

Video-conferencing systems can be transformed into video-surveillance units, using the equipment to snoop, record, or publicly broadcast presumably private video conferences.

These issues have raised security concerns with most service providers. The result is an increased investment on security equipments and security management solutions that can secure the telecom infrastructure. The security solutions can handle E2E security in the enterprise that includes simple firewall based software solutions to complex security gateway policy managers.

58. Service performance management: These applications help to monitor the service performance including the customer's experience using the service. These applications collect data from network performance management solutions and compute the values of service performance from the network data. These applications provide reports on the quality of service and variations in service key performance indicators.

    Service performance management provides both current performance data as well as the historical view of service performance. This helps in service and infrastructure planning.

    Resource performance is not the only feed for service performance management application. The service performance management application can perform tests to determine service performance either from the application or using external service test tools or applications.

59. Service problem management: These applications provide a service view of the customer problem. It interacts closely with customer problem management and network problem management applications. It is also common to have a single module for managing the three views for problem resolution. The problem reported by the customer is mapped to service level impacts and assigned to the appropriate team for resolution using this application.

    These applications first collect the problem details, consolidate the problem information, set appropriate priority, assign the problem to the appropriate department for resolutions, and manage the problem to closure. It also prepares a closure report that can be referred to in future for solving a similar or related problem. Interacting applications are informed on the status of the problem during change in status or when queried.

60. Service event monitoring: These applications monitor the service parameters in configuration and fault. It helps the operator to be aware of approaching problems or degradations to service. The impact analysis in event monitoring systems can be extended to predict the likely impact of service degradations on specific customers and map the cause to and appropriate network resources.

    Event monitoring systems can also be programmed to perform specific actions when a service level event occurs. For example when an event on possible service failure is generated, the event monitoring system can be programmed to notify the appropriate operator to take corrective action.

Rule-based event handling is a popular method of service operation support automation.

61. Service rating and discounting management: These applications calculate the customer specific charges and apply discounts in coming up with the final bill. It collects billings records, formats the information in the records, identifies usage, and applies the rate for the service consumed by the customer.

    Rating applications should be able to support different payment models and rate plans. Threshold setting and planning can also be performed using service rating and discounting applications. Predefined thresholds can be set on a user's spending limit and notifications generated when charges reach a specific percentage of the spending limit. IP-based services have a different rating model compared to normal voice based services. Current rating solutions are expected to integrate with existing network environments and enhance the framework to support real-time rating capabilities of complex and converged service offerings.
62. Planning management: Every telecom company has a set of product and service planning activities that can lead to change. Planning management solutions are used to handle planning operations and change management. These solutions provide the necessary orchestration between planning duties and network engineering activities. It maintains work flows for proper execution of planned activities.

    Planning solutions interface with financial control in order to authorize the expenditure required to purchase the required resources for execution of plan. There can be interactions with third party modules and supplier modules to communicate the plan to meet business needs. It also supports project management of building projects in the initial planning phases.
63. Supply chain management: These applications deal with daily production and distribution in the supply chain. For proper production the demand has to be forecasted and inventory for production needs to be planned. Supply chain management solutions work on historical data to understand the product demand and provide valuable inputs to the operator preparing the production plan. In the distribution framework the orders planned by the customers at the telecom retail stores and outlets need to be considered in developing the warehousing plan for direct store and warehouse distribution.

    Another important factor in supply chain management is transportation. Transportation drives both production and distribution costs. Transportation costs to source the inventory required for production can be high if the material requirements are not properly communicated to the suppliers. The distribution of finished goods should ensure minimal cost in transportation without significant increase in cost of warehousing. Saving on empty miles by clubbing distribution goods transportation with production inventory transportation can also be done effectively using supply chain management solutions.

64. Network troubleshooting tools: Fixing issues in the network using troubleshooting tools is a key activity performed by the network operator. There are many network troubleshooting tools that help the operator to identify network issues and bug fixing. These tools complement network trouble management and element management systems to effectively isolate and debug a problem.

    Some of the common tools include a telnet client to log into the remote machine, the SSH client to do secure login and debugging, the SNMP MIB browser to do a management information base walk through, and ping utility to check if a server is active. The tools can be used independently or can be integrated with other solutions.
65. Compliancy verification application: Adherence to standards is now a major requirement for most telecom solutions. Service providers want the network and service solutions to adhere to specifications defined by standardizing bodies. There are standards defined for most products from equipment, interfaces, and protocols to information model used in management products. Compliancy verification solutions can be used to check if a solution is compliant to a specific standard or set of standards.

    These are mostly test execution tools that have a predefined set of test cases. The test cases identify the solution compliancy to the standard specifications. The compliancy verification applications executes the test cases and identifies the number of specifications a product is compliant to. The test execution report can be used by the service provider to select vendors and their products by evaluating whether the product meets the standards required for a specific application.
66. Platform solutions: These are not complete applications that have stand-alone existence. It acts as a platform or framework that can be used for making complete applications. It can be developed for a variety of management domains. Some platforms are for data collection where the framework supports data collection using many protocols like SNMP, NETCONF, TL1, and so on. The solution using this platform can utilize one or more of the supported protocols for communication.

    Another common example of a platform is a service delivery platform that can be used for rapid creation and delivery of service. Development platform like an IDE (integrated development environment) can be used for effective programming. Similarly there can be event handling, performance monitoring, and other network and service level platforms.
67. Customer experience management (CEM): These solutions aims to satisfy the customer aspirations. While customer relation management (CRM) tries to achieve this by emphasizing their company's own goal, customer experience management keeps the customer as the focus for building a company's business strategy. In other words, a company's goals are used to determine effective methods of improving customer experience in CRM

and improving customer experience is used to identify the company strategy in CEM.

These applications obtain implicit and explicit feedbacks from the customer to identify the best way to improve business and relationships with the customer. Implicit feedbacks are obtained using algorithms that can derive the feedback based on customer behavior. Explicit feedbacks can be obtained from surveys and feedback forms that can be feed to the CEM application to generate experience reports.

68. Interface management: These solutions are mostly adapters that are used for conversion from one format to another. The interface adapters can be used when there is a protocol mismatch between the client and the server. For example, when the agent running on the network element supports legacy SNMP and when the server is based on XML, a protocol converter is required.

    Interface managers can also be used to make a product compliant to standards. For example, an adapter that can be used to convert legacy calls to standards based MTOSI calls is an interface manager. In this example when the adapter is integrated with the legacy solution, the product will then be compliant to MTOSI standards similar to next generation network management solutions.

69. Project management: Every development or enhancement activity in a service provider or service developer space is handled as a project that needs to be managed. As part of project management the milestones need to be tracked, resources need to be assigned based on skill set, software need to be procured to support development, estimates need to be prepared, and all activities to support the project from inception to completion need to be performed in project management.

    Project management solution is a collection of tools that can be used to manage the project effectively. These tools can be applied at different stages in the project life cycle like initiation, planning, requirements gathering, design, implementation, integration, testing, and support. The project management tool also maintains a repository of best practices and methodologies that can be referred by users of the application to get necessary templates and guidelines in preparing artifacts in the project.

70. Data synchronization: Synchronization of data is a major issue in telecom management. This has led to development of data synchronization solutions that check for discrepancies in data between application databases and synchronize the data. For example, the data available in the network management module might be different from the inventory management module.

    When a device goes down and is deallocated from an operation, the network management solution will get immediately notified. When the faulty element is shown as deallocated in the network management solution, this element might have an active status in the inventory management module when the data is not synchronized. The data synchronization solution routinely checks for discrepancies and will ensure that the network and inventory data base are synchronized.

71. Integration tools: These tools enable operators to quickly integrate diverse support systems and network elements. EAI tools like Vitria, WebM, and Tibco are examples of integration tools. Electronic integration of support systems is required for an automated back-office in large-scale operations. Tools like EAI creates an open data exchange platform optimized for heterogeneous interprocess communication.

    These tools are not limited to OSS and can be used to support communication between software systems in other domains also. Compliancy to standards can significantly reduce integration costs and minimize the integration effort.

## 23.3 Conclusion

With evolution in technology new and improved, networks are coming up in the market and there is an increased demand for management solutions that can manage these next generation networks or its elements by hiding the inherent complexity of the infrastructure. The capital and operational expenditure in managing the networks keeps on increasing when the OSS/BSS framework is not supporting the advances in network and service technology. The problem most service providers face is considerable effort to have an OSS/BSS framework that can manage a new network and effectively bill the customer to yield business benefits.

There are multiple OSS/BSS applications that are discussed in this chapter. These applications makes it easy to manage the network and services in a multiservice, multitechnology, and multivendor environment. Standardizing bodies are coming up with standards and guidelines in developing these management solutions. Current developments in telecom industry have lead to the addition of new networks and expansion of existing networks to support new technologies and products. The networks will continue to evolve and the only way to perform network management in a cost effective and operationally friendly manner is to do cost-effective OSS/BSS management.

This chapter gives the reader an overview of the management applications that make up the OSS/BSS stack from a holistic perspective. It considers the customer, service, and network layers in OSS/BSS and the applications that are used for managing interactions with third-party vendors, suppliers, stake holders, and partners. The intent is to meet the goal of this book, which is to give the reader a fundamental understanding of the concepts without diving into the details of a specific topic.

It is important to note that TeleManagement Forum (TMF) has come up with a Telecom Application Map (TAM), which is an application framework that can be referred to for understanding the role and the functionality of the various applications that deliver OSS and BSS capability. The TAM documents provide a comprehensive description of management applications and the components that make up the individual applications. TAM is part of NGOSS and has applications defined under groups similar to SID and eTOM to ensure that applications can be easily mapped to appropriate processes and information models.

## Additional Reading


1. www.tmforum.org
   GB929: Applications Framework Map (TAM). Latest release of TAM documents can be downloaded from the Web site.
2. Lawrence Harte and Avi Ofrane. *Billing Dictionary, BSS, Customer Care and OSS Technologies and Systems*. Fuquay Varina, NC: Althos Publishing, 2008.
3. Kornel Terplan. *OSS Essentials: Support System Solutions for Service Providers*. New York: John Wiley & Sons, 2001.
4. Andreas Molisch. *Wireless Communications*. New York: Wiley-IEEE Press, 2005.
5. Andrew McCrackan. *Practical Guide To Business Continuity Assurance*. Norwood, MA: Artech House Publishers, 2004.
6. James K. Shaw. *Strategic Management in Telecommunications*. Norwood, MA: Artech House Publishers, 2000.
7. Christos Voudouris, Gilbert Owusu, Raphael Dorne, and David Lesaint. *Service Chain Management: Technology Innovation for the Service Business*. New York: Springer, 2008.

# Chapter 24

# Information Models

This chapter is about information models. Two of the most popular information models in telecom domain, the Common Information Model (CIM) and Shared Information/Data Model are discussed in this chapter. TeleManagement Forum (TMF) and Distributed Management Task Force (DMTF) have developed a mapping between these two information models.

## 24.1 Introduction

Information models are important for application scalability. For example, consider a simple order management solution that uses two fields, first name and last name to represent the name of a customer. Another application for customer support uses three fields, first name, middle name, and last name to represent the name of a customer that contacts the support desk for assistance. The data collected and used by one of these applications cannot be used directly by the other application due to discrepancy in the number of fields used to collect the customer name. If the customer support application wants to check the order details of the customer by interacting with the order management module, then the customer name sent by the customer support application will not match the name used by the order management module. So standardization of information is required to ensure that an application and the information base are more scalable.

For the above example, the standardization of information is achieved by defining the number of fields to be used in defining a customer name. For a customer entity, there will be multiple attributes that can be used to define the entity. The number of fields to be used for each of these attributes, if defined, will improve interoperability between modules that work on the customer entity. So if a customer name is defined to have first name, middle name, and last name in all applications, then applications that need to exchange information relating to customer name will be able to interoperate without using an adapter.

The SID model was developed by TeleManagement Forum and the CIM was developed by Distributed Management Task Force. These models are widely adopted in current industry. While SID is mainly for telecom solutions, CIM is mostly used in enterprise solutions. With convergence, there was a need for a common model in telecom and enterprise industry. This led to the development of standards to map SID and CIM as a joint venture of TMF and DMTF. GB932 and GB933 from TMF and DSP2004 and DSP2009 are some of the documents that detail the alignment between the models.

This chapter explains the fundamentals of SID and CIM so that the reader can understand how an information model can be used in developing applications that are compliant to information standards. The common terminologies used in the standard documents on information models is discussed in this chapter so that the reader can easily interpret information while reading a SID or CIM document for developing a specific application. The choice of what information model needs to be used in developing an application is driven by business requirements. SID is a part of the NGOSS framework. Hence some of the terminologies and concepts of SID can be found to have similarities with the eTOM business. This is intentional as TMF wanted to have a mapping of business processes to information models and applications developed using the information model.

## 24.2 Shared Information/Data Model (SID)

The SID is a data centric framework similar to eTOM, which is a process centric framework of TMF. The SID information model represents business concepts, and the characteristics and relationships exhibited by the business concepts in an implementation independent manner. The cases and contracts of NGOSS can be created using SID.

The SID provides the reference models and common information/data catalogs that support eTOM business process models. Together SID and eTOM helps to ensure process to information alignment. The modeling language, UML is used for creating models in SID. The view of a business or system is created using UML class models and sequence diagrams in SID.

Common definitions for terms like resource, product, customer and service, and their relationships are provided by SID. This results in a common language for information modeling with service developers, service providers, and integrators. SID is used as a reference model for data design in telecom solution development.

The SID is composed of five building blocks:

1. Business entity: An item of business that follows a well-defined life cycle is represented as a business entity. It can be tangible things like a supplier, active things like inventory order, or conceptual things like customer account.
2. Aggregate business entity: A well-defined set of information and operations that helps to characterize a set of business entities is referred to as an aggregate business entity or ABE.
3. Attribute: This is a set of characteristics describing a business entity. Business entities can have relationships with other business entities and exhibit a set of characteristics.
4. Relationship: An association of business interests between two business entities is called a relationship. It should be noted that a business entity can establish an association with itself.
5. Domain: These are a collection of ABEs associated with a specific management area that are derived from analysis of process and information frameworks. It can be seen that domains making up the SID framework are consistent with eTOM level 0 concepts.

The SID framework comprises of business entities and their attributes represented using UML models with process mapping to eTOM. The SID specifications are grouped into domains, each of which has a set of SID business entities and attributes collected in aggregate business entities that can be directly used for data design by the service developers and service providers.

There is coupling between the different SID domains with a high degree of cohesion within the domain (see Figure 24.1). Use of SID leads to business benefits relating to cost, quality, timeliness, and adaptability of enterprise operations. With standard information and interface models, service providers can focus on value creation for their customers without worrying about the interoperability between applications.

The most compelling requirement in enterprises of achieving business and IT alignment can be achieved with the SID model apart from satisfying the NGOSS information and data needs. SID derives its roots from the eTOM that provides a business process reference framework and common business process vocabulary for the communications industry. The eTOM offers enterprises an effective way to organize their business processes and the communication between these processes. The companion model to the eTOM, the SID business view model provides an information reference model and a common vocabulary from a business entity perspective.

The SID uses the concepts of domains and aggregate business entities (or subdomains) to categorize business entities, thereby reducing duplication and promoting reuse. The SID scheme is categorized into a set of layers. This allows a specific team to identify and define specifications on a layer with minimal impact on other layers.

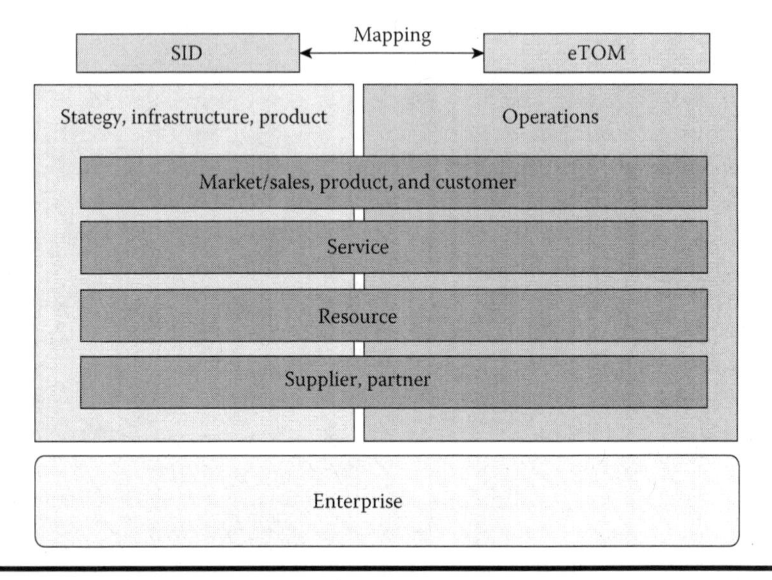


**Figure 24.1 SID domains mapped on eTOM level 0 model.**

This allows independent definition allowing distributed work teams. It should be considered that there is not necessarily a one–one relationship between the layers in SID.

The eTOM together with the SID model provides enterprises with a complete view of their business considering both process and information perspectives. SID provides the definition of the affect of information entities on the business processes defined in the eTOM. A combination of SID and eTOM provides the requisite materials that will help an enterprise to achieve its business goals.

The SID framework organizes information in a business view using business entities. The definitions using SID are made as generic as possible to promote reuse. The layered framework in SID helps to logically partition data relating to specific domains. The use of business oriented UML class models makes it easy to work with SID.

The domains of SID align with the business framework of eTOM. Creation of domains enables segmentation of the total business problem and allows TMF teams to be focused on a particular domain of interest. When the resultant business entity definitions within each domain are used in conjunction with the eTOM business process framework, it will provide a business view of the shared information and data. Aggregate business entities allow further partitioning of the information within each domain.

The business entities, attributes, and relationships are detailed with textual descriptions in each SID addendum. SID provides a consolidated Rational Rose UML based model that characterizes the entities to a model that can be easily understood for implementation of business applications.

The ABEs are organized using a categorization pattern. This ensures consistency of the ABE content and structure within each domain. The managed entity categories include:

- Strategy and plan specification: This is a description of the managed entity that is used to build the entity.
- Interaction: Any business communication with the managed entity is termed an interaction.
- Price: This signifies the cost of a managed entity.
- Configuration: The internal structure of a managed entity is its configuration.
- Test: This is the operation performed to identify the state of the managed entity.
- Trouble: Any problem like faults and alarms associated with the managed entity is referred to as trouble.
- Usage: This is the period of time during which a managed entity is in use.

Since a detailed discussion of SID is outside the scope of this book, let's look into the basic building block, the domains in SID. Domains have the following properties:

- Domains are a collection of enterprise data and associated operations.
- Business entities in domains encapsulate both operations and enterprise information.
- They provide for robustness of corporate/enterprise data formats.

The domains in SID are:

1. Market domain: This domain includes strategy and plans in the market space. Market segments, competitors, and their products, through to campaign formulation is covered in this domain.
2. Sales domain: This domain covers all sales related activity from sales contacts or prospects through to the sales force and sales statistics. Both the sales and marketing domain has data and contract operations that support the sales and marketing activities needed to gain business from customers and potential customers.
3. Product domain: The information and contract operations related to product life cycle are covered in product domain. The ABEs in product domain deal with product planning, performance, offerings, and usage statistics of the products delivered to the customer.
4. Customer domain: All data and contract operations associated with individuals or organizations that obtain products from a service provider are included here. It represents all types of interactions with the customer and management of customer relationship. The contract operations related to the

customer bills for products, collection of payment, overdue accounts, billing inquiries, and adjustments made as a result of inquiries are all accounted in the customer domain. The customer domain maps to the customer layer in the eTOM model.

5. Service domain: This domain deals with the definition, development, and operational aspects of services. Various eTOM processes that deal with the definition, development, and management of services offered by an enterprise are supported by the SID business entities in service domain. The domain also includes agreement on the service levels to be offered, deployment, and configuration of services. Information definition for other service related activities like management of service problems, service usage, quality of service, and service planning are covered in service domain.
6. Resource domain: This domain has ABEs managing the definition, development, and operational aspects resources. The components of the infrastructure and the products and services that use the infrastructure are covered in resource domain. Association of the resources to the products and services provides results in a detailed set of resource entities. The information models to perform planning, configuration, and monitoring are provided by SID in the resource domain.
7. Supplier/partner domain: The data and contract operations associated with a supplier/partner are included here. It includes planning of strategies, handling of all types of contact, management of the relationship, and the administration of supplier/partner data. Data and contract operations related to the supplier/partner bills, disputes, and inquiries are also handled in this domain.

Another important grouping used in SID is common business entities. These are business entities shared across two or more domains. As a result these business entities are not owned by any particular domain. The GB922 document of TeleManagement Forum covers the business view of SID and GB926 covers the system view of SID.

## 24.3 Common Information Model (CIM)

The CIM is an information model developed by DMTF. It uses object-oriented constructs and design in creating information models. The hierarchical, object-oriented architecture in CIM makes it easy to depict complex interdependencies. The CIM schema provides the actual model descriptions and captures notions that are applicable to all common areas of management, independent of implementations.

The CIM is a method for describing resources and their interactions in a heterogeneous IT environment. It is a technology neutral model for describing and accessing data across an enterprise and is not associated to a specific vendor. CIM can be used for modeling actual and virtual resources.

Another important standard, Web-based enterprise management (WBEM) also developed by DMTF is commonly used with CIM. While CIM defines the information model, WBEM defines the protocols used to communicate with a particular CIM implementation. CIM provides interoperability at the model level and WBEM provides interoperability at the protocol level. WBEM standards can be used to develop a single set of management applications for a diverse set of resources. DMTF has defined specifications on the schema and operations with CIM. This section is intended to give an overview on the CIM concepts.

Some of the basic building blocks in CIM are:

1. CIM specifications: The common information model specification describes an object-oriented meta-model based on the unified modeling language. The details for integration with other management models are defined in CIM specification. Data elements are presented as object classes, properties, methods, and associations in CIM specifications.

   CIM applies the basic structuring and conceptualization techniques of the object-oriented paradigm using a uniform modeling formalism that supports the cooperative development of the CIM schema. The specification defines the syntax and rules for each of the CIM meta-schema elements, the rules for each element and CIM naming mechanism. The specification also defines a CIM syntax language based on interface definition language (IDL) called managed object format (MOF).
2. Schema: The schema is a group of classes used for administration and class naming. Each class name is unique for a schema and has the format schemaname_classname. CIM schema provides model descriptions and captures notions that are applicable to all common areas of management, independent of implementations.

   CIM represents a managed environment as a collection of interrelated systems, each of which is composed of a set of discrete elements. The management schema consists of management platforms and management applications, like configuration management, performance management, and change management. The CIM schema supplies a set of classes with properties and associations thus creating a conceptual framework centered on the schema class. The core and common model discussed later in this chapter, make up the CIM schema.
3. Class: The properties and the methods common to a particular kind of object are defined in CIM class. Each CIM class represents a managed element and is scoped by the schema in which it belongs. The class name should be unique for the schema in which the class belongs.

4. Property: It contains the properties describing the data of the CIM class. CIM property will have a name, data type, value, and an optional default value. Each property must be unique within the class.
5. Method: Methods describe the behavior of the class. A method can be invoked and includes a name, return type, optional input parameters, and optional output parameters. Methods must be unique within the class, as a class can have zero or more methods. Methods with these characteristics are termed CIM methods. The method parameter and return type must be one of the CIM supported data types. The IN or OUT qualifier specifies the parameter type to be input or output.
6. Qualifier: Values providing additional information about classes, associations, indications, methods, method parameters, properties, or references are called a CIM qualifier. A qualifier can be used only with a qualifier type definition and the data type and value must match to that of the qualifier type. Qualifiers have a name, type, value, scope, flavor, and an optional default value. The scope of a qualifier is determined by the namespace in which it is present. The qualifier type definition is unique to the namespace to which it belongs.
7. Reference: This is a special property data type declared with REF key word. Use of REF indicates that the data type is a pointer to another instance. Thus defining the role each object plays in an association is done using a reference. The role name of a class in the context of an association can be represented by the reference.
8. Association: Associations are classes having the association qualifier. A type of class or separate object that contains two or more references representing relationships between two or more classes is called an association. Associations are used to establish relationship between classes without affecting any of the related classes.
9. Indication: Indications are types of classes having the indication qualifier. The active representation of the occurrence of an event that can be received by subscribing to them is called an indication. CIM indications can have properties and methods and can be arranged in a hierarchy.

   There are two types of indications; namely, life cycle indications and process indications. Life cycle indications deals with CIM class and instance life cycle events. Class events can be class creation, edit, or destroy and instance events can be instance creation, edit, destroy, and method invocation. Process indications deals with alert notifications associated with objects.
10. Managed object format (MOF): This is a language based on the IDL through which CIM management information could be represented to exchange information. Textual descriptions of classes, associations, properties, references, methods, and instance declarations, their associated qualifiers, and comments are the components of a MOF specification. A set of class and instance declarations make up the MOF file.

CIM is a collection of models. Some of the models in CIM are:

1. Core model: The core model is a set of classes, associations, properties for describing managed systems. The core model can be represented as a starting point to determine how to extend the common schema. It covers all areas of management with emphasis on elements and associations of the managed environment.

   The class hierarchy begins with the abstract-managed element class. It includes subclasses like managed system element classes, product related classes, configuration classes, collection classes, and the statistical data classes. From the classes, the model expands to address problem domains and relationships between managed entities.
2. Common models: Information models that capture notions that are common to a particular management area, but not dependent of any particular technology or implementation are called the common models. This offers a broad range of information including concrete classes and implementations of the classes in the common model. The classes, properties, associations, and methods in the common models are used for program design and implementation.
3. Application model: This model is intended to manage the life cycle and execution of an application. Information in application model deal with managing and deployment of software products and applications. It details models for applications intended for stand-alone and distributed environment.

   It supports modeling of a single software product as well as a group of interdependent software products. The three main areas in application model are the structure of the application, life cycle of the application, and transition between states in the life cycle of an application.

   The structure of an application can be defined using the following components:
   - Software element: It is a collection of one or more files and associated details that are individually deployed and managed on a particular platform.
   - Software feature: It is a collection of software elements performing a particular function.
   - Software product: It is a collection of software features that is acquired as a single unit. An agreement that involves licensing, support, and warrantee is required between the consumer and supplier for acquiring a software product.
   - Application system: It is a collection of software features that can be managed as an independent unit supporting a particular business function.
4. Database model: The management components for a database environment are defined in the CIM database model. This database model has a number

of supportive classes representing configuration parameters, resources, and statistics. There are three major entities in the database model:

- Database system: The software application aspects of the database environment makes up the database system.
- Common database: Organized data on logical entity is called a common database.
- Database service: It represents the process or processes performing tasks for the database.

5. Device model: The functionality, configuration, and state of hardware can be described using the device models. Management of the device depends on the configuration of the hardware and software as well as on the interconnections between the devices. A hardware component that can offer multiple functionalities is realized using multiple logical devices. The hardware that can be modeled ranges from a simple battery to complex storage volumes. Device models also describe relationships. The devices in the model are represented as components under CIM_System.
6. Event model: An event is an incident that can involve a change in the state of the environment or behavior of a component in the environment. An event usually results in a notification. The events present in the event model could be pervasive or infrequent events. The consequences of an event can last a long time affecting many things. Some events may require immediate action on the part of the observer whereas other events could be handled at a later point of time.
7. Interop model: The management components describing the WBEM infrastructure and how they interact with the infrastructure are defined as the CIM interop model. The WBEM infrastructure has:
   - CIM client: The client issues operation requests to the CIM server. It also receives and processes operation responses from the CIM server.
   - CIM server: The server issues operation responses after receiving and processing requests from the CIM client.
   - CIM object manager (CIMOM): It is a part of the CIM server that facilitates communication between server components.
8. Metrics model: This model consists of CIM classes specifying the semantics and usage of a metric and classes containing data values captured for a particular instance of the metric definition class. Metrics model facilitate dynamic definition and retrieval of metric information.

   Management of the transaction response time information generalized to the concept of unit of work and providing the additional metric information about the transaction can be done using metric definition.

## 24.4 Conclusion

The SID and CIM are two widely accepted frameworks for information modeling. There is considerable divergence between the two models as they were developed by different standardizing organizations. TeleManagement Forum the developers of SID and Distributed Management Task Force the developers of CIM have jointly worked toward harmonizing the two models. The TMF document GB932/DMTF document DSP2004 has the physical model mapping and TMF document GB933/DMTF document DSP2009 has the logical model mapping.

Using a standard information model is required for system interoperability. Ease of integration is a must in telecom management space where multiple operation support solutions interoperate to manage the operations in the Telco. The information models can be mapped to business process and applications where it needs to be used. In short, use of standard information models is mandatory in next generation telecom management solutions. After a brief overview of what an information model is expected to achieve, this chapter explains the basic concepts of the two most popular information models.

## Additional Reading

1. www.tmforum.org
   Latest release of the following documents can be downloaded from the Web site
   - GB922 and GB926: SID Solution Suite
   - GB932 and GB933: CIM-SID Suite
2. John P. Reilly. *Getting Started with the SID: A Data Modeler's Guide.* Morristown, NJ: TeleManagement Forum, 2009.
3. *Distributed Management Task Force (DMTF) CIM Tutorial at*: www.wbemsolutions.com/tutorials/CIM/
4. Chris Hobbs. *A Practical Approach to WBEM/CIM Management.* Boca Raton, FL: CRC Press, 2004.
5. John W. Sweitzer, Patrick Thompson, Andrea R. Westerinen, Raymond C. Williams, and Winston Bumpus. *Common Information Model: Implementing the Object Model for Enterprise Management.* New York: John Wiley & Sons, 1999.

# Chapter 25

# Standard Interfaces

This chapter is about OSS interfaces for interaction. Two of the most popular information models in telecom domain, Multi-Technology Operations System Interface (MTOSI) and OSS through Java Initiative (OSS/J) are discussed in this chapter. How these two standards can complement each other is also covered in this chapter.

## 25.1 Introduction

There are multiple operation support systems currently available in the telecom market. In a service provider space, multiple OSS solutions need to interoperate seamlessly to reduce operational costs and improve operational efficiency. To make this possible the interfaces for interaction need to be standardized. Most OSS solution developers are making their products compliant to MTOSI or OSS/J interface standards.

In current industry, compliancy to standards is more a necessity. Let us take a simple example of a simple fault management module. Now the network provisioning module might request active fault alarms using the interface call "getAlarmList ()," the inventory management module might request active fault alarms using the interface call "getActiveAlarms ()," and the event management module might request active fault alarms using the interface call "getFaultLogs ()." On the other end, the fault management module might be exposing the interface getActiveFault () for other modules to query and get the active alarm list as response.

One solution to fix this problem is to write an adapter that translates the call between modules, like getAlarmList () from network provisioning module to getActiveFault () that the fault management module can understand. In the example we are discussing, the solution of developing adapters will result in three adapters for the fault management module alone. So in a service provider space where there will be many OSS solutions, the number of adapters required would make interoperability costs huge. Again when a new OSS solution is introduced, a set of adapters needs to be developed for the modules it is expected to integrate with. So adapters are not a viable solution that would be cost effective to the service provider.

Another option to fix the problem of interoperability is to have standard interfaces. Let us look into how this works for the example we are discussing. If it is decided by the standardizing body that getActiveFault () is the standard interface for getting active alarms from the fault management module, then the service provider can request compliancy to standard for modules that interact with the fault management module. That is the fault management module that will respond only to getActiveFault () request and any module compliant to standard that interacts with the fault management module is expected to use the standard interface. When a new module is introduced that is compliant to standards interoperability with fault management module, this will not be an issue.

This chapter explains the fundamentals of MTOSI and OSS/J so that the reader can understand the significance of interface models and how an interface model can be used in developing applications that are compliant to standards. The common terminologies used in the standard documents on interface models are also discussed in this chapter.

## 25.2 MTOSI Overview

Before we discuss MTOSI, let us first give a brief overview on MTNM the predecessor of MTOSI. MTNM was formed by combining the ATM information model (ATMIM) team that was formed to address ATM management and SONET/SDH information model (SSIM) team. MTNM was one of the most deployed interface suites of TM Forum. It includes a business agreement document titled TMF513, an information agreement titled TMF608, a CORBA solution set titled TMF814, and an implementation statement titled TMF814A. Popular NMS solutions like AdventNet Web NMS provide support for TMNF MTNM interfaces. MTNM was mainly focusing on management of connectionless networks with an emphasis on Ethernet and control plane-based networks.

When most OS suppliers started converting MTNM IDL to XML, TMF started the MTOSI group. This group took a subset of MTNM and generalized the procedures for OS-OS interactions (see Figure 25.1) on an XML-based interface that was most required for the telecom market. A major MTOSI goal is to make the XML interface independent of the transport and gradually supports more transports.

Use of MTOSI has many benefits to different players in the telecom value chain. To equipment vendors, MTOSI results in reduction of development costs and provides a faster time to market. It also lowers deployment risks and enables

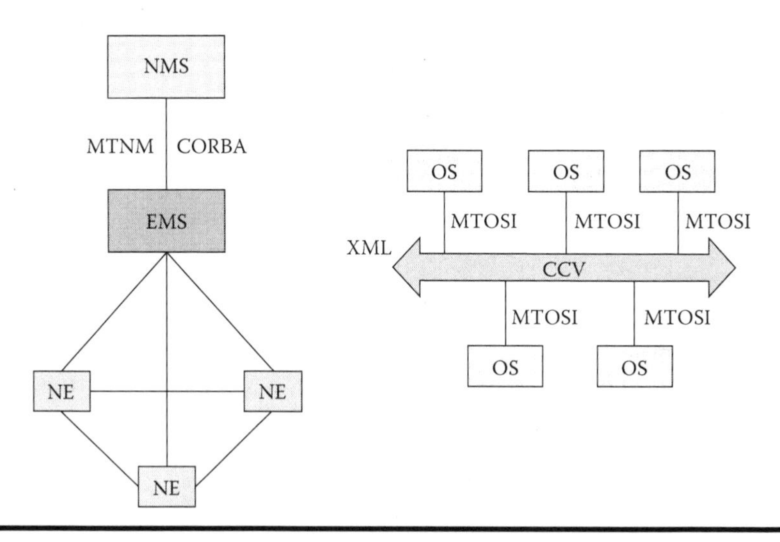


**Figure 25.1 MTNM (point to point) and MTOSI (bus) structure.**

differentiation. Service providers benefit by a reduction of integration costs and faster time to market for new service rollout. Use of MTOSI helps in reducing development cost for software vendors. It reduces relationship costs and provides easier maintenance and testing that enables product differentiation. For system integrators, MTOSI simplifies interoperability testing and provides faster time to market for integrated solutions. The technical skills and training requirements are also reduced with the use of MTOSI.

MTOSI leverages the knowledge gained from MTNM deployments and artifacts. It supports multiple synchronous and asynchronous communication styles to satisfy the interaction needs. It supports bulk inventory and alarm operations, and replaces many fine grain operations that were advocated in previous interface specifications. In XML, MTOSI supports most Web service standards including WDSL, SOAP, and XSD. It can be seen that MTOSI offers support for service oriented architecture (SOA) and enterprise service bus (ESB).

In MTOSI, the requirements of network management or service resource management have been specified in terms of the entities visible across the interface and the service requests that may be invoked across the interface. The interface agreement document contains the data model describing the entities and associated service requests while the business agreement document contains requirements for the visible data entities and the associated service requests.

The information exchange infrastructure through which OSS solutions interact is called a common communications vehicle or CCV. The CCV is used to publish new entities. The OSS has direct access to the CCV. The existence of the entity instance on the CCV is announced by the publishing. Storage and naming of the entity instance is done by naming OS. The entity name could be the name used to announce the entity. The discovering OS and the naming OS need not be the same OS for any particular entity.

The following attributes of the entities are visible across the interface:

- Name: The entity instance on the CCV can be uniquely identified using the "name" attribute. Once the naming OS sets the attribute, its value does not vary for the life of the entity.
- Discovered name: In the case where the OS that publishes the entity on the CCV is not the naming OS, the name of the entity is known as the discovered name. This attribute may be left empty if the naming OS publishes the entity first on the CCV. The value of the name attribute and the discovered name attribute of the entity can be set to the same value or different value.
- Naming OS: The OS that is responsible for setting the "name" of the entity also sets up the naming OS attribute and this attribute represents an identifier for the steward of the entity. There is a unique naming OS for each entity published on CCV.
- Alias name list: The attribute is a list of name-value pairs for the entity. The name refers to the type of alias and the value component holds the alias itself.
- User label: It is an optional attribute that represents the "user friendly" name for the entity.
- Owner: It is an optional attribute that is used to identify the owner of the entity.
- Vendor extensions: This attribute is used by vendors to further qualify and extend the entity characteristics.

MTOSI provides steps to discover new network entities such as managed elements and topological links in cases where two different suppliers are providing the network inventory management OS and the network discovery/activation OS. Another interesting feature in MTOSI is backward and forward compatibility.

- Backward compatibility: This feature allows a new version of a client OS to be rolled out in a way that does not break existing server OSs. An older version of a message can be sent to a client OS (by server OS) that understands the new version and still have the message successfully processed.
- Forward compatibility: This feature allows a new version of a server OS to be rolled out in a way that does not break existing client OSs. The server OS can send a newer version of an instance and still have the instance successfully processed.

MTOSI allows a client OS to perform multiple operations on multiple objects using a single request to a server OS. In this approach, the transactional logic is delegated to the server OS facilitating in performance tasks like configuration of a managed element and all associated equipment and ports. Also the same operation can be performed on a large number of objects of the same type in an efficient manner. These requests may contain multiple actions and/or objects and the requests may be of long duration.

## 25.3 MTOSI Notifications

In this section, the notification feature is taken as a sample MTOSI feature for discussion. Some of the event notifications provided by MTOSI include:

- Creation of a particular object
- Attribute value changes for attributes of an object
- Conveying all the change information in single notification

The interface offers capability for both single event inventory notifications like the creation of a particular entity or attribute value changes for several attributes of a given entity, as well as multievent inventory notifications, which is a direct method for getting bulk inventory updates from the source OS to other interested OSs.

To get a feel of notification generation like object creation notification (OCN) and attribute value change (AVC) notification, let us discuss two methods by which a new network entity is announced on CCV. The first method requires the discovery OS to keep track of the planned entities, and the network entities to be planned and named at the interface before they are discovered. The steps in the first method are:

1. A network entity is entered into the inventory OS.
2. The network entity is represented by the object created by inventory OS.
3. This object is transferred to the planning resource state. The network entity has not been installed now.
4. The "Object Creation Notification" is sent from the inventory OS to the discovery OS.
5. The network entity is actually installed later.
6. The discovery OS detects the installed entity and issues an "Attribute Value Change" notification on the resource state of the network entity.
7. This results in change from planning to installing.

In the second method the new network entity is first announced on the CCV by the discovery OS before it is given a unique name. Here we will deal with object discovery notification (ODscN) and object creation notification (OCN). The steps in the second method are:

1. A network entity that has not been announced by the inventory OS at the interface is directly detected by the discovery OS.
2. The discovery OS then sends an object discovery notification to the notification service.

3. The notification is received by the inventory OS.
4. The new object is given a name by the inventory OS.
5. Inventory OS issues an object creation notification.
6. Other modules like discovery and activation OSs receives and processes the OCN.

Service interfaces and data entities define interface implementation. The service interfaces consist of operations and notifications that have parameters. Attributes make up data entities. Additional service interfaces like operations and notifications are needed as the interface is extended. Interface elements like operations will not be deleted when passing from one version to the next unless an error with an existing element is identified. Instead of "delete" term, "Deprecate" is used with respect to parameters for operations and notifications, since it may be easier to leave the operation/notification signature alone and just deprecate usage of a particular parameter.

## 25.4 OSS/J Overview

While the NGOSS program of TMForum focuses on the business and systems aspects, essentially the problem and solution statement of OSS-solution delivery, the OSS/J program focuses on the implementation, solution realization, and deployment aspects using Java technology. The OSS through the Java (OSS/J) initiative defines a technology specific set of API specifications. In addition to API specifications, OSS/J also includes reference implementations, technology compatibility kits, and multitechnology profiles for OSS integration and deployment.

The core business entities (CBE) model of shared information/data (SID) model that helps to provide a technology neutral implementations view of the NGOSS architecture is used by the OSS/J API specifications. The Java 2 Platform Enterprise Edition and EJB-based Application Architecture are the enabling technologies in OSS/J, used as tools for developers to easily develop applications. The J2EE platform offers faster solution delivery with a reduced cost of ownership. The Enterprise JavaBean (EJB) component architecture is designed to hide complexity and enhance portability by separating complexities inherent in OSS applications.

Rapid integration and deployment of a new OSS application can be done by OSS through Java initiative's framework due to standard EJB components that are already equipped with connectors. Application integration capabilities are provided for server-side applications by APIs that are defined within the initiative using J2EE platform and EJB architecture.

A standard set of APIs using Java technology for OSS and BSS can be defined and implemented by using OSS through Java initiative. Members of OSS/J are leveraging standards from 3rd Generation Partnership Project (3GPP), Mobile Wireless Internet Forum (MWIF), and the TeleManagement Forum (TMF) to implement common Java APIs and provide a source code that can be made freely available to the industry. OSS/J implementation groups include telecom equipment manufacturers, software vendors, and systems vendors that are formed to develop and implement interoperable and interchangeable components to build OSS applications.

Addressing the lack of interoperable and interchangeable OSS applications for the OSS industry is the goal of the initiative. Developing and implementing common APIs based on an open component model with well-defined interoperable protocols and deployment models serve to accomplish the goal.

OSS/J initiative is performed under the open Java community process (JCP). Using JCP the initiative will develop the following for each application area:

- Specification documents: It details the interface APIs.
- Reference implementation (RI): This is a working version that facilitates use of the APIs.
- Technology compatibility kit (TCK): A suite of tests, tools, and documentation that will allow third parties to determine if their implementation is compliant with the specification.

Production of demonstration systems to validate the OSS/J APIs against suitability for supporting actual field scenarios is a major goal of OSS through the Java initiative. These systems should also meet performance and scalability for usage in actual field scenarios. Defining OSS through Java APIs and maturing APIs from other sources results in the addition of more complete business scenarios to the demonstration systems.

The OSS applications are divided into three fundamental parts by the J2EE application model. The three parts are:

- Components: The application developed is comprised of a set of components. This makes components the key focus area for application developers.
- Containers: It provides service transparently to both transaction support and resource pooling. Using containers, component behaviors can be specified at deployment time, rather than in the program code. Containers also provide security in addition to transaction support.
- Connectors: Connectors define a portable service API to plug into existing OSS vendor offerings. It is a layer beneath the J2EE platform and promotes flexibility by enabling a variety of implementations of specific services. System vendors can conceal complexity and remote portability by implementing containers and connectors.

Use of OSS/J has many benefits to different players in the telecom value chain. To equipment vendors, it facilitates integration of equipment with more vendor products. Service providers are able to develop agile solution and improve interoperability using OSS/J component architecture. To software vendors it provides a common API support for across-the-board integration. This accelerates the delivery of new products, lowering integration and maintenance costs. The use of OSS/J will reduce integration costs for system integrators.

Various commercial off the shelf (COTS) and legacy applications, both network and customer facing, are integrated by the OSS systems. OSS/J initiative works on specific application areas in OSS domain and uses J2EE, EJB, and XML technologies for its development. APIs and message structures are defined for each application by the implantation group so that the OSS components can be developed and assembled into complete solutions. APIs are specified and implemented as EJB components. OSS applications are developed using these APIs.

OSS/J is emerging as a widely accepted OSS/BSS standard that helps Telco to realize standardization of interface, benefiting the service providers with vendor independence. OSS systems developed the integration approach and construction approach using J2EE technology. Integration approach is intended to integrate existing applications. Construction approach develops EJB-based applications, where both application and its serving components are in the container.

Easier portability and rapid application development are possible through EJB. Based on the standardized APIs, it is possible to develop vendor independent EJB connectors for various protocols like CMIP, SNMP, TL1, and the knowledge can be reused for multiple OSS/J systems. Developing EJB connectors itself as COTs components offers many opportunities and the knowledge can be reused for developing connectors for proprietary protocols also.

OSS/J model can address the impact of the emergence of rapid service creation environments like IP multimedia subsystems (IMS) and service delivery platforms (SDP) on traditional OSS/BSS systems. OSS/J delivers standard functional OSS APIs, each including specifications, reference implementations, conformance test suites, and certification process and source codes. Open IT standards form the basis of OSS/J technology.

Use of OSS/J leads to a market of interoperable and interchangeable OSS components. OSS/J APIs include EJB and XML/JMS profiles, WS profiles, and WSDL interfaces for the common API. Vendors can be compliant to OSS through Java API by supporting any one of these profiles. Though OSS/J is generic it can be specialized to be model and technology specific. It provides coverage throughout the eTOM processes. OSS/J can be extended to meet specific operational requirements by use of a disciplined extension mechanism. It also supports metadata and queries.

The best practices and design patterns implemented in each of the API back up OSS/J deliverables. eTOM can define a common map for communications industry business processes and a reference point for internal process reengineering needs, partnerships, alliances, and general working agreements with other providers as a result of endorsement by a large majority of the world's leading service providers. When applied to the full range of Telco operations, maximum flexibility is provided by the technologies of OSS/J.

OSS/J has the following characteristics:

- OSS/J uses J2EE as middleware. This is an open standard middleware. J2EE can provide tightly coupled system implementation. It comes with implementation conformance test tools and certification process.
- OSS/J uses XML over JMS that enables loose coupling among systems and interfaces.
- OSS/J components are Web services-enabled using WSDL specifications.
- OSS/J offers open standards for its APIs. The standards come with a source code. They are certifiable at implementation level and come with conformance test tools.

Before we conclude the discussion let's look into OSS/J certification. The OSS/J API specification and reference implementation are accessible for free download. OSS solution developers can use the API specification and reference implementation code to develop OSS/J API in products that are planned to be certified. To certify the developed API, the solution developer has to download the technology compatibility kit within the Java community process, which is a test suit that is available for free.

The test suite needs to be run against the API developed by the service provider. The test execution will generate a test report that will specify if the test run was successful. On obtaining a successful test run in the report, the solution developer needs to send the test execution report to the OSS/J certification committee. The certification committee will validate the report and the process used to create it. Once the certification committee acknowledges the certification of the product, the certification can then be published. The published certificates will also have a reference to the test execution report. The publication of the test report eases the procurement of certified products for faster integration.

## 25.5 MTOSI and OSS/J

MTOSI focuses on reducing the cost of integration in a commercial environment by providing a full XML interface specification detailing the model, operations and communications enabling out of the box interoperability. The focus of OSS/J is also reduction of the integration cost in a partner/open-source environment. OSS/J provides a Java interface specification that offers basic operations but does not contain the model or interaction as specific as in MTOSI. Instead, OSS/J allows sharing of reference implementations to enable interoperability. Thus it becomes evident that MTOSI and OSS/J are complementary technologies. Using a mapping/mediation approach MTOSI and OSS/J can interoperate. For a closed partner engagement, OSS/J is best suited where as for an out-of-the-box commercial environment MTOSI is the best.

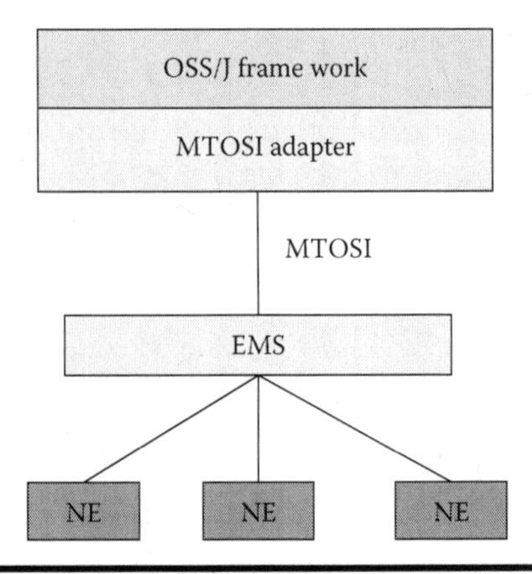


**Figure 25.2 MTOSI and OSS/J can interoperate.**

OSS/J offers technology independent APIs. There are separate APIs for various OSS functions. MTOSI gives OSS interfaces for a multitechnology environment and is based on specific models. MTOSI models are defined in TMF 608. OSS/J, being associated to a generic set of operations, can have an adapter written when a specialized set of operations needs to be executed with MTOSI. This is shown in Figure 25.2 where the generic OSS/J framework has an MTOSI adapter for interaction wherever a defined set of operations are to be executed.

MTOSI and OSS/J has some marked differences. While OSS/J is Java based, MTOSI is XML based with emphasis on Web services. MTOSI however has guidelines on using JMS, though its usage is not mandatory for implementing MTOSI compliant APIs. OSS/J also supports XML, but the XML API is wrapped with Java API.

The MTOSI back end implementation is not based on a specific technology as compared to OSS/J where the back end implementation is in Java. As MTOSI evolved from MTNM most of the issues in interface APIs were fixed and the definitions were to offer a service oriented interface on a heterogeneous platform environment. So when multitechnology, multinetwork support is required MTOSI would be a good choice, while in open source based environment OSS/J would be the popular choice. Support to both these interface standards can be implemented in the same OSS solution.

## 25.6 Conclusion

An application programming interface (API) is an interface that defines the ways by which an application program may request services from another application. APIs provide the format and calling conventions the application developer should use to work on the services. It may include specifications for routines, data structures, object classes, and protocols used to communicate between the applications.

The APIs are mostly abstract interface specifications and controls the behavior of the objects specified in that interface. The application that provides the functionality described by the interface specification is the implementation of the API. This chapter gives the reader an overview of the management interface standards used in OSS/BSS development. It considers MTOSI and OSS/J for developing applications in the customer, service, and network layers of OSS/BSS. The basic concepts of these two popular interface standards are discussed without diving into the details of a specific topic.

## Additional Reading

1. www.tmforum.org
   (Latest version of the documents can be downloaded from the Web site)
   TMF608: MTOSI Information Agreement
   TMF517: MTOSI Business Agreement
   TMF854: MTOSI Systems Interface: XML Solution Set
   TMF608: MTNM Information Agreement
   TMF513: MTNM Business Agreement
   TMF814: MTNM IDL Solution Set
   TMF814A: MTNM Implementation Set and Guidelines
   OSS APIs, Roadmap, Developer Guidelines, and Procurement Guidelines
2. Software Industry Report. *Sun & Leading Telecom Companies Announce Intention To Start Next Generation OSS Through Java.* Washington, DC: Millin Publishing Inc., 2000.
3. Kornel Terplan. *OSS Essentials: Support System Solutions for Service Providers.* New York: John Wiley & Sons., 2001.
4. Dr. Christian Saxtoft. *Convergence: User Expectations, Communications Enablers and Business Opportunities.* New York: John Wiley & Sons., 2008.
5. John Strassner. *Policy-Based Network Management: Solutions for the Next Generation.* San Francisco, CA: Morgan Kaufmann, 2003.

# IMPLEMENTATION GUIDELINES

# Chapter 26

# Socket Communication

This chapter is about the basics of network programming. It introduces the reader to the concept of writing client–server code. The client usually acts as the agent sourcing data to the server that listens for data from multiple clients. Out of the different techniques that are used for client–server programming, this chapter specifically deals with socket programming. Programming for TCP and UDP transport between client and server is also discussed.

## 26.1 Introduction

To complete this book on fundamentals of EMS, NMS, and OSS/BSS, it is essential to discuss the implementation basics for developing solutions using the concepts discussed in preceding chapters.

Some of the important aspects in developing management applications are:

- These applications mainly deal with the application layer. Management protocols usually lie in this layer.
- There will be multiple agents or clients running on the network elements that will collect data and send to a central server. Even for upper layers, there will be business and service level applications that take feed from multiple NMS clients.
- There are different techniques of communication based on environment. Socket-based programs, RPC-based programs, and web-based programs are quite common in implementing application interactions for management solutions.

- The specific application layer protocol is more a matter of choice based on data to be sent. For usual management operations like get, set, and trap functionalities, management protocols like SNMP or NETCONF could be used. While SNMP is the choice in most legacy solutions, the more recent applications use the power of XML and use XML-based protocols like SOAP. For other management functions like file transfer, FTP is still the most popular choice, and for alerts as mail SMTP is used.

This part of the book is intended to cover all these topics for developing management applications including a case study on the design approach to be followed in developing an NGNM solution and then customizing the solution for a specific network. The coverage still limits the discussion to the fundamentals with information on additional reading to get detailed information on specific topics.

This chapter gives details on the application layer and client–server programming. The socket interface is then discussed in detail how TCP and UDP commutation is achieved with socket programming. The programming language used in this chapter is "C" considering most initial implementations of socket programming were done with socket APIs (application programming interface) in C programming language. It is important to note that C programming language was initially developed for the UNIX operating system by Dennis Ritchie.

## 26.2 Application Layer

The application layer is the top layer in the OSI model and TCP/IP model. The application layer manages communication between applications and handles issues like network transparency and resource allocation. The layers below the application layer provide the framework for communication between application programs in the application layers.

The important aspects of the application layer programs are:

- Client–server paradigm: The interaction between applications is based on a request response model. When an element connected to a network wants to communicate with another element in the network, it raises a request message identifying itself as the source and specifying the destination to which the message needs to be sent. Upon receiving the service request, the destination element returns a response message to the source.

  In a client–server model, there will be one or more clients connected to a server. The management application sends a request to the agents to perform a service like getting management data, setting the data, or notifying when an event occurs. The agents send a response message to the management application on completing a service request. The response message can be synchronous or asynchronous message.

- Services: There are different types of services that can be offered in the application layer. These make use of the client–server paradigm for message exchange. Some of the popular protocol–based services are:
  - Services using simple mail transfer protocol (SMTP): The SMTP service allows transfer of message between two users on the Internet via electronic mail.
  - Services using file transfer protocol (FTP): The FTP service allows the exchange of files between two elements.
  - Services using hypertext transfer protocol (HTTP): The HTTP transfer protocol facilitates the use of the World Wide Web.
- Addressing: The client message includes the address of the server as the destination address and its own address as the source address for communication. Similarly for messages from the server, the server address is used as the source address and the address of the client is the destination address. The application layer has its own addressing format, which is different from the address format used by the underlying layers. For example, electronic mail service has a host address format like editor@crc.com and the world wide web service has a host address format like www.crcpress.com.

  The application program uses an alias name as the address rather than an actual IP address. The main part of the alias is the address of the remote host and the remaining part is related to the port address of the server and directory structure where the server program is located. Instead of using the IP address, the application program uses an alias name with the help of another entity called DNS (domain name system).
- Capabilities: There are a standard set of capabilities that are expected in any application layer program. Three of the capabilities from the standard set that are discussed here are reliability of data transfer, high throughput, and minimal delay in communication. Not all application programs implement these capabilities even though these are implied requirements in commercial grade management solutions.
  - Reliability of data transfer: Some application programs include reliability as part of its protocol implementation or use the services of a reliable transport layer protocol like TCP. There are applications that do not require reliable data transfer as in intermittent data transfer over UDP.
  - High throughput: Having a high throughput ensures transmission of a maximum amount of data in a unit of time. In most cases a high throughput will require high bandwidth. When the volume of data to be transferred is high as in transmission of video files, high throughput becomes a necessary requirement.
  - Minimum delay: Some applications are very sensitive to delay. For example, communication delay can adversely affect an interactive real-time application program.

## 26.3 Monitoring

Networking protocols helps to manage the network and makes it easier to work on the network. Some of the utility protocols may not be required for every network application though it might be possible to add unique features to solutions using these protocols. Applications based on DNS, ICMP (Internet control message), TELNET, and other utility protocols in the TCP/IP suite like ARP (address resolution protocol) or RIP (routing information protocol), can be effective tools for an operator performing network troubleshooting. Some of these methods are discussed here (see also Figure 26.1).

Domain name system: This is a client–server application that identifies each host on the Internet with the help of a unique user-friendly name referred to as a domain name. This functionality is useful as most people find it difficult to remember strings of numbers like an IP address of a host. The DNS helps to convert domain names into the IP address. This is done by mapping an easily recognizable domain name with IP addresses. A worldwide network of DNS stores the list of domain names against IP addresses.

Internet control message protocol: This is also referred to as PING. It is a protocol that is used to report broken network connections or other router-level problems that end hosts might need to know. The PING utility available in most computers helps to identify if another computer in the network is switched on and how much delay there is over the connection to send messages to it.

TELNET: This protocol is used for debugging servers and for investigating new TCP-based protocols. To debug a server, a TCP connection is made to the server

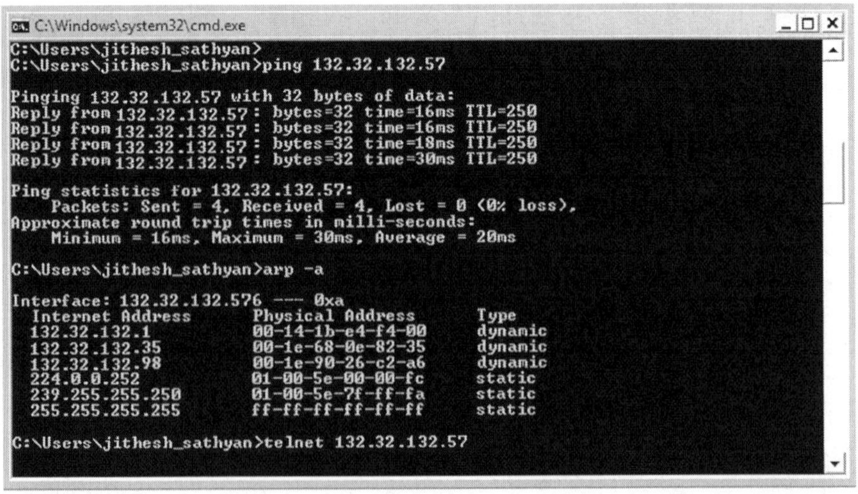

```
C:\Windows\system32\cmd.exe
C:\Users\jithesh_sathyan>
C:\Users\jithesh_sathyan>ping 132.32.132.57

Pinging 132.32.132.57 with 32 bytes of data:
Reply from 132.32.132.57: bytes=32 time=16ms TTL=250
Reply from 132.32.132.57: bytes=32 time=16ms TTL=250
Reply from 132.32.132.57: bytes=32 time=18ms TTL=250
Reply from 132.32.132.57: bytes=32 time=30ms TTL=250

Ping statistics for 132.32.132.57:
    Packets: Sent = 4, Received = 4, Lost = 0 (0% loss),
Approximate round trip times in milli-seconds:
    Minimum = 16ms, Maximum = 30ms, Average = 20ms

C:\Users\jithesh_sathyan>arp -a

Interface: 132.32.132.576 --- 0xa
  Internet Address      Physical Address      Type
  132.32.132.1          00-14-1b-e4-f4-00     dynamic
  132.32.132.35         00-1e-68-0e-82-35     dynamic
  132.32.132.98         00-1e-90-26-c2-a6     dynamic
  224.0.0.252           01-00-5e-00-00-fc     static
  239.255.255.250       01-00-5e-7f-ff-fa     static
  255.255.255.255       ff-ff-ff-ff-ff-ff     static

C:\Users\jithesh_sathyan>telnet 132.32.132.57
```

**Figure 26.1 Working with PING, ARP, and TELNET in MS-DOS (Microsoft disk operating system).**

that is to be debugged using telnet utility. Since most telnet clients facilitate the connection to ports other than default port 23, the telnet utility can also be used to investigate new TCP-based protocols. Most operating systems include telnet clients, so telnet is rarely used programmatically.

Address resolution protocol resolves IP addresses into their equivalent MAC addresses and routing information protocol helps to identify the number of hops to a destination. There are many more utility applications that can aid in network troubleshooting.

## 26.4 Client–Server Model

The client-server model is the most common way by which an element requests services from another network element (see Figure 26.2). In a typical network like two computers connected through the Internet, the local computer (client) runs a program that requests a service from another program on the remote computer (server). There can be multiple clients in the network that request for service from a server. The communication channel between server and client could always be connected or the connection can terminate upon completion of service. Using the IP address of the client and the port address of the specific server program, a communication channel is established between the client and the server.

Once a connection is established, the channel remains active and open for communication. After the completion of the service from the server, the communication channel between the client and the server can be closed. The server program keeps running on the remote computer, open to connection requests from clients. A client program runs only when it is needed while a server program runs all the time because it does not know when its services will be needed.

Clients can be run on a machine either iteratively or concurrently. When the clients are run iteratively, the machine allows one client to start, run, and

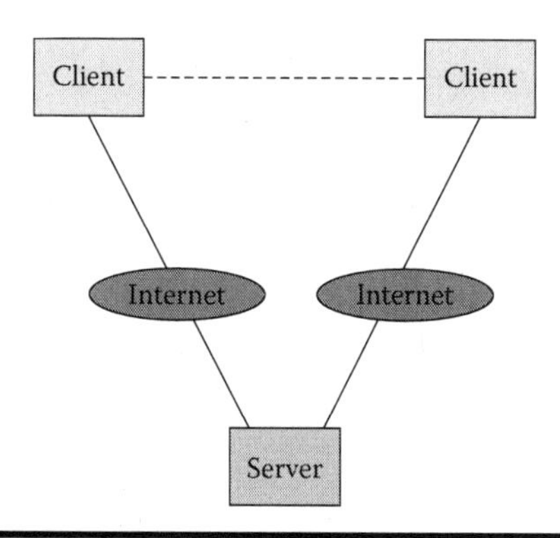


**Figure 26.2 Client server model.**

terminate before another client is started by the machine. When the clients are run concurrently, the machine allows running two or more clients at the same time.

Similarly the servers can be run on a machine either iteratively or concurrently. An iterate server accepts, processes, and sends the response to the client before accepting another request; that is, it processes a single request at a time. A concurrent server processes a large number of requests at the same time. It shares its time between a large numbers of requests. The transport layer protocol and the service method facilitate the server operation. UDP—connectionless transport layer protocol or TCP—connection oriented transport layer protocols are used commonly.

Based on the number of requests the server can handle and the type of connection, servers can be classified into four types:

- Connectionless iterative: The connectionless iterative server processes a single request at a time and uses a connectionless transport layer protocol like UDP.
- Connectionless concurrent: The connectionless concurrent server processes many requests at the same time and uses a connectionless transport layer protocol like UDP.
- Connection oriented iterative: The connection oriented iterative server processes a single request at a time and uses a connection oriented transport layer protocol like TCP.
- Connection oriented concurrent: The connection oriented concurrent server processes many requests at the same time and uses a connection oriented transport layer protocol like TCP.

## 26.5 Socket Interface

Sockets are the communication structures used in socket programming. They are a connection independent method of creating communication channel between processes. It acts as an interface between the application process and transport layer. The socket acts as an end point providing communication between two processes. Socket interfaces are a set of system calls.

Most operating systems define socket as a structure with the following fields (see Figure 26.3):

- Family: This field defines the protocol group like IPV4, IPV6, or UNIX domain protocols.
- Type: This field defines the type of the socket. Socket type can be raw socket, stream socket, or datagram socket.
- Protocol: This field defines the protocol used. It can be TCP for connection oriented communication and UDP for connectionless communication.

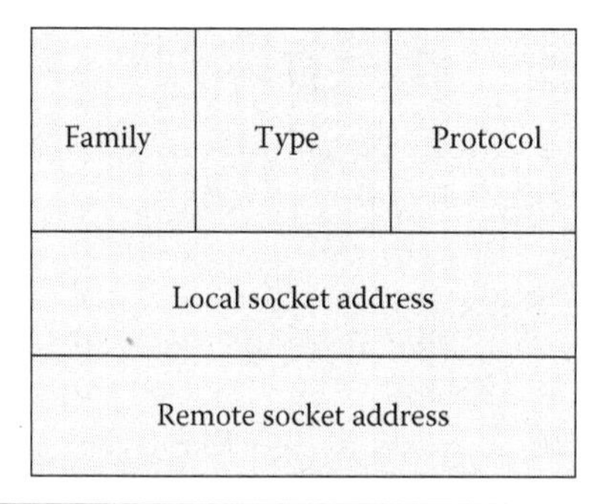


**Figure 26.3 Socket structure.**

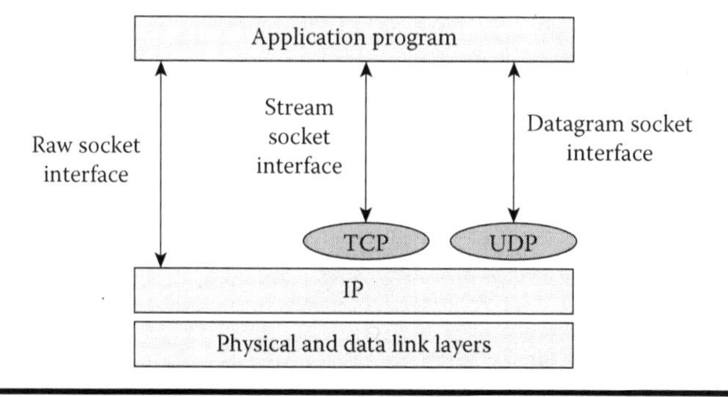


**Figure 26.4 Types of socket.**

- Local socket address: This field is a combination of the local IP address and the port address. It is the socket address of the local application program.
- Remote socket address: This field defines the combination of the remote IP address and the port address. It is the remote socket address of the remote application program.

There are three types of sockets (see Figure 26.4):

- Raw socket: Sockets that directly use the IP services are referred to as raw sockets. Neither the stream socket nor the datagram sockets are used by some protocols like ICMP or OSPF that directly uses the IP services of raw socket.
- Stream socket: Stream sockets are used with connection oriented protocol like TCP. It uses pairs of stream sockets to connect an application program with another application program on the network.
- Datagram socket: Datagram sockets are used with connectionless protocol like UDP. It sends messages from one application program to another application program on the network using pairs of datagram sockets.

The two types of transport protocols commonly used for socket communication:

- TCP provides a reliable, connection oriented, and flow controlled sequenced channel of communication. Stream sockets are used with connection oriented protocols like TCP.
- UDP is used for intermittent transmission. It is performed in an unreliable connectionless channel. Though UDP preserves the message boundaries, they cannot perform sequencing and delivery functions. Datagram sockets are used with connectionless protocols like UDP. They deliver datagram to other endpoints. To add reliability, specific measures have to be taken in programming to check for packet loss, to resend packets when required.

In programming, protocols are represented as numbers and not usually taken as constants. Below TCP or UDP, Internet protocol (IP) is used in the network layer for communication.

## 26.6 Basic Socket System Calls

1. socket: This function is used to create the socket of a given domain, type, protocol.

```
#include <sys/types.h>
#include <sys/socket.h>
int socket (int domain, int soc_type, int protocol);
```

// "domain" can be AF_INET or AF_UNIX.
// "soc_type" is the socket type. It can be SOCK_STREAM for stream socket and SOCK_DGRAM for datagram socket.
// "protocol" is usually set as zero. So the type defines the connection protocol within domain.

2. bind: This function is used to associate the socket id with an address to which other processes can connect.

```
#include <sys/types.h>
#include <sys/socket.h>
int bind (int socID, struct sockaddr *addrPtr, int
length);
```

// "socID" is the id of the socket.
// "addrPtr" is a pointer to the address family. Each domain has a different address family as defined in structure sockaddr.
// "length" is the size of *addrPtr.

3. listen: This function is used by the server to specify the maximum number of connection requests allowed. That is the server queue size for connections from client.

```
#include <sys/types.h>
#include <sys/socket.h>
int listen (int socID, int size);
```

// "socID" is the id of the socket.
// "size" is the number of pending connection requests allowed at the server.

4. accept: This function is used to identify the socket id and address of the client connecting to the server socket. Waits for an incoming request and when received creates a socket for it.

```
#include <sys/types.h>
#include <sys/socket.h>
int accept (int socID, struct sockaddr *addrPtr, int
*lengthPtr);
```

// "socID" is the id of the socket.
// "addrPtr" is a pointer to the address family.
// "lengthPtr" is the maximum size of address structures that can be called.

Based on the method of accepting connections, servers are classified into:

- Iterating server: In this type only one socket is opened by the server at a time. After completion of processing in that connection, the next connection is accepted. Having only a single connection usually leads to poor performance and utilization.
- Forking server: In this type, after the connection is accepted from a client, a child process is forked off to handle the connection. Forking allows multiple clients to be connected to the server with a separate forked process handling a client connection. If data sharing is required then forking should be implemented with multithreading.
- Concurrent single server: In this type, the server can simultaneously wait on multiple open socket ids. The server process wakes up only when new data arrives from the client. This reduces the complexity of threads implementations.

5. Send: This function is used to send data. The return value of the function indicates the number of bytes sent. The send function is used in connection oriented implementation.

```
#include <sys/types.h>
#include <sys/socket.h>
int send (int socID, const char *dataPtr, int length,
int flag);
```

// "socID" is the id of the socket.
// "dataPtr" is a pointer to the data.
// "length" is the number of data octets sent.
// "flag" is to specify if it is a normal message (value 0) or requires special priority.

6. Recv: This function is used to receive data. The return value of the function indicates the number of bytes received. The receive function is used in connection oriented implementation.

```
#include <sys/types.h>
#include <sys/socket.h>
int recv (int socID, char *dataPtr, int length, int flag);
```

// "socID" is the descriptor of the socket from which the data is to be received.
// "dataPtr" is a pointer to the data.
// "length" is the buffer size for data received.
// "flag" allows the caller to control details.

7. Sendto: This function is used to send data. The return value of the function indicates the number of bytes sent. The sendto function is used in a connectionless protocol implementation.

```
#include <sys/types.h>
#include <sys/socket.h>
int sendto (int socID, const void *dataPtr, size_t
dataLength, int flag, struct sockaddr *addrPtr,
socklen_t addrLength);
```

// "socID" is the id of the socket to which data is to be sent.
// "dataPtr" is a pointer to the data.
// "dataLength" shows the number of data octets.
// "flag" allows the sender to control details.
// "addrPtr" is a pointer to the address family.
// "addrLength" is the size of address structure.

8. Recvfrom: This function is used to receive data. The return value of the function indicates the number of bytes received. The recvfrom function is used in connectionless protocol implementation.

```
#include <sys/types.h>
#include <sys/socket.h>
int recvfrom (int socID, void *dataPtr, int dataLength,
int flag, struct sockaddr *addrPtr, int *addrLength);
```

// "socID" is the id of the socket to which data is to be sent.
// "dataPtr" is a pointer to the data.
// "dataLength" is the buffer size for data received.
// "flag" allows the caller to control details.
// "addrPtr" is a pointer to the address family.
// "addrLength" is the size of address structure.

9. Shutdown: This function is used to prevent sending or receiving data between sockets.

```
#include <sys/types.h>
#include <sys/socket.h>
int shutdown (int socID, int flag);
```

// "socID" is the socket on which send or receive needs to be disabled.
// "flag" is used to set the type of operation to be disabled.

10. Connect: This function is used by the client to connect to a server that is listening for connection.

```
#include <sys/types.h>
#include <sys/socket.h>
int connect (int socID, struct sockaddr *addrPtr, int
addrlen);
```

// "socID" is the id of the socket.
// "addrPtr" is a pointer to the address family.
// "addrlen" is the size of socket address.

11. close: This function is used to close connection corresponding to the socket descriptor and free the socket descriptor.

```
#include <sys/types.h>
#include <sys/socket.h>
int close (int sockID);
```

// "socID" is the id of the socket to be closed. It can be called by both the client and the server.

## 26.7 Client Server Implementation for Connection Oriented Communication

The following activities are performed by the client and server in connection oriented communication (see also Figure 26.5):

- Server creates a server socket.
- Server binds with the socket.
- Server listens for connection requests from client.
- Client creates a client socket.
- Client sends a connection request to server.
- Server accepts the request from client and connection is established between the client and server.
- Client and server exchange (send and receive) messages.
- Client closes the socket connection with server.
- Server can wait for connection request for another client or close the server socket.

The server can use only one well-known port at a time for communication, but there are several client connections open at the same time that requires the presence of several ports. To overcome this problem the server issues several ephemeral ports along with the well-known port. The connection made to the well-known

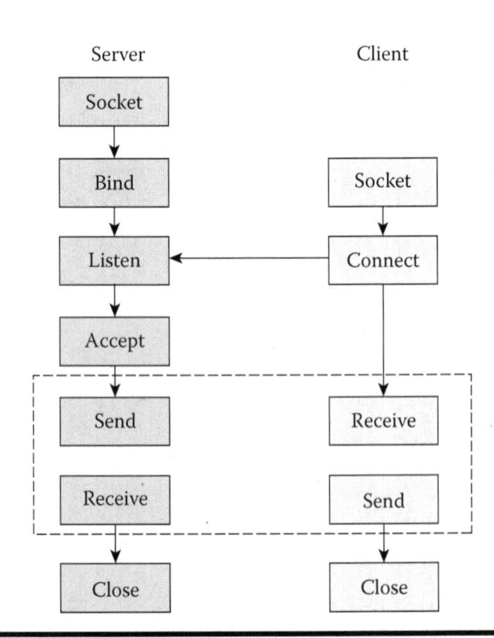


**Figure 26.5 Connection oriented communication.**

port is assigned to the temporary port so that the data transfer takes place between the temporary port at the client site and the temporary port at the server site. This helps to free the well-known port so that the server can make a connection with another client.

```
/* ===================================== *
* Connection Oriented (TCP) Server Program *
* ===================================== */


/* Include Header files */

#include <sys/types.h>
#include <sys/socket.h>
#include <netdb.h>
#include <netinet/in.h>
#include <stdio.h>
#include <string.h>

void main(void)
{

   int responseListenID;
   int responseAcceptID;
   socklen_t clientAddrLen;

   struct sockaddress serverAddr;
   struct sockaddress clientAddr;

      /* Create Socket */
   responseListenID = socket(AF_INET,SOCK_STREAM,0);

   memset (&serverAddr, 0, sizeof (serverAddr));

      /* define port (selectedPort) to use for listening */
   serverAddr.sin_family = AF_INET;
   serverAddr.sin_port = htons(selectedPort);
   serverAddr.sin_addr.s_addr = htonl(INADDR_ANY);

      /* Bind the socket */
   bind (responseListenID, &serverAddr, sizeof (serverAddr));

      /* Listen for connect request from client */
   listen (responseListenID, 1);
   clientAddrLen = sizeof(clientAddr);

      /* define infinite "for" loop to accept connections
      from client */
   for(; ;)
   {
      /* accept response from client */
```

```
    responseAcceptID = accept(responseListenID,&clientAddr,
    &clientAddrLen);
      /* fork to handle multiple client request as separate
      process*/
    pid = fork();
      if (pid>0)
      {
        /* this is the parent process of fork operation */
        close(responseAcceptID);
        continue;
      }
      else
      {
          /* this is the child process and needs to handle the
          specific client*/
        close(responseListenID);
         /*Implement read and write, for communication with
         client */
      close(responseAcceptID);
      }    /* if-else end here */
    }    /* for loop end here */
}
```

```
/* ==================================== *
* Connection Oriented (TCP) Client Program *
* ==================================== */

/* Include Header files */

#include <sys/types.h>
#include <sys/socket.h>
#include <netdb.h>
#include <netinet/in.h>
#include <stdio.h>
#include <string.h>

/* Define size of buffer */
#define DATABUF 512

void main(void)
{
    int commSocket;
    struct sockaddress hostAddr;
```

```
    char buf[DATABUF];
    struct hostent *hptr;

       /* Create Socket */
    commSocket = socket (AF_INET, SOCK_STREAM, 0);
    memset (&servAddr, 0, sizeof(hostAddr));
    hostAddr.sin_family = AF_INET;

       /* define port (selectedPort) to use connection */
    hostAddr.sin_port = htons(selectedPort);

       /* define domain name (address) */
    hptr = gethostbyname ("address");
    memcpy((char*)&hostAddr.sin_addr.s_addr,hptr->h_addr_
    list[0],hptr->h_length);

       /* connect to server */
    connect (commSocket, hostAddr, sizeof struct sockaddress);
    memset (buf, 0, DATABUF);

       /* Have a loop to send and receive information */
       while (gets(buf))
       {

       /* Implement read and write, for communication with
       server */

       }
 close (commSocket);
}
```

## 26.8 Client Server Implementation for Connectionless Protocol Communication

The connectionless iterative server processes a single request at a time and uses a connectionless transport layer protocol like UDP. During the process of accepting, processing, and sending the request the server does not attend to other packets.

The following activities are performed by the server in a connectionless protocol communication (see Figure 26.6):

- Server creates a server socket.
- Server binds with the socket.
- Server sends and receives messages with the client.
- Server closes the connection.

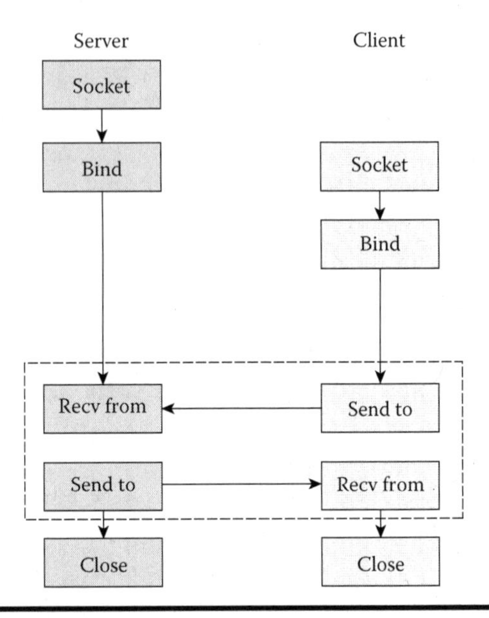


**Figure 26.6 Connectionless protocol.**

The following activities are performed by client in connectionless protocol communication:

- Client creates a client socket.
- Client binds with the socket.
- Client sends and receives messages with the server.
- Client closes the socket connection with server.

```
/* ==================================== *
* Connectionless (UDP) Server Program *
* ==================================== */


/* Include Header files */

#include <sys/types.h>
#include <sys/socket.h>
#include <netdb.h>
#include <netinet/in.h>
#include <stdio.h>
#include <string.h>
```

```
/* Define size of buffer */
#define DATABUF 512

void main(void)
{
   char buf [DATABUF];
   int socketID;
   socklen_t addrLen;

   struct sockaddress serverAddr;

      /* Create Socket */
   socketID = socket (AF_INET, SOCK_DGRAM, 0);
   memset (&serverAddr, 0, sizeof (serverAddr));

   serverAddr.sin_family = AF_INET;

      /* define port (selectedPort) to use connection */
   serverAddr.sin_port = htons (selectedPort);

   serverAddr.sin_addr.s_addr = htonl (INADDR_ANY);

      /* Bind the Socket */
   bind (socketID, &serverAddr, sizeof (serverAddr));

   addrLen = sizeof (serverAddr);
   memset(buf, 0, DATABUF);

      /* Define infinite loop to send and receive data from
      client
   for(; ;)
   {
      /* The while loop is invoked when data is received */
      while(recvfrom (socketID, buf, MAXBUF, 0, &clientAddr,
      &addrLen)>0)
      {
         /* Implement function to process the received
         information */

         /* Send data to the client */
         sendto (socketID, buf, MAXBUF, 0, &clientAddr, &addrLen);
         memset (buf, 0, DATABUF);

      } /* end of while loop */

   } /* end of for loop */

/* Close the socket connection */
close(socketID);

}
```

```
/* ====================================== *
* Connectionless (UDP) Client Program *
* ====================================== */

/* Include Header files */

#include <sys/types.h>
#include <sys/socket.h>
#include <netdb.h>
#include <netinet/in.h>
#include <stdio.h>
#include <string.h>

/* Define size of buffer */
#define DATABUF 512

void main (void)
{
   char buf [DATABUF];
   int socketID;

   socklen_t remoteAddrLen;
   struct sockaddress remoteAddr;
   struct hostent *hptr;

      /* Create Socket */
   socketID = socket (AF_INET, SOCK_DGRAM, 0);
   memset (&remoteAddr, 0, sizeof (remoteAddr));

   remoteAddr.sin_family = AF_INET;

      /* define port (selectedPort) to use connection */
   remoteAddr.sin_port = htons (selectedPort);

      /* define domain name (address) */
   hptr = gethostbyname ("address");

   memcpy ((char*) &remoteAddr.sin_addr.s_addr, hptr->h_addr_
   list[0], hptr->h_length);

   memset (buf, 0, DATABUF);
   remoteAddrLen = sizeof (remoteAddr);

/* No need to bind if communication is not peer-to-peer */

   while (gets(buf))
   {
```

```
        /* send information to server */
        sendto (socketID, buf, sizeof (buf), 0, &remoteAddr,
        sizeof(remoteAddr));
        memset (buf, 0, sizeof (buf));

        /* receive information from server */
        recvfrom (socketID, buf, DATABUF, 0, &remoteAddr,
        &remoteAddrLen);
        printf ("%s\n",buf);
        memset (buf, 0, sizeof(buf));

    } /* end of while loop */

/* Close the socket connection */
close(socketID);

}
```

## 26.9 Conclusion

Most legacy developments on NMS are client–server based and these solutions usually use socket communication for interaction. In the client–server model only when intermittent communication is involved the transmission mechanism is UDP based and when a continuous reliable connection is required, TCP is used.

This chapter gives the reader a basic understanding of socket programming by introducing the reader to the basic terminologies. The sample code given in this chapter is intended to bridge the gap between theory and implementation, by first explaining the program and then supplementing the discussion with code. Comments have been added in the code, so that the reader can easily follow the program execution.

## Additional Reading

1. Lincoln D. Stein. *Network Programming with Perl.* United Kingdom: Addison-Wesley Professional, 2001.
2. W. Richard Stevens, Bill Fenner, and Andrew M. Rudoff. *Unix Network Programming, Volume 1: The Sockets Networking API.* 3rd ed. United Kingdom: Addison-Wesley Professional, 2003.
3. Elliotte Rusty Harold. *Java Network Programming.* 3rd ed. California: O'Reilly Media Inc., 2009.
4. Michael Donahoo and Kenneth Calvert. *TCP/IP Sockets in C: Practical Guide for Programmers.* San Francisco, CA: Morgan Kaufmann, 2000.
5. Richard Blum. *C# Network Programming.* Sybex, California: Sybex, 2002.

# Chapter 27

# RPC Programming

This chapter is about the remote procedure call (RPC) programming. It introduces the reader to the concept of writing RPC client–server code. Remote procedure call defines a powerful technology for creating distributed client–server programs. The RPC runtime libraries manage most of the details relating to network protocols and communication. With RPC, a client can connect to a server running on another platform. For example, the server could be written for Linux and the client could be written for Win32. The RPC is for distributed environment.

## 27.1 Introduction

Programs communicating over a network need a paradigm for communication. The RPC is used in a distributed environment where communication is required between heterogeneous systems. It follows the client–server model of communication. In a remote procedure call based communication a client makes a procedure call to send a request to the server. The arrival of the request causes the server to dispatch a routine, perform the service the client has requested, and send back the response to the client.

The machine implementing the set of network services is called the server and the machine that requests for the service is the client. The server can support more than one version of a remote program. The control moves through two processes in the remote procedure call model. They are the caller process and server process. The caller process sends a message containing the procedure parameters to the server process and waits for a reply message that contains the procedure results. On arrival of the result, the caller process resumes execution. The server side process is dormant until it gets a call message. When the server process gets a request, it will process the same and send a response. It can be seen that only one of the two processes is active at any given time in the RPC model.

The RPC call model can support both synchronous and asynchronous calls. Asynchronous calls permit the client to perform useful work while waiting for the reply from the server. To ensure effective utilization of the server, the server can create a task to process an incoming request, so that it will be free to receive other requests. A collection of one or more remote programs, implementing one or more remote procedures such that the procedures, their parameters, and results are documented in the protocol specification of the program is called a RPC service. The network client will initiate remote procedure calls to access the services.

In order to be forward compatible with changing protocols, a server may support more than one version of a remote program. The caller places arguments to a procedure in a well-specified location and transfers control to the procedure in the local model. As the control is gained by the caller it extracts the results of the procedure from the well-specified location and continues execution. This chapter gives an overview on RPC with emphasis on programming. The XDR (eXternal Data Representation) has been used in examples as the presentation layer to wrap data in a format that both the client and server can understand with the RPC model that involves data transfer between heterogeneous systems.

## 27.2 RPC Overview

The RPC is a powerful technique for constructing distributed, client–server based applications that are based on extending the notion of conventional local procedure calling. In RPC, the called procedure need not exist in the same address space as the calling procedure. The two communicating processes may be on the same system or on different systems with a network connecting them. RPC allows programming in a distributed environment avoiding the details of the interface with the network.

The RPC is transport independent and can be used in a variety of applications facilitating isolation of the application from the physical and logical elements of the data communications mechanism. The client–server modeling with RPC is easier to program and offers powerful computing (see Figure 27.1). Clients make remote calls through a local procedure interface when RPC is combined with the ONC RPCGEN protocol compiler. RPC, being analogous to function calls, passes the calling arguments to the remote procedure and the caller waits for a response to be returned from the remote procedure.

During an RPC call, the client program first calls the RPC function. The server will execute the request and call the dispatch routine for the specific service the client has requested. The server will then execute the service and once the execution is complete, the response is sent to the client. So the client makes a procedure call sending a request to the server and waits. The client thread is blocked from

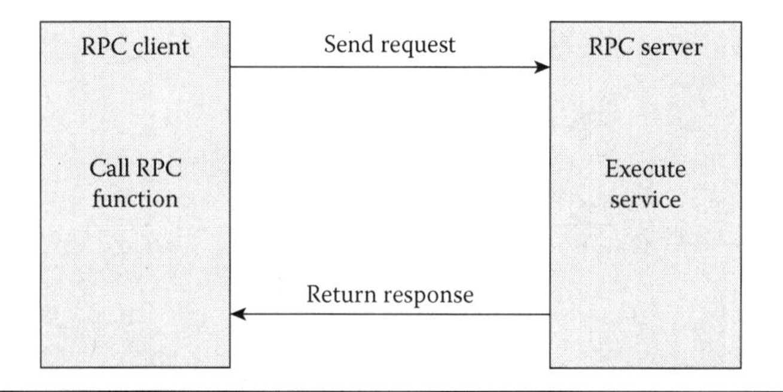


**Figure 27.1 RPC Client–server model.**

processing until either a reply is received or the request times out and the client program continues as soon as the RPC call is completed.

A remote procedure is identified by a combination of program number, version number, and procedure number. In a group of related remote procedures, each remote procedure is identified by a unique procedure number in the program. The program is further identified using the program number, which is unique. Each version of a program consists of a collection of procedures that are available to be called remotely. Multiple versions of an RPC protocol can be made available using the version numbers.

In order to develop an RPC application, first the protocol for the client–server communication needs to be defined. This is followed by the development of the client program and the server program. The programming aspects to be looked into for RPC development are discussed in this chapter.

## 27.3 RPC Concepts

### *27.3.1 Transport*

RPC protocol is a message protocol that is independent of the transport protocol and is specified using XDR language. The RPC protocol concerns only with the specification and interpretation of messages. Reliability of communication is not guaranteed with RPC and this is dependent on the transport protocol. Most of the issues in reliability can be eliminated if the application runs on a connection oriented transport such as TCP/IP. When an unreliable connectionless transport like UDP is used, the application needs to implement its own retransmission and timeout policy. The RPC protocol does not attach specific semantics to the remote procedures or their execution so as to ensure transport independence.

Consider RPC running on top of an unreliable transport and application retransmits RPC messages after short timeouts and receives no reply, then it can be inferred that the procedure was executed zero or more times. If a reply is received, it can be inferred that the procedure was executed at least once. The transaction ID is used primarily by the client RPC layer to match replies to requests and the client application may choose to reuse its previous transaction ID when retransmitting a request.

The server cannot examine this ID in any other way except as a test for equality. A server takes advantage of the transaction ID that is packaged with every RPC request to ensure that the procedure is executed at least once. The transaction ID can also ensure that a previously granted request from a client does not get granted again. If the application runs on a reliable transport, it can infer that the procedure was executed exactly once from the reply message. If no reply message is received, it cannot assume the remote procedure was not executed. An application needs timeouts and reconnection to handle server crashes even with a connection oriented protocol like TCP.

## 27.3.2 Presentation

The XDR is a popular presentation layer in implementing RPC applications. XDR can be used to transfer data between diverse computer architectures and enables communication between diverse machines like Sun Solaris, VAX, or AIX from IBM. For communication between heterogeneous systems, a common format that is understood by both systems is required. XDR enables RPC to handle arbitrary data structures, in a machine independent manner regardless of the byte order or structure layout conventions used. This machine dependency is eliminated by converting the data structures to XDR before sending (see Figure 27.2).

Using XDR, any program running on any machine can create portable data by translating its local representation into the XDR representation and vice versa. Data conversion with XDR involves two tasks. They are:

- Serializing: This is the process of converting from a particular system format to XDR format.
- De-serializing: This is the process of converting from a XDR format to a particular system format.

Intricate data formats can also be described using XDR language. Using a data description language like XDR provides less ambiguous descriptions of data.

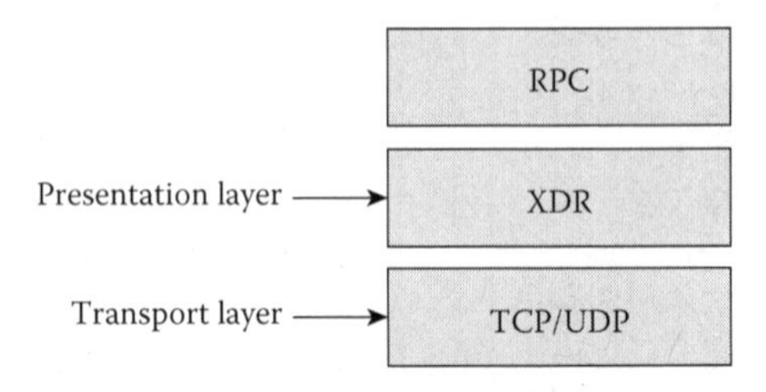


**Figure 27.2 Sample protocol stack for RPC implementation.**

Moreover XDR is easy to understand and implement. There is a close analogy between XDR types and types defined in other high level languages. On-the-fly data interpretation can be performed across machines as the language specification itself is an ASCII string.

### 27.3.3 RPCGEN Compiler

The rpcgen compiler automates the process of writing RPC applications. With rpcgen the overall development time involved in coding and debugging low-level routines is reduced. This way the programmer can work on the functionality to be implemented in the remote procedures without worrying about the network interface code. The rpcgen accepts remote program interface definitions written in the RPC language and converts them into C language output.

The output of rpcgen includes:

- A header file for the common definitions.
- A remote proxy of the client routines called the stub.
- A remote proxy of the server routines called the skeleton.
- XDR filter routines for parameters and results.
- Stub routines prototyped in C programming language.

In the absence of rpcgen, the programs written by the programmer has to make local procedure calls to the client skeletons using remote program. The rpcgen compiler allows programmers to mix low-level code with high-level code. After writing the server procedures, the main program can be linked with the server skeletons to produce an executable server program. Handwritten routines can be easily linked with the rpcgen output.

The client skeletons interface with the RPC library hides the network from its caller. The server skeleton hides the network from server procedures that are to be invoked by remote clients. The most important file for conversion when using rpcgen is the protocol definition file. This file has a description of the interface of the remote procedures. It also maintains function prototypes and definition of any data structures used in the calls. These definitions will include both argument types and return types that need to be converted. The protocol definition file can also include "C" (programming language) code shared by client and server.

```
/* ===================
     Before Conversion
=================== */

type1 PRO1(operands1)
type2 PRO2(operands2)
```

```
/* ===================
     After Conversion
       Client Stub
=================== */

type1 *pro_1(operands1 *, CLIENT *);
type2 *pro_2(operands2 *, CLIENT *);

/* ===================
     After Conversion
     Server Procedure
=================== */

type1 *pro_1_svc(operands1 *, struct svc_req *);
type2 *pro_2_svc(operands2 *, struct svc_req *);
```

### 27.3.4 RPC Procedure Identification

An RPC call message identifies the procedure to be called using the three fields:

- Program number: There can be multiple programs running on the remote host and each of the programs is identified by a unique number. Once a program number is administered by some central authority it can be implemented as a remote program.
- Program version number: The version field of the call message identifies the version of the RPC protocol being used by the caller. Presence of protocol version makes it possible to speak old and new protocols through the same server process.
- Remote procedure number: The procedure number documented in the specific program's protocol specification identifies the procedure to be called.

```
/* =============
     Example
============= */

program TEST_PROG
{
      version TEST_VERSION
      {

      type1 PRO1(operands1) = 1;
      type2 PRO2(operands2) = 2;

      } = 3;

} = 2000055;
```

```
// In this example:
// >> the program with name "TEST_PROG" has program number
2000055
// >> the version with name "TEST_VERSION" has version
number 3
// >> the procedure with name "PRO1" has program number 1
```

### 27.3.5 Port Mappers

Servers that convert RPC program numbers into universal addresses (IP port numbers) are called port mappers. When an RPC server is started, it tells the port mapper the port number it is listening to and what RPC program numbers it is prepared to serve. When a client wants to make an RPC call to a given program number, it first checks the port mapper on the server machine to determine the port number where RPC packets should be sent for accessing the specific service. The port mapper program maps RPC program and version numbers to transport-specific port numbers, which enables dynamic binding of remote programs.

This mapping is desirable because the range of reserved port numbers is very small and the number of potential remote programs is very large. The port mapper also aids in broadcast RPC. A given RPC program will usually have different port number bindings on different machines, so there is no way to directly broadcast to all of these programs. To broadcast to a given program, the client actually sends its message to the port mapper that has a fixed port number. The port mapper picks up the broadcast and calls the local service specified by the client. When the port mapper gets the reply from the local service, it sends the reply back to the client.

### 27.3.6 Broadcasting

In broadcast RPC-based protocols, the client sends a broadcast packet to the network and waits for numerous replies. Broadcast RPC uses unreliable, packet-based protocols like UDP as its transport. Servers that support broadcast protocols respond only when the request is successfully processed without any specific pointers on the errors. Broadcast RPC uses the port mapper to convert RPC program numbers into port numbers.

When normal RPC expects one answer, the broadcast RPC expects many answers from each responding machine. A major issue with broadcast RPC is that when there is a version mismatch between the broadcaster and a remote service, the broadcast RPC never detects this error.

Also services need to first register with the port mapper to be accessible using the broadcast RPC mechanism. Broadcast requests are limited in size to the maximum transfer unit (MTU) of the local network.

## 27.3.7 Batching

Batching allows a client to send a sequence of call messages to a server. It uses reliable byte stream protocols like TCP for transport. The client never waits for a reply from the server, and the server does not send replies to batch requests. A sequence of batch calls is usually terminated by a RPC to flush the pipeline. Since the server does not respond to every call, the client can generate new calls in parallel, with the server executing previous calls. In addition, the reliable connection oriented implementation can buffer many calls and send them to the server in a single write system call.

The RPC architecture is designed so that a client sends a call to a server and waits for a reply that the call has succeeded. Clients do not compute while servers are processing a call, which is inefficient if the client does not want or need an acknowledgment for every message sent. The RPC batch facilities make it possible for clients to continue computing while waiting for a response.

## 27.3.8 Authentication

Identification is the means to present or assert an identity that is recognizable to the receiver. Authentication provides the actions to verify the truth of the asserted identity. Once the client is identified and verified, access control can be implemented. Access control is the mechanism that provides permission to allow the requests made by the user to be granted, based upon the user's authentic identity. Access control is not provided in RPC and must be supplied by the application. Several different authentication protocols are supported in RPC. A field in the RPC header indicates what protocol is being used. The RPC protocol provides the fields necessary for a client to identify itself to a service and vice versa. The call message has two authentication fields, the credentials and verifier. The reply message has one authentication field, the response verifier.

The RPC protocol specification defines all three fields as the following *opaque* type:

```
enum auth_flavor {
         AUTH_NONE = 0,
         AUTH_UNIX = 1,
         AUTH_SHORT = 2,
         AUTH_DES = 3,
         AUTH_KERB = 4
         /* and more to be defined */
};

struct opaque_auth {
auth_flavor flavor;
.....
};
```

Here *opaque_auth* structure is an *auth_flavor* enumeration followed by bytes that are opaque to the RPC protocol implementation. The interpretation and semantics of the data contained within the authentication fields is specified by individual, independent authentication protocol specifications. When authentication parameters are rejected, the response message contains information stating why they were rejected.

In scenarios where the caller is not aware of the authentication parameters and the server does not want to perform any specific authentication, then AUTH_NULL (0) can be used. In this case, the *auth_flavor* value of the RPC message credentials and verifier, and response verifier correspond to null authentication. The caller of a remote procedure may want to identify itself as it is identified on a trusted UNIX system. In this case AUTH_UNIX (1) is used.

## 27.4 Defining the Protocol

The easiest way to define and generate the protocol is to use a protocol complier such as rpcgen. For the protocol you must identify the name of the service procedures, and data types of parameters and return arguments. The protocol compiler reads a definition and automatically generates client and server stubs. The rpcgen uses its own language (RPC language or RPCL), which looks very similar to preprocessor directives. The rpcgen exists as a stand-alone executable compiler that reads special files denoted by ".x" extension. To compile a RPCL file, simply do

```
>> rpcgen example.x
```

This will generate four possible files:

- example_clnt.c: This is the client proxy file.
- example_svc.c: This is the server stub file.
- example_xdr.c: This is XDR file with filters.
- example.h: This is the header file needed for any XDR filters.

The external data representation (XDR) is a data abstraction needed for machine independent communication. The client and server need not be machines of the same type. Simple NULL-terminated strings can be used for passing and receiving the directory name and directory contents. The program, procedure, and version numbers for client and servers can be set using rpcgen or relying on predefined macros in the simplified interface.

Program numbers are defined in a standard way:

- 0*x*00000000 - 0*x*1*FFFFFFF*: Defined by Sun
- 0*x*20000000 - 0*x*3*FFFFFFF*: User Defined

- 0*x*40000000 - 0*x*5*FFFFFFF*: Transient
- 0*x*60000000 - 0*xFFFFFFFF*: Reserved

## 27.5 Interface Routines

Some of the RPC interface routines are:

- rpc_reg(): This routine will register a procedure as an RPC program on all transports of the specified type.
- rpc_broadcast(): This routine will broadcast a call message across all transports of the specified type.
- rpc_call(): This routine is used to make remote calls to a procedure on the specific remote host.
- clnt_create(): This is the client creation routine, called with details of where the server is located and transport type to use.
- clnt_tp_create(): This routine is used to create the client handle for the specified transport.
- clnt_call(): This routine is used by the client to send a request to the server.
- clnt_create_timed(): This routine allows the programmer to specify the maximum time allowed for each type of transport tried during the creation attempt.
- svc_create(): Server handles for all transports of the specified type are created using this interface routine.
- svc_tp_create(): This routine is used to create server handle for the specified transport.
- rpcb_set(): This routine is used to set a map between an RPC service and a network address.
- rpcb_unset(): This routine is used to unset a map between the RPC service and network address created using set routine.
- rpcb_getaddr(): This routine is used to get the transport addresses of specified RPC services.
- svc_reg(): The program and version number pair are associated with the specified dispatch routine using this interface routine.
- svc_unreg(): The association set using svc_reg() is deleted using svc_unreg().

## 27.6 RPC Implementation

1. Files that are required for executing RPC server and client:
   - Interface file: Defining the interface details. These include the input/output structure and the program, version, and procedure number.

- Client proxy file and server stub file: These can be generated using protocol compilers, mibdl.exe for windows and rpcgen for unix.
- Interface header: Also generated by a protocol compiler from the interface file. Add this header in both client and server code.
- Client and server code: To make a handle for communication, registering procedures, and so on.
- XDR file: –Unix uses XDR (eternal data representation) in the presentation layer for RPC. This file is also created by protocol compiler.

2. First let us create the interface file (msg.x):

```
// Structure for input
struct nodeRequest{
short nodeNumber;
};
//Structure for output
struct nodeResponse{
short nodeState;
};
//RPC details
program MESSAGEPROG {
     version PRINTMESSAGEVERS {
     nodeResponse getState (nodeRequest) = 1;
     } = 1;
} = 0x20000055;
// The procedure "getState" takes "nodeRequest" as the
input parameter and gets "nodeResponse" as the output
parameter
```

3. Using rpcgen: Protocol compiler have tags to perform additional functions. With the options used, all the basic files are created.

```
>>rpcgen -N -a msg.x
```

Output files:

```
-rw-r--r--   1 jithesh root   210 Aug 16 23:38 msg.x   // IDF
-rw-r--r--   1 jithesh root   597 Aug 16 23:38 msg_xdr.c // XDR File
-rw-r--r--   1 jithesh root   4398 Aug 16 23:38 msg_svc.c // Server stub
-rw-r--r--   1 jithesh root   671 Aug 16 23:38 msg_clnt.c // Client Proxy
-rw-r--r--   1 jithesh root   606 Aug 16 23:38 msg.h // IDF header
```

-rwxr-xr-x 1 jithesh root 1063 Aug 16 23:38 makefile.msg* // Make file
-rw-r--r-- 1 jithesh root 1148 Aug 17 00:20 msg_client.c // Client code
-rw-r--r-- 1 jithesh root 459 Aug 17 00:20 msg_server.c // Server code

4. Modify client code (msg_client.c):

```
nodeResponse *result_1; // already available code
nodeRequest getstate_1_arg1;     // already available code

// Add the code below after the code specified above
which is already available in the output of rpcgen

getstate_1_arg1.nodeNumber = 3;

     switch(result_1->nodeState)
     {
     case 1: printf("%s Node State is Active \n");break;
     case 2: printf("%s Node State is Busy \n");break;
     default: printf("%s Node State is Faulty \n");break;
     }

// The response message from server is of type short.
// Each state of the node of type short maps to an
actual state.
```

5. Modifying server code (msg_server.c)

```
static nodeResponse result; // already available code

// Add the code below after the code specified above
which is already available in the output of rpcgen

result.nodeState = 1;

// Here the server is always returning the node state as
"1" corresponding to Active
```

6. Compile/linking:

```
/usr/local/bin/gcc -c msg_xdr.c
/usr/local/bin/gcc msg_client.c msg_clnt.c msg_xdr.o -o
client -lnsl
/usr/local/bin/gcc msg_server.c msg_svc.c msg_xdr.o -o
sever -lnsl
```

Output:
-rwxr-xr-x 1 jithesh root 8096 Aug 17 00:20 client*
-rwxr-xr-x 1 jithesh root 11720 Aug 17 00:20 sever*

7. Some points to be noted:
   - Just like socket programming here connection handling is involved also.

- We need to create handle, register rpc, destroy handle after use. The protocol compiler generates code for this or the programmer can write code for the same.
- The programmer needs to update xdr_free to release memory, though this is not mandatory for the program to execute.

**8.** Testing

```
# ./sever

# ps -ef| grep sever
   root 2894   1 1 10:42:26?        0:00 ./sever
   root 2994 10839 0 10:42:33 pts/4   0:00 grep sever

// This shows that the server is running with process id
2894
# ./client
usage:  ./client server_host

# ./client 122.102.201.11
 Node State is Active

// 1. Client connects to server
// 2. Calls the getState procedure
// 3. Gets response as "1"
// 4. Prints string "Node State is Active" corresponding
to "1" in switch condition
```

## 27.7 Conclusion

Remote procedure call defines a powerful technology for creating distributed client-server programs. The RPC runtime libraries manage most of the details relating to network protocols and communication. With RPC, a client can connect to a server running on another platform. For example, the server could be written for Linux and the client could be written for Win32. This makes RPC programming different from normal socket based communication.

It is to be noted that RPC in not quite common in working on Web pages when using the Internet. One of the most important concepts to understand about the Web services framework (WSF) is that it is not a distributed object system. Web services communicate by exchanging messages, more like JMS than RMI. The WSF doesn't support remote references, remote object garbage collection, or any of the other distributed object features developers have come to rely upon in RMI, CORBA, DCOM, or other distributed object systems. Web communication deals with a browser type of client process and Web server type of server process (socket programming). The RPC is for distributed environment. This concluding paragraph is intended to give the reader an understanding on how RPC is different from socket programming that is discussed in the preceding chapter and Web communication that will be discussed in the next chapter.

## Additional Reading


1. Per Brinch Hansen. *The Origins of Concurrent Programming: From Semaphores to Remote Procedure Calls.* New York: Springer, 2002.
2. John Bloomer. *Power Programming with RPC.* California: O'Reilly Media, Inc., 1992.
3. David Gunter, Steven Burnett, Gregory L. Field, Lola Gunter, Thomas Klejna, Shankar Lakshman, Alexia Prendergast, Mark C. Reynolds, and Marcia E. Roland. *Client/Server Programming With RPC and DCE.* United Kingdom: Que, 1995.
4. Guy Eddon. *RPC for NT Building.* United Kingdom: CMP, 2007.

# Chapter 28

# Web Communication

This chapter is about Web-based programming. It introduces the reader to the concept of writing programs for the World Wide Web (www). It is quite common in current telecom management industry to have Web-based NMS, EMS, and OSS/BSS solutions. Programming for the distributed environment on the Internet needs an understanding of the underlying concepts. The basic socket programming principles are reused though the communication framework is changed.

## 28.1 Introduction

The World Wide Web (WWW) is a flexible, portable, user-friendly service that is a repository of the information spread all over the world yet linked together using the Internet framework. The World Wide Web was initiated by CERN (European Laboratory for Particle Physics) for handling the distributed resources for scientific research. It has now developed into a distributed client-server service where a client can access the server service using a browser. Web sites are the locations where the services are distributed.

Information published on the Web or displayed on a Web client like a browser is called a page. The Web page is a unit of hypertext or hypermedia available on the Web. The hypertext stores the information as document sets (text) that are linked through pointers. A link can be created between documents, so that the user browsing through a document can move to another document by clicking the link to another document. Hypermedia contains graphics, pictures, and sound. Homepage is the main page for an individual or organization. A server can provide information on a topic in one or more Web pages. Being a single server, this is an undistributed environment. Multiple servers can also provide the information on a topic in multiple pages. When multiple servers are involved it is a distributed environment.

Solutions published on the Web give the user the flexibility of accessing information anytime and anywhere. The Web server will host the EMS, NMS, or OSS/BSS application. Any user who is authorized to work on the application can access the management application using a Web browser. This makes it easy for the user, as there need not be a custom management client on the user's machine to access the application on the server. Web-based solutions are becoming increasingly popular in the current industry. Some of the big players in network management like IBM, HP, and AdventNet have all offered Web-based NMS solutions.

This chapter provides the reader an understanding of the World Wide Web. The HTTP for communication and the browser architecture for the Web client are discussed in detail. This chapter ends with a sample implementation of a Web server. The Web server implementation is written in C# programming language. C Sharp (C#) programming language was developed by Microsoft as part of the .Net initiative. The program is supplemented with sufficient comments to ensure that a person who is not familiar with this programming language understands the steps in developing a Web server.

## 28.2 Hypertext Transfer Protocol (HTTP)

The data on the World Wide Web can be accessed using the HTTP. This protocol transfers data in plain text, hypertext, audio, video, and other formats. HTTP follows the client server model. The data is transferred between client and server similar to a socket-based communication. HTTP messages are read and interpreted by the HTTP server and HTTP client.

The HTTP can be considered a combination of SMTP and FTP based on the properties it shares with these communication protocols. In HTTP the request messages from the client to the server and the response messages from the server to the client carry data in a letter form with a MIME-like format as used in SMTP. However, HTTP messages are delivered quickly compared to SMTP messages that are stored and then forwarded. HTTP is similar to FTP in that it transfers files and uses TCP services. However HTTP uses a single TCP connection, which is not the case in FTP.

In HTTP, data is transferred between client and server and there is no separate control connection. It is a stateless protocol although it uses the TCP services. The HTTP request and response messages follow the same format. The client sends a request message to the server thereby initiating the transaction. The server sends the response message back to the client (see Figure 28.1).

The HTTP request message has a request line (see Figure 28.2). The request line has the request type, uniform resource locator (URL), and HTTP version. The

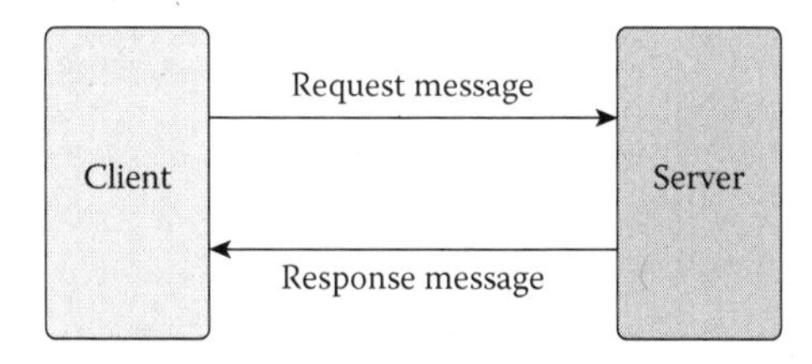


**Figure 28.1 HTTP request-response model.**

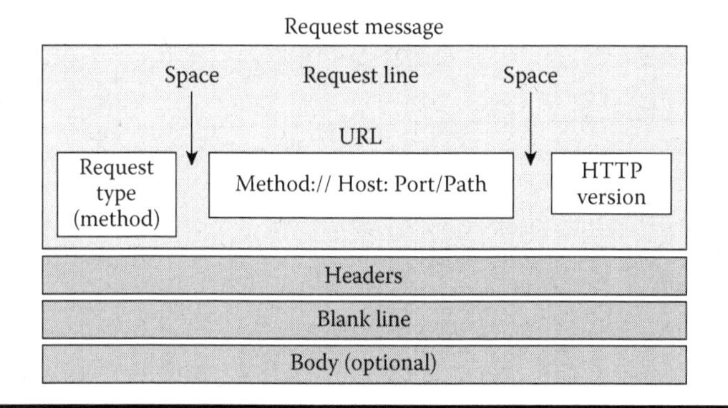


**Figure 28.2 HTTP request message.**

request type categorizes the request into various methods, the URL is a standard for specifying any information on the Internet facilitating document access from the World Wide Web, and version indicates the HTTP version used. Version 1.1 is the latest version of HTTP. The older versions of HTTP like the 1.0 and 0.9 are also in use.

The URL is not limited to use in HTTP. It is a Web address defining where a resource is available. The URL in the request line defines the following:

- Method: The method protocol is used to retrieve a document. Some of the example method protocols are FTP and HTTP.
- Host: The Web server that stores the Web pages or the information to be published is called the host. The hosts are given alias names that usually begin with characters "www."
- Port: The Port number of the server is optional in the URL. If present, it is placed between the host and the path such that it is separated from the host with a semicolon. For example, request line "http://ww.crmapp:8050" has 8050 as the port number.
- Path: This is the path to the file where the information is located. The path has details of the directories including the subdirectories and files to access the file.

The request type field present in the request message defines different messages (methods). Some of these methods are:

- GET: The GET method is the main method for retrieving a document from the server. When a GET request is issued by the client, the server responds with the document contents in the body of the response message.
- POST: Information can be provided by the client to the server using the POST method.
- HEAD: By using HEAD method it is possible to get information about a document. The server response to the HEAD method does not contain a body.
- LINK: Document can be linked to another location by specifying the file location in the URL part of the request line and the destination location in the entity header using the LINK method.
- UNLINK: Links created by the LINK method are deleted by the UNLINK method.
- PATCH: This method is a request for a new or replacement document. The PATCH method contains the list of differences that should be implemented in the existing file.
- MOVE: File location can be moved by specifying the source file location in the URL part of the request line and the destination location in the entity header using the MOVE method.
- OPTION: Information on the available options is requested by the client to the server using the OPTION method.
- DELETE: Document on the server can be deleted using the DELETE method.
- COPY: A file can be copied to another location by specifying the source file location in the URL part of the request line and the destination location in the entity header using the COPY method.

The HTTP response message (see Figure 28.3) has a status line that includes HTTP version, status code, and the status phrase. The HTTP version in the status line and the request line are the same. A three digit code representing the status is the status code and the status code is explained in text form by the status phrase.

Information can also be exchanged between the client and the server using headers. The client can request the document be sent in a special format or the server can send additional information about the document using headers. One or more header lines make up the header and each header line consist of the header name, a colon, a space, and a header value. A header line belongs to the following four categories:

- General header: General information about the message are provided by the general header. The general header can be present in both request and response.

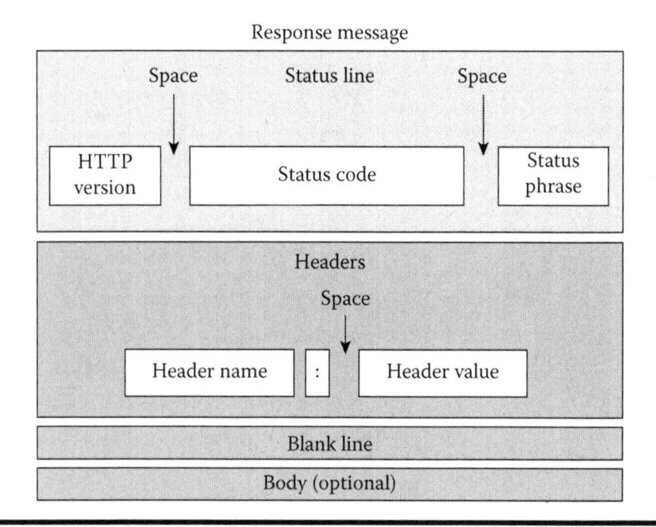


**Figure 28.3 HTTP response message.**

- Request header: The client's configuration and its preferred document format are specified by the request header. The request header is present only in the request message.
- Response header: The server's configuration and its special information about the request are specified by the response header. The response header is present only in the response message.
- Entity header: Information about the document body is provided by the entity header. It is usually present in response messages. Request messages like POST method that has a body can also uses this header.

The request messages contain only general header, request header, and entity header and the response messages contain only general header, response header, and entity headers.

An important aspect of HTTP interaction is the proxy server. It is a server that keeps the copies of responses to recent requests. When a request is received from the HTTP client, the proxy server first checks its cache and sends the request to the corresponding server only if the response is not stored in the cache. This helps to reduce the traffic and the load on the original server as well as to improve the latency. The proxy server thus collects the incoming response and stores it for future requests from other clients. The client can use the proxy server only if it is configured to access the proxy.

The HTTP connection can be a persistent or nonpersistent connection. Let us discuss the difference between these connections and the HTTP version that supports the specific connection.

- Persistent connection: This connection is terminated on the basis of a client request or if a timeout has been reached. So here the connection remains open for more requests after the server sends the response. The server sends the data requests with each response. In cases where the data is created dynamically, the server does not know the data length. To handle this scenario, the server closes the connection after sending the data facilitating the client to understand that the end of the data has been reached. Persistent connection is the default in HTTP version 1.1.
- Nonpersistent connection: In this connection, the server closes the connection after sending the response to the client, the data is read until the client encounters the end-of-file marker. A single TCP connection is opened for each request or response in the nonpersistent connection. So a TCP connection is opened by the client when a request is to be sent. The nonpersistent connection is supported in HTTP version 1.0. These connections impose high overhead on the server and needs a slow start procedure each time the connection is opened.

## 28.3 Web Client and Server

The browser is a generic Web client application used to retrieve, present, and traverse information on the World Wide Web, private information hosted on a Web server, or data resources in file servers. A browser has three parts: the controllers, client programs, and interpreters. The inputs from the keyboard/mouse are received by the controller. The documents are accessed through client programs and the documents are displayed on the screen using interpreters. Some of the most popular Web browsers are Windows Internet Explorer, Mozilla Firefox, Google Chrome, and Opera.

The following are the Web-based document types:

- Static documents: The documents that are created and stored inside the server are called static documents. The server can change the document contents but the user cannot change it. A document copy is sent to the client through document access and the document is displayed using a browsing program by the user.
- Dynamic documents: The Web server runs an application program for the creation of a dynamic document on the arrival of a request. Here the server checks the URL to find if it defines a dynamic document. If so, the program is executed by the server and the program output is send to the client. These documents do not exist in a predefined format. The program output is returned to the browser that requested the document as a response. The dynamic document content differs from one request to another as it creates fresh document for each requests.

- Active documents: Programs that must be run at the client site comes under active documents. For a program that creates an animated graphics on the screen or a program that interacts with the user, requires the program to be run at the client site where the animation or interaction takes place. Active documents are not run on the server. The server only stores the active document in the form of binary document.

Next let us look into the programming aspect of each of these document types.

1. Programming for static documents
   The HTML is the language used for creating Web pages that can be viewed using a browser. The browser can format the data on a Web page for interpretation. Through HTML, the formatting instructions can be embedded in the file itself allowing any browser to read the instruction and format the text according to the workstation being used. The data present in the main text and the formatting instructions used to format the data are both represented as ASCII characters, so that every type of server can receive and interpret the document.
   The Web page has two parts: the head and the body. The head is the first part of the Web page that contains the page title and other parameters that the browser will use. The text in the body makes up the actual content of the Web page. The appearance of the text can be modified by embedding tags into the text (see Figures 28.4 through 28.9). The tags that define the document appearance are present in the body of the Web page. A tag comes in pairs—the beginning tag bearing the tag name and the ending tag with a slash followed by the tag name. Both the tags are enclosed in symbols < and >.

| | | |
|---|---|---|
| <HTML> | </HTML> | Defines HTML document |
| <HEAD> | </HEAD> | Defines document head |
| <BODY> | </BODY> | Defines document body |

**Figure 28.4 HTML skeletal tags.**

| | | |
|---|---|---|
| <B> | </B> | Boldface |
| <I> | </I> | Italic |
| <U> | </U> | Underlined |
| <SUB> | </SUB> | Subscript |
| <SUP> | </SUP> | Superscript |

**Figure 28.5 HTML text formatting tags.**

| <TITLE> | </TITLE> | Defines document titles |
|---|---|---|
| <Hn> | </Hn> | Defines headers (n is an integer) |

**Figure 28.6 HTML title and header tags.**

| <CENTER> | </CENTER> | Centered |
|---|---|---|
| <BR> | </BR> | Line break |

**Figure 28.7 HTML dataflow tags.**

| <OL> | </OL> | Ordered list |
|---|---|---|
| <UL> | </UL> | Unordered list |
| <LI> | </LI> | An item in a list |

**Figure 28.8 HTML list tags.**

| Image | | |
|---|---|---|
| <IMG> | </IMG> | Defines an image |
| Hyperlink | | |
| <A> | </A> | Defines an address (hyperlink) |
| Executable contents | | |
| <APPLET> | </APPLET> | Document is an applet |

**Figure 28.9 HTML image, hyperlink, and Applet tags.**

2. Programming for dynamic documents
   Consider a dynamic document example that involves collection of the time stamp on the server to map logs obtained in the client to the time it was send from the server. Here the information is dynamic and changes from moment to moment. In this scenario, the client can request the server to run the system command that generates the data and sends the results to the client. The dynamic documents can be handled by the common gateway interface (CGI) technology that contains a set of rules and terms that allows the programmer to define how the dynamic document should be written, how input

data should be supplied to the program, and how the output result should be used.

The term *common* in CGI indicates that the set of rules are common to any language like C, C++, Unix Shell, or Perl. The term *gateway* in CGI indicates that a CGI program is a gateway used to access other resources like databases and graphic packages. The term *interface* in CGI indicates that there exists a set of predefined terms like variables and cells that can be used in any CGI program. Codes that are written in one of the languages that support CGI are known as a CGI program. A programmer can write a simple CGI program by encoding a sequence of thoughts in a program using the syntax of any of the languages specified above.

3. Programming for active documents

The active documents are compressed at the server site so that when the client requests the active document, the server sends a document copy in the form of a byte code. Thus the document is run at the client site or decompressed at the client site. The conversion to the binary form saves the bandwidth and transmission time. The client can also store the retrieved document so that it can run the document again without making a request to the server.

The steps from the creation to the execution of an active document are:

- The programmer writes the program to be run on the client.
- The program is compiled and the binary code is created.
- The compressed document is stored in a file at the server.
- The client sends a request for the program.
- The requested copy of the binary code is transported in a compressed form from the server to the client.
- The client changes the binary code into executable code.
- The client links all the library modules and prepares it for execution.
- The program is run and the result is presented by the client.

Java is an object-oriented language that is popular for developing Applet programs to be run on the client as active documents. Java has a class library allowing the programmer to write and use an active document. An active document written in Java is known as an Applet. Java is a combination of high level programming language, a run time environment, and a class library allowing the programmer to write an active document that can be run on a browser. It should be noted that Java program does not necessarily need a browser and can be executed as a stand-alone program.

Private data, public methods, and private methods can be defined in the Applet. An instance of this applet is created by the browser. The private methods or data are invoked by the browser using the public methods defined in the Applet. In order to execute the Applet, first a Java source file is created using an editor. From the file, the byte code is then created by the Java compiler.

This is the Applet that can be run by a browser. An HTML document is then created and the Applet name is inserted between the <APPLET> tags. Java is just one of the programming languages that can be used for creating active documents, though Applet is used in the Java context. The client applications will execute within a container application such as the browser.

## 28.4 Implementing Web Server

The server implementation is similar to the socket server program discussed in the Chapter 26 on socket programming. C# is used in this example, so that the reader can identify how different the implementation is when different programming languages are used. The example program is given sufficient comments and description to help the reader develop a program in the language of their choice.

1. At the base of a HTTP server is a TCP server. The server has to be multithreaded, so an array list of sockets is declared to handle multiple connections.

```
private ArrayList listSockets;
...
```

2. Every HTTP server has an HTTP root, which is a path to a folder on the hard disk from which the server will retrieve Web pages. This path needs to be set. Then initialize the array list of sockets and start the main server thread.

```
// Set the path;
String path = " < path > ";

listSockets = new ArrayList();
Thread threadListener = new Thread(new
ThreadStart(newConnection));
threadListener.Start();
```

3. The newConnection function manages new incoming connections, allocating each new connection to a new thread for handling client requests. HTTP operates over port 80. In this program we have set the port as 9080 and this can be customized.

   Start the TcpListener on the port specified. This thread runs in an infinite loop and waits for a socket connection for invoking AcceptSocket method. Once the socket is connected a new thread called the socketHandler function.

```
public void newConnection()
{
// Specify port to use
int port = 9080;

TcpListener tcpListener = new TcpListener(port);
tcpListener.Start();

        while(true)
        {
                Socket socketHandler = tcpListener.
                AcceptSocket();

                if (handlerSocket.Connected)
                {
                        listSockets.Add(socketHandler);
                        ThreadStart thdstHandler =
                        new ThreadStart(handlerThread);
                        Thread thdHandler =
                        new Thread(thdstHandler);
                        thdHandler.Start();

                }

        }
}
```

4. The first task this thread must perform, before it can communicate with the client, is to retrieve a socket from the top of the ArrayList. Once this socket has been obtained, it can then create a stream to this client by passing the socket to the constructor of a NetworkStream. A StreamReader is used to read from the incoming NetworkStream. The incoming line is assumed to be HTTP GET in this example and the same can be extended to handle other HTTP POST.

   The physical path also needs to be resolved. It can be read from disk and sent out on the network stream. Then the socket is closed. The response fileContents needs to be modified to include the HTTP headers so that the client can understand how to handle the response.

```
public void handlerThread()
{
Socket socketHandler = (Socket)listSockets[listSockets.
Count-1];

String streamData = "";
String fileName = "";
String[] input;
```

```
StreamReader reader;
NetworkStream networkStream =
new NetworkStream(socketHandler);
reader = new StreamReader(networkStream);
streamData = reader.ReadLine();
input = streamData.Split(" " .ToCharArray());

// The input line is assumed to be: GET <some URL path>
HTTP/1.1
// Parse the filename using input

// Add the HTTP root path
fileName = path + fileName;

FileStream fs = new FileStream(fileName, FileMode.
OpenOrCreate);
fs.Seek(0, SeekOrigin.Begin);
byte[] fileContents = new byte[fs.Length];
fs.Read(fileContents, 0, (int)fs.Length);
fs.Close();

// Modify fileContents to include HTTP header.

socketHandler.Send(fileContents);
socketHandler.Close();

}
```

5. To test the server:
   - Create an HTML page.
   - Save the page as index.htm at HTTP root path.
   - Run the server.
   - Open a browser and type in http://localhost:9080. Localhost should be replaced by the IP address of the server when server is running on a different machine from where the browser is launched.
   - The index.htm page is displayed.

## 28.5 Conclusion

In a typical client-server model, each application has its own client program and had to be separately installed on the client machine. Only after the client application is installed on the machine will the user interface be available. Any upgrade of client application would require an upgrade of solution running on individual user workstation also. This resulted in a huge maintenance cost and decreased productivity. Also with the wide spread use of Internet, there was an increased demand for making applications available anytime and anywhere.

The advantage of Web applications is that it uses a generic client application. Web documents are written in a standard format such as HTML that can be presented on any Web browser and does not require a specific client application. So any change in the server will not require an upgrade on the client workstation. During the session, the Web browser interprets and displays the pages and acts as the universal client for any Web application. Most of the current EMS, NMS, and OSS/BSS applications are Web based. The intent of this chapter is to introduce the reader to Web application development, so that the fundamentals can be supplemented with additional reading to develop scalable Web-based operation support solutions.

## Additional Reading


1. Ralph F. Grove. *Web-Based Application Development.* Sudbury, MA: Jones & Bartlett Publishers, 2009.
2. Leon Shklar and Rich Rosen. *Web Application Architecture: Principles, Protocols and Practices.* 2nd ed. New York: Wiley, 2009.
3. Susan Fowler and Victor Stanwick. *Web Application Design Handbook: Best Practices for Web-Based Software.* San Francisco, CA: Morgan Kaufmann, 2004.
4. Ralph Moseley. *Developing Web Applications.* New York: Wiley, 2007.
5. Wendy Chisholm and Matt May. *Universal Design for Web Applications: Web Applications That Reach Everyone.* California: O'Reilly Media, Inc., 2008.

# Chapter 29

# Mail Communication

This chapter is about the mail communication. It introduces the reader to the concept of writing programs for mail service. It is quite common in current telecom management applications to have alerts and summary reports sent by mail from the EMS, NMS, and OSS/BSS solutions to the concerned operator or business manager. SMTP (simple mail transfer protocol) has been used as the mail service protocol to explain the concepts of communication by mail.

## 29.1 Introduction

The standard mechanism for electronic mailing on the Internet for sending a single message that includes text, video, voice, or graphics to one or more recipients is known as simple mail transfer protocol or SMTP. The electronic message called mail has an envelope containing the sender address, the receiver address, along with other information. The message has a header part and a body. The header has information on destination and source, while the body of the message holds the content to be read by the recipient.

The header has:

- The sender address: Details of the address from which the message is sent.
- The receiver address: Details of the recipient address to which the message is to be sent.
- Message subject: An identifier for the message.

The Internet user having an e-mail ID receives the mail to his mailbox that is periodically checked by the e-mail system and informs the user with a notice. If the user is ready to read his mail, a list containing the information about the mail like the sender mail address, the subject, the time when the mail was sent or received are displayed. The user can select the message of his choice and the contents of the message is displayed on the screen.

E-mail addresses are required for delivering mail. The addressing system used by SMTP consists of the local part and the domain name that are separated by an @ symbol (local_identification@domain_name). Some of the terminologies that the reader will encounter in this chapter are:

- User agent: The user mailbox stores the received mails for a user for retrieval by the user agent.
- Email address: The name of the user mailbox is defined in the local part of the email address.
- Mail exchangers: These are the hosts that receive and send e-mail. The domain name assigned to each mail exchanger may be a DNS database or a logical name.

This chapter has details on the user agent, mail transfer agents, and the mail delivery. SMTP is used for explaining the concepts. Perl programming language is used to show a programming example on implementing mail communication. Standard SMTP libraries are available in most of the high level programming languages. These libraries make client and server programming for mail communication easy. This chapter also uses an SMTP library in Perl programming language to code the example program for mail communication.

## 29.2 Mail Delivery Process

The SMTP is the TCP/IP protocol that defines e-mail services on the Internet. The client–server is used for exchange of mail messages. Commands are sent from the client to the server while responses are sent from the server to the client. Both the commands and the responses carry messages across client and server (see Figure 29.1). A keyword followed by zero or more arguments makes up a command. There are 14 commands defined by SMTP.

There are three phases of mail transfer:

1. Connection establishment
2. Message transfer
3. Connection termination

The SMTP server starts the connection phase after the client has established a TCP connection to the well-known port 25. Exchange of a single message takes

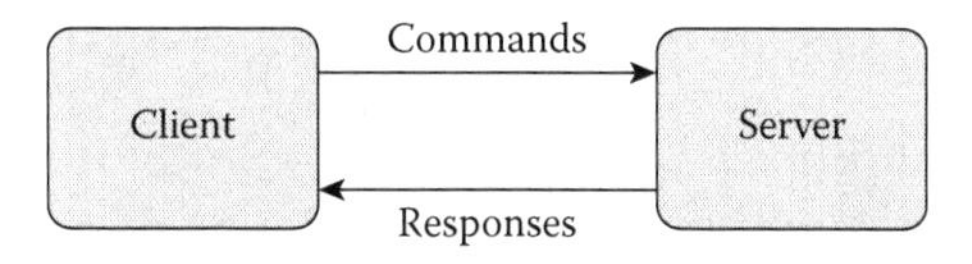


**Figure 29.1 Client–server model of mail transfer.**

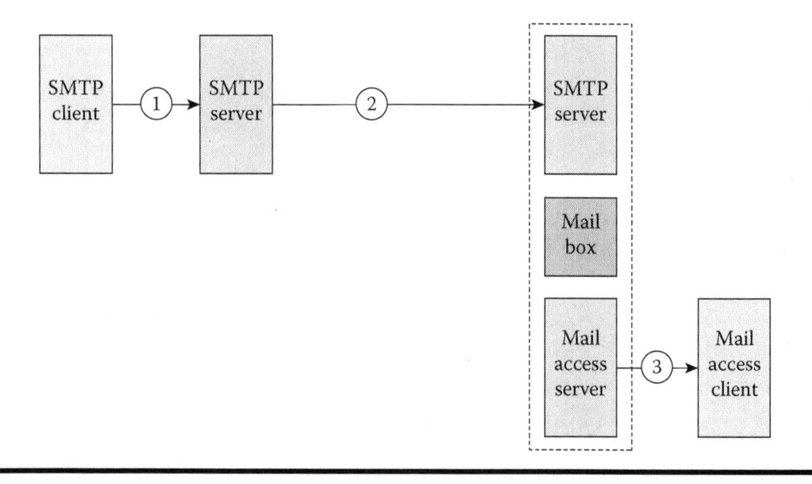


**Figure 29.2 Mail delivery process.**

place between the sender and one or more recipients after the establishment of a connection between the SMTP client and server. The connection between the SMTP client and server is terminated by the client after the successful transfer of the message between them.

The transfer of mails from the sender to the receiver involves the following stages (see Figure 29.2):

a. Stage 1: The mail is transferred from the user agent that uses the SMTP client software to the local server that uses the SMTP server software. There would be another remote SMTP server as the connectivity to the remote server may not be available at all times from the SMTP client. The mail is stored in the local server before it is transferred to the remote server.
b. Stage 2: The local server now acts as the SMTP client to the remote server that is now the SMTP server. Data needs to be sent from the local server to the remote server that is now the SMTP server. This mail server receives the mail and stores the mail in the user mailbox for later retrieval.
c. Stage 3: In this stage the mail access client uses mail access protocols like POP3 (Post Office Protocol, version 3) or IMAP4 (Internet Mail Access Protocol, version 4) to access the mail so as to obtain the mail.

## 29.3 Mail Protocols

a. Simple mail transfer protocol (SMTP): This protocol pushes the messages to the receiver without the server requesting a transfer. So SMTP is a push-based protocol. It is used to send data from client to a server where the mails are stored. Some other access protocol is to be used by the recipient to retrieve the messages from the mail server mailbox where it was stored.
b. Post office protocol, version 3 (POP3): This is a mail access protocol. The recipient computer has client POP3 software and the mail server has the server POP3 software. The client can access mail by downloading e-mail from the mail server mailbox. A connection is opened between the user agent or the client and the server on TCP port 110. The user name and password are sent for authentication.

   The delete mode and the keep mode are the two modes of POP3. After the retrieval of the mails, they are deleted from the mailbox through the delete mode. The mails are stored in the user's local workstation where they can save and organize the received mails after reading or replying. In the keep mode, the mails remain inside the mailbox even after retrieval and are employed. The keep mode is used when the user accesses the mails through a system away from the local workstation, so that the mail can be read and maintained for later retrieval and organizing. However POP3 has several limitations.
c. Internet mail access protocol, version 4 (IMAP4): It is a mail access protocol. IMAP4 is a powerful and complex mail access protocol used to handle the transmission of email. Using IMAP4 the user can create the mailbox hierarchy in a folder for e-mail storage or even create, delete, or rename the mailboxes on the mail server. In cases where the bandwidth is low and the e-mail contains multimedia with high bandwidth requirements, the user can partially download the e-mail through IMAP4. It is also possible to check the e-mail header or search for a string of characters in the e-mail content before downloading.

## 29.4 User Agent

The user agent is a component of the electronic mail system. It is a software package that performs the following basic functions:

- Composing messages: The e-mail messages to be sent out are composed by the user agent. Some user agents provide a template on the screen that has to be filled in by the user while other user agents have a built-in editor that can perform sophisticated tasks like spell check, grammar check, and so on. The user can create his message in a word processor or text editor that can be imported or cut and pasted to the user agent.

- Reading messages: The incoming messages are read by the user agents. A user agent (when invoked, checks the mail in the incoming mailbox) referred to as an inbox. For each of the received mail, the user agent shows a one-line summary with fields like the number field, a flag indicating whether the mail has been read and replied to or read but not replied to, sender name, subject field if subject line in the message is not empty, and message size.
- Replying to messages: The user agent allows the user to send a reply message to the original sender of the message or to all the recipients such that the reply message contains the original message for quick reference and the new message.
- Forwarding messages: The user agent allows the message to be send to a third party apart from the original sender or the recipients of the copy with or without adding any extra comments.

There are two types of mailboxes, called the inbox and outbox. These are actually files with a special format that are created and handled by the user agent (see Figure 29.3). The received mails are kept inside the inbox and all the sent e-mails are kept inside outbox. The messages received in the inbox remain there until the user deletes the messages.

User agents can be of two types:

- Command driven user agents: The command driven user agents were used in the earlier days but are still present as underlying user agents in servers. It performs its task by accepting a one-character command from the keyboard. By typing a predefined character at the command prompt the user can send a reply to the sender of the message and by typing another predefined character at the command prompt the user can send reply to the sender and all the recipients.
- GUI-based user agents: The GUI (graphical user interface) based user agents contain user interface components like icons, menu bars, windows that allows

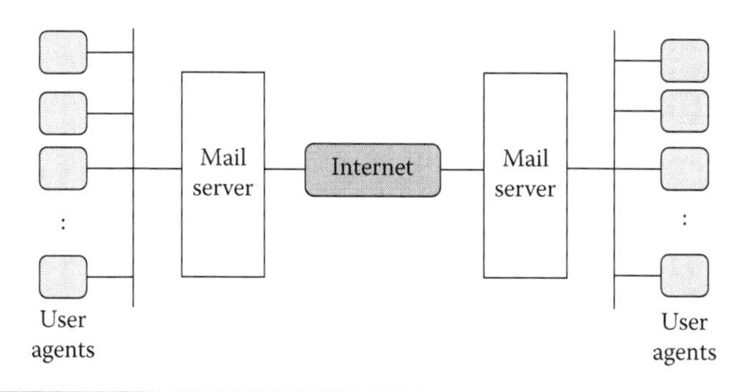


**Figure 29.3 Role of user agents in E2E mail transfer.**

the user to interact with the software by using both the keyboard, and the mouse thereby making the services easy to access.

There are user agents from multiple vendors that can be used for day-to-day mail communication. Some of the most popular application based user agents include Microsoft Outlook, Mozilla Thunderbird. Some vendors offer Internet mail by hosting mail server on their server and providing the user with Web-based user agent like Gmail, Yahoo Mail, Microsoft MSN Mail, Rediff Mail, and so on.

In Web-based mail, the mails are transferred to the receiving mail server from the sending mail server through SMTP and HTTP is employed in the transfer of messages to the browser from the Web server. The user sends a message to the Web site to retrieve the mails. The Web site asks the user to specify the user name and password for authentication. The transfer of messages from Web server to the browser happens only after user authentication. The messages are transferred in HTML format from the Web server to the browser that loads the user agent.

## 29.5 Multipurpose Internet Mail Extensions

Multipurpose Internet mail extensions (MIME) is an extension to SMTP that allows non-ASCII data to be sent through SMTP (see Figure 29.4). SMTP has several limitations; such as it cannot be used for languages that are not supported by 7-bit ASCII characters and it cannot be used to send binary files, video, or audio. The MIME acts as a supplementary protocol. The non-ASCII data at the sender site is transformed to ASCII data through MIME and are delivered to the client SMTP to be sent through the Internet. The ASCII data are received by the server SMTP and are delivered to MIME to be transformed into original data.

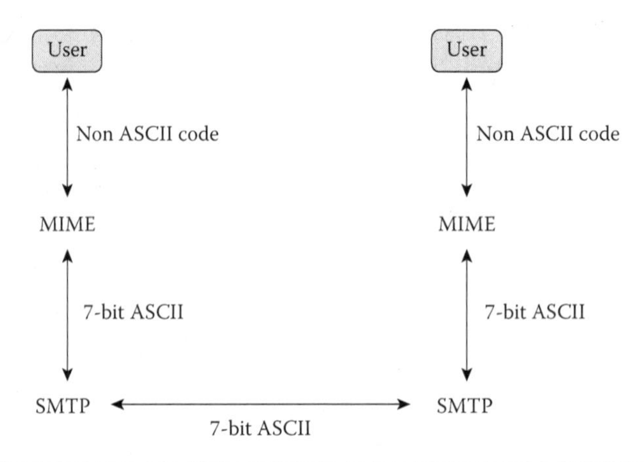


**Figure 29.4 Role of MIME.**

MIME-Version, Content-Type, Content-Transfer-Encoding, Content-Id, Content-Description are the five headers defined by MIME that can be added to the original SMTP header sections to define the transformation parameters. Let us look into each of these headers:

1. MIME-Version => The version of the MIME used, is defined by this header.

```
MIME-Version: 1.1
```

2. Content-Type => The data type that are used in the body of the message are defined by this header. The type and the subtype of the contents are separated by a slash. The header may contain other parameters depending on the subtype (see Figure 29.5).

```
Content-Type: < type / subtype; parameters >
```

| Type | Subtype | Description |
|---|---|---|
| Text | Plain | Unformatted text |
| Multipart | Mixed | Body contains ordered parts of different data types |
| | Parallel | Body contains parts of different data types |
| | Digest | Similar to mixed but default type is message/RFC822 |
| | Alternative | Parts are different versions of the same message |
| Message | RFC822 | Body is an encapsulated message |
| | Partial | Body is a fragment of a bigger message |
| | External body | Body is a reference to another message |
| Video | MPEG | Video is in MPEG format |
| Audio | Basic | Single channel encoding of voice at 8KHZ |
| Image | JPEG | Image is in JPEG format |
| | GIF | Image is in GIF format |
| Application | PostScript | Adobe PostScript |
| | Octect stream | General binary data of 8-bit bytes |

**Figure 29.5 Content-Type => Type and subtype used in MIME.**

3. Content-Transfer-Encoding => The methods to encode the messages into zeros and ones to facilitate transport are defined by this header.

```
Context-Transfer-Encoding: < type >
```

The encoding can be:

- 7-bit: This encoding handles ASCII characters and short lines. It is a 7-bit ASCII encoding where the line length should not exceed 1000 characters.
- 8-bit: This encoding handles non-ASCII characters and short lines. It is an 8-bit encoding and sends non-ASCII characters but the line length should not exceed 1000.
- Binary: This encoding handles non-ASCII characters of unlimited length. It is an 8-bit encoding where non-ASCII characters can be sent and the line length can exceed 1000 characters.
- Base64: In this encoding, 6-bit blocks of data are encoded into 8-bit ASCII characters. It transforms the data made of bytes into printable characters that can be sent as ASCII characters or any character set type that is supported by the underlying mail transfer mechanism.

Base 64 Alphabet Table

| Value | Code | Value | Code | Value | Code | Value | Code | Value | Code | Value | Code |
|---|---|---|---|---|---|---|---|---|---|---|---|
| 0 | A | 11 | L | 22 | W | 33 | H | 44 | S | 55 | 3 |
| 1 | B | 12 | M | 23 | X | 34 | I | 45 | T | 56 | 4 |
| 2 | C | 13 | N | 24 | Y | 35 | J | 46 | U | 57 | 5 |
| 3 | D | 14 | O | 25 | Z | 36 | K | 47 | V | 58 | 6 |
| 4 | E | 15 | P | 26 | A | 37 | L | 48 | W | 59 | 7 |
| 5 | F | 16 | Q | 27 | B | 38 | M | 49 | X | 60 | 8 |
| 6 | G | 17 | R | 28 | C | 39 | N | 50 | Y | 61 | 9 |
| 7 | H | 18 | S | 29 | D | 40 | O | 51 | Z | 62 | + |
| 8 | I | 19 | T | 30 | E | 41 | P | 52 | 0 | 63 | / |
| 9 | J | 20 | U | 31 | F | 42 | Q | 53 | 1 | | |
| 10 | K | 21 | V | 32 | G | 43 | R | 54 | 2 | | |

- Quoted-Printable: This encoding is used when the data consist mainly of ASCII characters with a small portion of non-ASCII characters. The ASCII characters are sent as it is, but the non-ASCII characters are sent as three characters—the first character being an equal sign (=), the second and the third are the hexadecimal representation of the byte.

4. Content-Id => The whole message is identified in a multiple message environment through the Content-Id header.

```
Content-Id: id = <content-id>
```

5. Content-Description => The nature of the body; that is, whether the body is an image, an audio, or a video are defined by the Content-Description header.

```
Content-Description: <description>
```

The Base 64 and Quoted-Printable are preferred over 8-bit and binary encoding. MIME is an integral part of communication involving SMTP client and SMTP server.

## 29.6 Implementation

In this example, the Perl programming language is used for implementing SMTP server and client. The implementation is not tied to a specific programming language and the SMTP client and server can be implemented in other programming languages as well. This example is just to illustrate the necessary components required to implement the SMTP server and client.

Net::SMTP library has been used in this example as it implements the SMTP functions required for implement SMTP applications in the Perl programming language. Most high level programming languages have similar libraries for implementing applications that require mail communication.

a. SMTP Client using Net::SMTP

```
# ----------------- #
# SMTP Client #
# ----------------- #
# Include Net::SMTP library
use Net::SMTP qw(smtp);

# Create a Net::SMTP object
my $mailer = Net::SMTP->new("smtp.crcpress.com") ordie$@;

# SMTP commands can now be accessed with "mailer" object
# The SMTP server is smtp.crcpress.com

# Send a sample message to postmaster at the SMTP Server
$mailer->mail("user\@crcpress.com");
$mailer->to ("postmaster\@crcpress.com");

# Initiate the sending of data
$mailer->data();
```

```
# Mention the sender the way it needs to appear in the
message
# The actual sender is user@crcpress.com
$mailer->datasend("From: jithesh\@crcpress.com\n");

# Mention the receiver the way it needs to appear in the
message
# The actual receiver is postmaster@crcpress.org
$mailer->datasend("To: ohanley\@crcpress.com"\n\n\n");

# Specify the body of the message
$mailer->datasend("Test Mail\n");

# End the message
$mailer->dataend();
$mailer->quit or die "mail sending failure";
```

b. SMTP Server implementation using Net::SMTP::Server

```
# ------------------ #
# SMTP Server #
# ------------------ #

# Include libraries
use Carp;
use Net::SMTP::Server;
use Net::SMTP::Server::Client;
use Net::SMTP::Server::Relay;

# Create a Net::SMTP::Server object
$receiver = new Net::SMTP::Server('localhost', 25);
# Specify address and port for running the Server
# SMTP commands can now be accessed with "receiver" object

# Accept Connection from Client
while($conn = $receiver->accept()) {

# Handle the client's connection
my $sender = new Net::SMTP::Server::Client($conn);

# A better option is to spawn off a new parser for
handling the client
# This can be implemented as a new thread or using
fork( )

# Process the client.
$sender->process || next;
# This will ensure that the connecting client completes
the SMTP transaction.
```

```
# Relaying everything to a server
my $relay = new Net::SMTP::Server::Relay($sender->
{FROM},$sender->{TO},$sender->{MSG});
# In an ideal implementation the mails needs to be
stored in the server
# Storage ensures that mails can be accessed later using
an access protocol
}
```

## 29.7 Conclusion

The intent of this chapter is to introduce the reader to the basic concepts of mail communication. The chapter started with a basic overview of mail messaging and mail delivery process. Then some of the mail protocols for data transfer and access were discussed. MIME, which is an important concept in mail delivery, was also handled. The chapter concludes with an implementation example of SMTP server and client using the Perl programming language. Event handling is an essential activity in operation support solution and mail communication is a way of notifying operators and business managers on critical events.

## Additional Reading


1. Pete Loshin. *TCP/IP Clearly Explained.* 4th ed. San Francisco, CA: Morgan Kaufmann, 2003.
2. John Rhoton. *Programmer's Guide to Internet Mail: SMTP, POP, IMAP, and LDAP.* Florida: Digital Press, 1999.
3. Rod Scrimger, Paul LaSalle, Mridula Parihar, and Meeta Gupta. *TCP/IP Bible.* New York: John Wiley & Sons, 2001.
4. Laura A. Chappell and Ed Tittel. *Guide to TCP/IP.* Kentucky: Course Technology, 2001.
5. Candace Leiden, Marshall Wilensky, and Scott Bradner. TCP/IP for Dummies. 5th ed. New York: For Dummies, 2003.

# Chapter 30

# File Handling

This chapter is about the file transfer that is a common method of data exchange. Performance and other data records are collected in the network elements at regular intervals and stored in files. These files in the network elements needs to be transferred to the OSS solutions for analysis and plotting the data in the records in a user-friendly manner. File transfer protocol (FTP) is the most common protocol used for transfer of files across systems. This chapter introduces the reader to file handling concepts and the use of FTP in file transfer.

## 30.1 Introduction

The transfer of files from one system to another is provided by the application layer protocol called FTP. This means that FTP can be used in telecom management solutions for file transfer between agents running on network elements and management server. FTP involves a client–server based interaction. FTP avoids the common problems present during a file transfer and favors easy file transfer by taking care of:

a. The difference in the file name conventions between two systems.
b. The difference in directory structures.
c. The difference in representing text and data.

Using TCP services, FTP establishes one connection in port 20 for data transfer known as the data connection. This connection can be used to transfer different types of data. It establishes another connection in port 21 for the control information like commands and responses known as the control connection. A line of

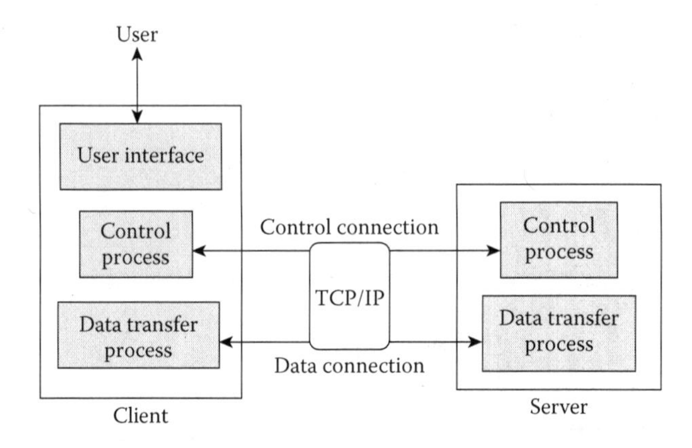


**Figure 30.1 Client–server model in file transfer.**

command or response line is transferred at a time using the control connection. User interface, client control, and client data transfer process are the three components of the client. Server control process and server data transfer process are the two components of the server (see Figure 30.1).

The data connection is established between the client data transfer process and the server data transfer process. If many files are being transferred, the data connection can be opened and closed many times. The control connection is established between the client control process and the server control process. The control connection is opened and maintained for the entire interactive FTP session such that for each file transfer, the command for file transfer opens the connection and the connection is terminated after the file transfer. The presence of a separate channel for data transfer and control helps to achieve maximum throughput in data transfer and minimum delay in communication of control signals. This interactive connection between the user and the server remains open for the entire process.

File handling is a frequent activity in most operation support solutions. The most common example, where file transfer protocols like FTP and SFTP (secure transfer) is used, is for the management server to collect performance records from the network elements. The network elements generate performance reports at predefined intervals. These records are to be transferred as files to the management server, where the files are parsed to analyze the performance of the network elements. In this chapter, the basics of working with file transfer protocol are discussed along with implementation examples on working with files and file transfer.

## 30.2 Communication with FTP

The client and server between which files needs to be transferred may have different operating systems, file structures, file formats, and different character sets. Both these systems must communicate with each other to facilitate file transfer. FTP is the protocol that enables file transfer between two systems that may or may not process the same properties.

FTP uses the following communication methods:

- Control connection: Communication is achieved through commands and responses in the case of control connection. FTP uses an ASCII character set to communicate across a control connection (see Figure 30.2). The file type, transmission mode, and data structure are defined by the client to solve the heterogeneity problem during data transfer. The file is prepared for transmission through control connection before transferring through the data connection.
- Data connection: File transfer is done using the data connection (see Figure 30.3). The client decides the file type, data structure, and transmission to be used in the data connection.

Files of different types can be transferred across the data connection. The original file can be transformed into ASCII characters at the sender site and the ASCII characters are retransformed back to its own representation at the receiver site, in the case of ASCII files, which is the default format for transferring the text files. The

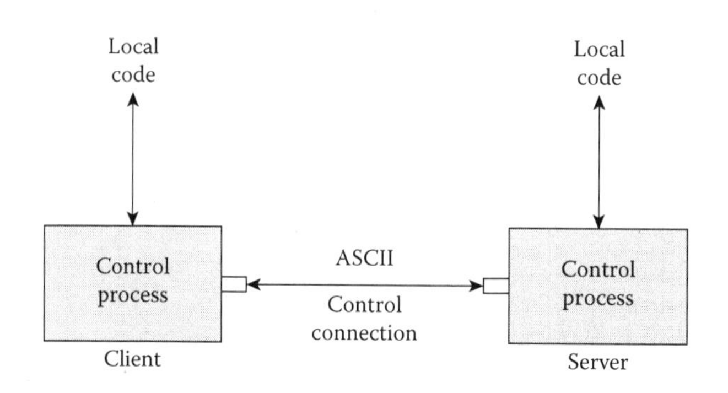


**Figure 30.2 Control connection.**

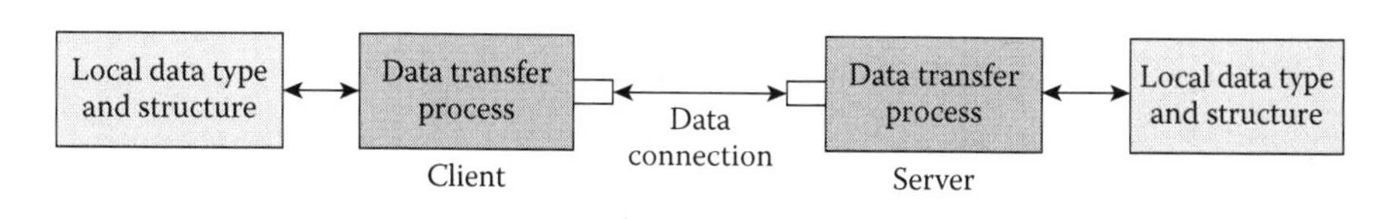


**Figure 30.3 Data connection.**

files can also be transferred by EBCDIC encoding if either one end or both the ends of the connection use EBCDIC encoding. Binary files like compiled programs or images encoded as zeros and ones can be transferred as continuous streams of bits without any interpretation or encoding.

Printability of the file encoded in ASCII or EBCDIC can be defined by adding the following attribute:

- Nonprint format is the default format used for transferring a text file that will be stored and processed later as the file lacks any character to be interpreted for the vertical movement of the print head.
- Files in TELNET format contain ASCII vertical characters like NL (new line), VT (vertical feed), CR (carriage return), and LF (line feed) indicating that the file is printable after transfer.

The different data structures that can be used for transferring files through data connection are:

- Record structure: Text files can be divided into records by using record structure.
- Page structure: This divides the file into pages such that each page has a page number and a page header facilitating the pages to be stored and accessed sequentially or randomly.
- File structure: In this structure, the file is a continuous stream of bytes having no structure.

The different transmission modes that can be used for transferring files through data connection are:

- Block mode: In the block mode the files are delivered in the form of blocks to TCP such that each block is preceded by a 3-byte header. The first byte provides a description of the block and they are called block descriptor. The block sizes are defined by the second and third bytes in the form of bytes.
- Compressed mode: In compressed mode the big files can be compressed by replacing the data units that appear consecutively by one occurrence and the repetition number. Blanks are compressed in text files and null characters are compressed in binary files. Run-length encoding method is used for compression.
- Stream mode: In the stream mode the data reaches the TCP as a continuous stream of bytes that is then split into segments of appropriate size. If the data consist of many records then each recorder will have a termination character that terminates the data connection by the sender. There is no need for a termination or end-of-file character in cases where the data is simply a stream of bytes. The stream mode is the default mode.

## 30.3 File Transfer with FTP

File retrieval, file storage, or transfer of directory list/file names from the server to the client, can be considered as file transfer in the case of FTP (see Figure 30.4).

- File retrieval: This is the process of copying the file from the server to the client.
- File storage: This is the process of copying the file from the client to the server.
- Directory list or file names: These are treated as a file that can be sent from the server to the client over data connection.

In file storage, the control commands and responses are exchanged across the control connection (see Figure 30.5). Then each of the data records is transferred across

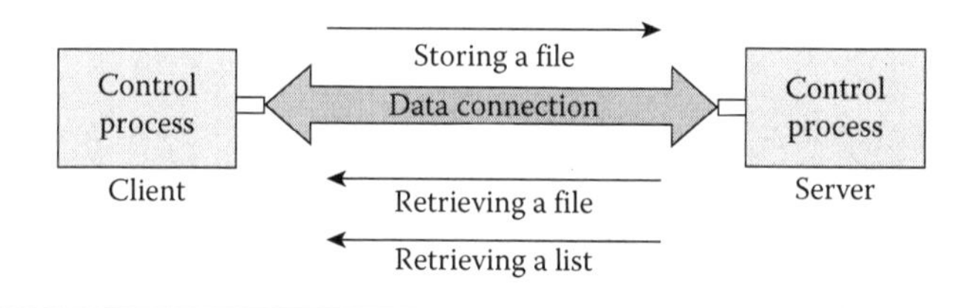


**Figure 30.4 File transfer.**

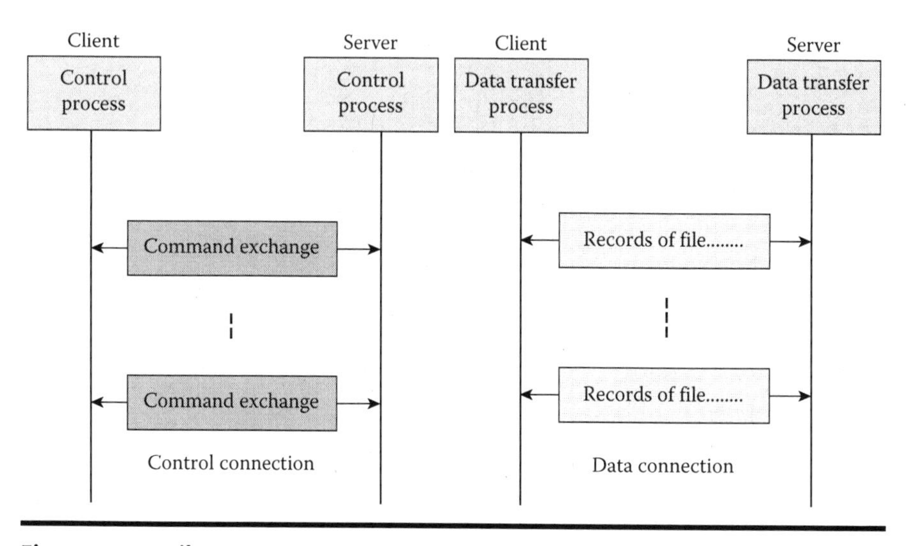


**Figure 30.5 File storage.**

the data connection. Finally the connection is terminated by exchanging certain commands and responses.

The FTP services can be accessed through user interfaces provided by the operating systems. After receiving the input line from the user, the FTP interface reads it and converts it to the corresponding FTP command. The user must have the user name and password on the remote server in order to use FTP. However, the user does not require having the user name and password to access the files present in some sites. Here the user gives the user name as anonymous and the password as guest. These are anonymous FTPs and only a subset of command is allowed to the anonymous users by certain sites. Anonymous login provides limited user access.

A knowledge on the files and subfolders present in each folder in a remote computer's file system helps to navigate through the folders. This data is returned on the data socket just like files. The folder listings are received from the server when the client issues a LIST command. The server that is waiting for created data sockets responds to the client request.

```
In Windows, the directory listing style with DOS will be of
the format:

08/12/2009  09:18 PM    <DIR>         pics
12/11/2009  08:10 PM    <DIR>         presentation
08/24/2009  10:31 AM    <DIR>         Program Files
05/14/2009  10:51 PM    <DIR>         ringtone
11/12/2009  11:51 AM       114,010   Proposal.pdf
11/04/2009  09:24 AM        32,877   submit.pdf
```

```
In UNIX FTP servers, the directory listing style will be of
the format:

d---rw-rw- 1 user group 0 Aug 24 2009 .
d---rw-rw- 1 user group 0 Aug 24 2009 ..
----rw-rw- 1 user group 564 Aug 24 14:20 Arch.bmp
----rw-rw- 1 user group 623 Aug 24 22:40 submit.pdf
----rw-rw- 1 user group 2132 Aug 24 15:00 File1.txt
d---rw-rw- 1 user group 0 Aug 24 2009 Test
```

## 30.4 File Commands

Some of the FTP commands and the command description are discussed in this section.

| *FTP Command* | *Description* |
|---|---|
| **STOR** | This FTP command is used for uploads. |
| **STOU** | The command is also used for uploads, where the server chooses the remote file name that is specified in the 250 response. |
| **ABOR** | The command ABOR is used for aborting the current data transfer. |
| **RETR** | This FTP command is used for downloads. |
| **DELE** | Command DELE is used for deleting the specified file. |
| **REST** | This FTP command is used for restarting the file transfer at a specified position. |
| **APPE** | The APPE command is used for appending. |
| **RNFR/RNTO** | Used to rename files (RNFR <old name>) (RNTO <new name>). |
| **RMD** | Command RMD is used for deleting the specified folder. |
| **PWD** | Command PWD is used for responding with the current working directory. |
| **LIST** | When the client issues a LIST command, the server responds with the contents of the current working directory. |
| **SYST** | FTP command SYST is used for responding with the name of the operating system or the OS being emulated. |
| **TYPE** | Indicates the format of the data, either A for ASCII, E for EBCDIC, I for Binary. |
| **MKD** | A new folder can be created by using MKD command. |
| **STAT** | FTP command STAT is used for responding with the status of a data transfer. |
| **HELP** | FTP command HELP is used for text information about the server. |
| **USER/PASS** | The username and password of the user can be specified by the FTP command USER and PASS, respectively. |

## 30.5 File Sharing

Files can be accessed across systems using methods other than FTP. This is done by sharing the files rather than transferring the files. Two file sharing methods are discussed here as examples:

1. Microsoft file sharing
   Common internet file (CIF) system is the native file-sharing protocol of Windows 2000 and XP that was developed by Microsoft. Due to NTLM encryption, CIF is more secure than FTP and generally faster and is used for providing the network drive functionality and print sharing. The terms NETBIOS and NETBEUI are the more correct names for Microsoft file and print sharing.
   Windows file sharing is widely used in enterprises where employees share office resources like printer or a central repository for files. For intranet a central repository using network share would be better than FTP, which would be slower and less secure. For external sharing of files, FTP would be more suitable because of its interoperability and ease of deployment.
2. Netware file sharing
   Netware file sharing is one of the fastest file transfer protocols over internal networks. It is nonroutable because they are built on top of the Internetworking packet exchange or Sequenced packet exchange (IPX/SPX) protocols. These packets can be converted to TCP/IP using translators. The performance factor is lost in this case. The Netware system, also referred to as IntranetWare is centered on a central Netware server that runs the Novell operating system started from a bootstrap DOS application. The authentication and privileges can be controlled by Netware Directory Service or NDS, which is hosted by the server and can replicate among other servers. Instead of NDS the older Novell servers (3.x) use binary and the bindery cannot replicate. By using the Novell core protocol (NCP), the clients locate the server and when a remote file server is found, it is mapped to a local drive on the client's machine.

## 30.6 Implementation

Four example programs are discussed in this section. The implementation of all the programs in this section is using the Perl programming language.

1. Read file: In this example, the contents of a file is read into a data structure and the data structure is printed to display the contents read from the file.

```
# --------------------------- #
# Read Contents of File #
# --------------------------- #

# Specify the file that needs to be read
$fileName = '/root/test.txt';
```

```
# Open the File for reading
open(FILE, $fileName);

# Read file contents into an array
@lines = <FILE>;

# Close file after reading to array
close(FILE);

# Print contents of the array
print @lines;
```

2. Write data to file: In this example, data is written to a file opened in write mode. After write operation with this program, the contents of the file can be verified using the program to read file.

```
# --------------------------- #
#    Write Data to File    #
# --------------------------- #

# Define the data to be written
$str = "Write this to file";

# Open the file in write mode
open FILE, ">/root/test.txt";

# Write the pre-defined string to file
print FILE $str;

# Close the file after writing
close FILE;
```

3. Manipulating files: In this example file manipulation with PERL is discussed. Every high level programming language has its own set of APIs for file manipulation. This example is just to introduce the reader to general file manipulations that are done in programming.

```
# --------------------------- #
#    File Manipulation    #
# --------------------------- #

# All files are assumed to be in the current directory
# remove file with name "test" using rm command
rm test;

# rename  file "old_name" to "new_name" using rename
command
```

```
rename "old_name", "new_name";# change file permissions
on the file
chmod 777 new_name;

# create link to a file
link "test1", "test2";

# unlink a created link
unlink test2;
```

4. FTP client: This example shows a FTP client implementation in PERL programming language. It uses Net::FTP library in PERL for implementation. The example shows an FTP client that gets a file from a remote server.

```
# --------------------------- #
#          FTP Client         #
# --------------------------- #

# Create an instance using new method
# It defines a connection to a remote server
$ftpHandle = Net::FTP -> new ("ftp.crcpress.com");

# Login to the server
$ftpHandle->login();

# Specify the user id and password
$ftpHandle->login("jithesh","password");
# Move to the directory that contains required file
$ftpHandle->cwd("/root/home/shared_files")

# Retrieve the file
$ftpHandle->get("test.txt");

# Disconnect after file transfer
$ftpHandle -> quit();
```

## 30.7 Conclusion

The intent of this chapter is to introduce the reader to the basic concepts of file transfer. The chapter started with a basic overview of file transfer using FTP. Then some of the details on communication with FTP are discussed along with other methods for file sharing. The chapter concludes with implementation examples using the Perl programming language. File handling and transfer is an essential activity in an operation support solution.

## Additional Reading

1. Candace Leiden, Marshall Wilensky, and Scott Bradner. *TCP/IP for Dummies*. 5th ed. New York: For Dummies, 2003.
2. Pete Loshin. *TCP/IP Clearly Explained*. 4th ed. San Francisco, CA: Morgan Kaufmann, 2003.
3. Rod Scrimger, Paul LaSalle, Mridula Parihar, and Meeta Gupta. *TCP/IP Bible*. New York: John Wiley & Sons, 2001.
4. Laura A. Chappell and Ed Tittel. *Guide to TCP/IP*. Kentucky: Course Technology. 2001.

# Chapter 31

# Secure Communication

This chapter is about implement security in management applications. The different types of encryption and security protocols are discussed in this chapter. The chapter ends with an implementation example for encryption and decryption of messages.

## 31.1 Introduction

The study of transforming messages in order to make them secure and immune to attacks is known as cryptography. The original text message without any transformation is said to be in plaintext format. After applying transformation on the original message to make it secure, the message is called ciphertext. The plaintext is transformed to the ciphertext by the user through the encryption algorithm and the ciphertext is transformed back to the plaintext by the receiver through the decryption algorithm (see Figure 31.1).

The process of ciphering (encrypting or decrypting) is performed using a key called the cipher, which is actually a number or a value. The key is kept secret and requires protection. In the encryption process the encryption key along with the encryption algorithm transforms the plaintext to the ciphertext and in the decryption process the decryption key along with the decryption algorithm transforms the ciphertext to the plaintext (see Figure 31.2). Symmetric key or secret key cryptography algorithms and the public key or asymmetric cryptography algorithms are the two groups of cryptography algorithms.

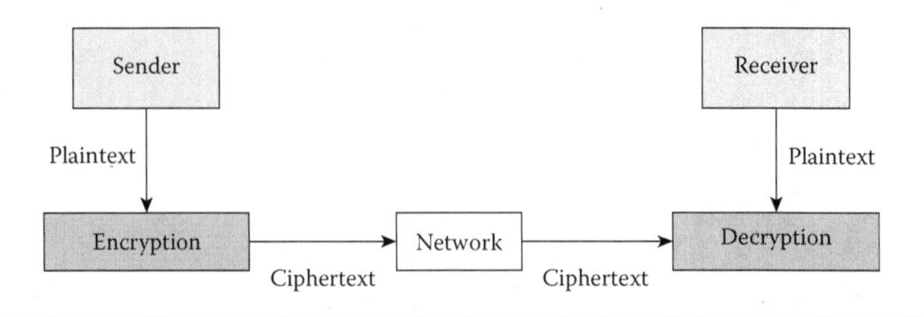


**Figure 31.1 Transformation of message from plaintext to ciphertext.**

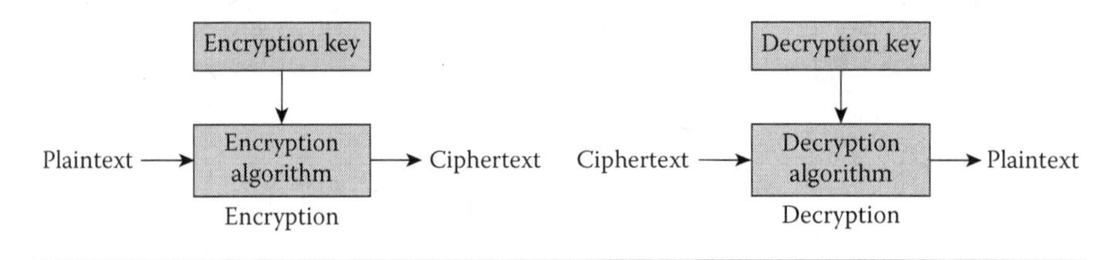


**Figure 31.2 Encryption and decryption process.**

## 31.2 Symmetric Key Cryptography

The symmetric key (secret key cryptography) is named so because the same key is used by both the sender and the receiver to perform the encryption and decryption process. The sender encrypts the data using the symmetric key along with the encryption algorithm. The receiver decrypts the data using the same symmetric key using the decryption algorithm. The algorithm used for decryption is the inverse of the algorithm used for encryption in symmetric key cryptography. Symmetric key cryptography is used to encrypt and decrypt long messages because the key is smaller. It is efficient and takes little time in encryption process. However it should be noted that the sender as well as the receiver must have the same unique symmetric key and the distribution of the keys between the two parties can be difficult.

Some of the cipher types are:

A. Substitution cipher: Using the substitution method, the substitution cipher substitutes a symbol with another. The characters present in the plaintext can be replaced with another character.

   It includes monoalphabetic and polyalphabetic substitution.

   a. Monoalphabetic substitution: A character in the plaintext is always converted to the same character in the ciphertext regardless of its position in the text. For example, if the algorithm says character C in the plaintext must be changed to character F, then every character C is changed to character F regardless of its position in the text. There is a one-to-one relationship existing between the character in the plaintext and the character in the ciphertext. Monoalphabetic substitution is simple but the code can be attacked easily.

  b. Polyalphabetic substitution: A character in the plaintext can be converted to many characters in the ciphertext. There is a one-to-many relationship existing between the character in the plaintext and the character in the ciphertext. The code can be broken through trial and error method.

B. Transpositional cipher: In this method the transpositional cipher retains the characters in their plaintext form but changes their position to make the ciphertext. The text is organized to form a two-dimensional table. The key defines what columns should be swapped and results in an interchanged column pattern. Here also the plaintext can be found out through a trial-and-error method.

C. Block cipher: The block ciphers take a number of bytes and encrypt them as a single unit. Two common implementations are P-box and S-box. P-boxes and S-boxes can be combined to obtain a more complex cipher block called the product block.

  a. P-box: Transposition at the bit level can be performed by the P-box where P stands for permutation. P-box implementation in hardware is faster than in software. The same number of ones and zeros are present in both plaintext and ciphertext. The encryption algorithm, decryption algorithm, and the key are embedded in the hardware.

  b. S-box: Substitution at the bit level can be performed by the S-box where one decimal digit is substituted by another. Here S stands for substitution.

Let us take DES and Triple DES as examples of symmetric key cryptography.

I. Data encryption standard (DES) is a complex block cipher where using a 56-bit key the algorithm encrypts a 64-bit plaintext (see Figure 31.3). A 64-bit ciphertext is created by putting the text through complex swap and transposition procedures. Using complex encryption and decryption algorithms DES can substitute eight characters at a time. Encryption and the key are the same for each segment, implying that, if the data are four equal segments then the result will also have four equal segments.

II. Triple DES is an algorithm compatible with DES that uses three DES blocks and 56-bit keys and operates on eight character segments. The encryption process in triple DES uses an encryption-decryption-encryption combination of DES while the decryption process in triple DES uses a decryption-encryption-decryption combination of DES (see Figure 31.4).

There are many other symmetric key algorithms like AES (Advanced Encryption Standard), RC4, IDEA (International Data Encryption Algorithm), Twofish, Serpent, Blowfish, and CAST5.

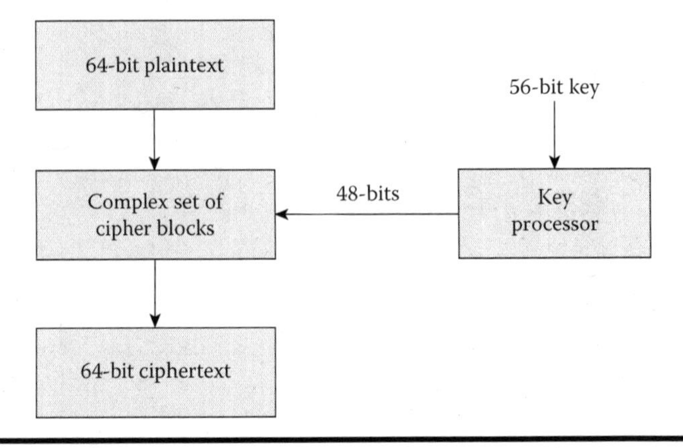


**Figure 31.3 Block view of DES.**

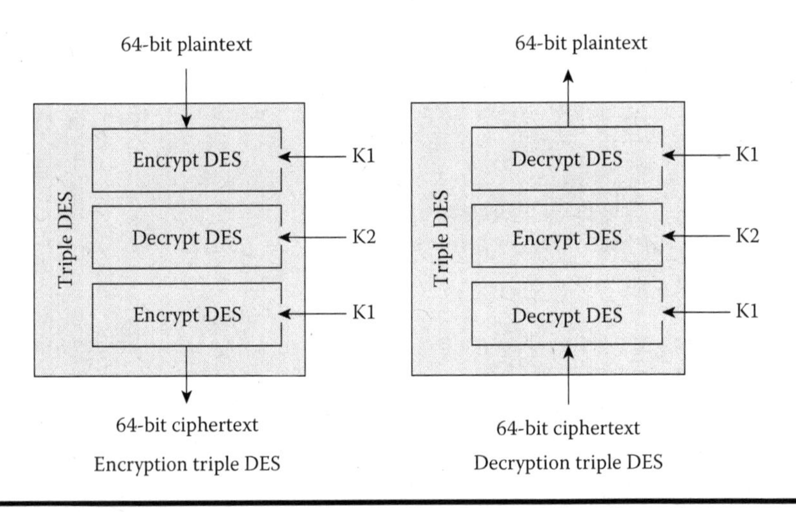


**Figure 31.4 Block view of triple DES.**

## 31.3 Public Key Cryptography

In this technique there are two keys: the private key is kept by the receiver and the public key is announced to the public. In public key cryptograghy, different keys are used for encryption and decryption. So the methodology uses asymmetric key algorithms. In the public key encryption or decryption process, each entity creates a public key and a private key so that communication to a third party is possible through the public key. The numbers of keys that are used in the public key cryptography are far smaller when compared to symmetric key cryptography. The complexity of the public key cryptography algorithm restricts its use for large

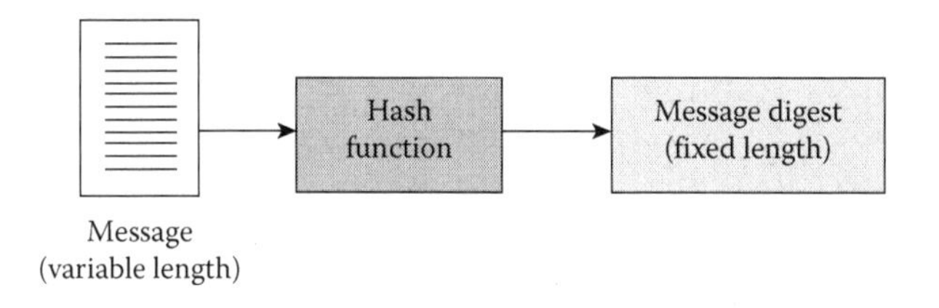


**Figure 31.5 Making message digest.**

amount of texts. Also the relationship between an entity and its public key must be verified in the public key method.

The digital signature that is used for authenticating the sender of documents and messages uses public key cryptography. Either the whole document or a digest (condensed version of the document) can be signed. Public key encryption can be used for signing an entire document where the private key is used by the sender for encryption and the public key of the sender is used by the receiver for decryption. Signing the whole document using public key encryption is inefficient for long messages. To overcome the inefficiency in signing the whole document using the public key encryption, a digest of the document is created using hash function and then signed (see Figure 31.5). Digest is a condensed version of the document. Message Digest 5(MD5) that produces a 120-bit digest and Secure Hash Algorithm 1(SHA-1) that produces a 160-bit digest are the two common hash functions.

Hash operation is one-way operation; that is, the digest can only be created from the document while the other way is not possible. Also hashing is a one-to-one function producing little probability that two messages will create the same digest. The digest so created is encrypted by the sender using the private key and attached to the original message and sent to the receiver who separates the digest from the original message and applies the same hash function to the message to create a second digest.

The receiver uses the public key of the sender to decrypt the received digest. If the two digests are the same then message authentication and integrity are preserved. The digest, which is a representation of the message, is not changed indicating that integrity is provided. The message is secure as the message and its digest comes from a true sender. Also the sender cannot deny the message as the sender cannot deny the digest.

There is an access to every user's public key in the case of public key cryptography. So as to prevent the intruder attack, the receiver can ask a certification authority or CA, which is a federal or state organization that binds a public key to an entity and then issues a certificate. This will help the receiver to give his public key to the users so that no user can accept a public key forged as that of the receiver. The problem behind public key certification is that one certificate may have the public key in one format and another certificate may have the public key in another format. To overcome this ITU has devised the X.509 protocol that describes the certificate in a structural way. When we use public key universally many DNS servers are required to answer the queries, such that the servers are arranged in a hierarchical relationship that is called public key infrastructure (PKI).

## 31.4 Message Security

The following services are offered under security:

- Privacy: It is a security service in which the transmitted message makes sense only to the intended receiver. Privacy can be achieved either by using symmetric key cryptography or public key cryptography.
- Message authentication: It is a security service in which the sender of the message is verified to be authentic and having the required privileges for sending a message to and requesting services from a specific receiver. Once the sender is authenticated, then the sender is authorized to perform a set of operations based on security rules.
- Message integrity: It is a security service in which the data reaches the receiver as they were sent without any accidental or malicious changes that may occur during transmission.
- Nonrepudiation: It is a security service in which the receiver must be able to prove that the message received by the receiver came from a specific sender.

Before starting a communication, the sender identity is verified using the authentication service. User authentication can be done by both symmetric key cryptography and public key cryptography.

- User authentication with symmetric-key cryptography: Here the identity of one entity (person/process/client/server) is verified for another entity. For the entire process of system access, the identity of the user is verified only once.

  In one approach the sender sends the identity and password in an encrypted message using the symmetric key. Here the intruder does not have any idea about the identity and password because it does not know the symmetric key. But the intruder can cause replay attack by intercepting both the authentication message and the data message and storing and resending them to the receiver again.

  In another approach, replay attack can be avoided by using a large random number called nonce that is used only once so as to distinguish a fresh authentication from a repeated one. Here the sender identity in plaintext is sent to the receiver who sends a nonce in plaintext. The sender then encrypts it using the symmetric key and sends back the nonce to the receiver. The intruder cannot replay the message as the nonce is valid only once.

- User authentication with public key cryptography: Here the sender encrypts the message with its private key and the receiver can decrypt the message using the sender's public key and authenticate the sender. But the intruder can either announce its public key to the receiver or encrypt the message containing a nonce with its private key, which the receiver can decrypt with the intruder's public key (assumed to be that of the sender).

## 31.5 Key Management

Key management includes both symmetric key distribution and public key certification. Symmetric keys are extremely useful when they serve a dynamic function between two parties such that they are created for one session and destroyed when the session is over. These keys are called session keys.

Some of the key management protocols are:

- Diffie–Hellman method: In the Diffie-Hellman protocol two parties are provided with a one-time session key, so that they can exchange data without remembering or storing it for future use. Without meeting, the two parties can agree on the key through the Internet. The disadvantage is that an intruder can easily intercept the messages sent.
- Kerberos: This is a very popular authentication protocol that is also a key distribution center (KDC). The KDC is a secret key encryption. It is a trusted third party that shares a symmetric key with the sender and another symmetric key with the receiver so as to prevent the flaw of the Difflе–Hellman method.

  Kerberos is not used for person-to-person authentication. The authentication server (AS), ticket-granting server (TGS), and the real or data server that provides services to others are the different servers involved in Kerberos. The authentication server (AS) is the key distribution center in Kerberos protocol. Each user that registers with the AS is provided with a user identity and password. These identities and passwords are stored in the AS database. After verifying the user, AS issues a session key to be used between the sender and TGS and then sends a ticket for TGS.

  The ticket-granting server (TGS) issues a ticket for the real server and a session key between the sender and receiver. User verification is separate from ticket issuing so that the sender can contact TGS several times for obtaining tickets for different real servers, although the sender ID is verified just one time with AS. The real or data server provides services to other users. Global distribution of ASs and TGSs are allowed by Kerberos where each system is called a realm. The user gets a ticket for the local server. The user can ask local TGS to issue a ticket that is accepted by a distant TGS, and the ticket is sent provided that the distant TGS is registered with the local TGS. The user can then use the distant TGS in order to access the distant real server.

- Needham-Schroeder protocol: This key management protocol uses multiple challenge response interactions between parties so as to achieve a flawless protocol.

## 31.6 Security Protocols

The internet engineering task force (IETF) developed a collection of protocols to provide security for a packet at the IP level called IPSec (Internet protocol security). Selection of encryption, authentication, and hashing methods are left to the user. IPSec simply provides the framework and mechanism. A logical unidirectional (simplex) connection can be established between the source and the destination by an IPSec signaling protocol called security association (SA). Two SA connections are required for a bidirectional (duplex) connection, one SA connection in each direction.

The following elements define an SA connection:

- SPI or 32-bit Security Parameter Index acts as a virtual circuit identifier in connection oriented protocols like Frame Relay or ATM.
- Protocol type used for security.
- Source IP address.

The transport mode and the tunnel mode are the two modes that define where the IPSec header is added to the IP packet. In transport mode, the IPSec header is added between the IP header and the rest of the packet and in tunnel mode, the IPSec header is added before the original IP header and a new IP header is added before the IPSec header. IPSec header, IP header, and the rest of the packet constitute the payload of the new IP header.

Authentication header (AH) protocol and the encapsulating security payload (ESP) protocol are the two protocols defined by IPSec.

- Authentication header (AH) protocol: Source host authentication and payload integrity carried by the IP packet are carried out by the AH protocol. But it does not provide privacy. Using a hashing function and a symmetric key, it calculates a message digest and inserts the digest in the authentication header. AH is placed in the correct location depending on the transport or tunnel mode.
- Encapsulating security payload (ESP) protocol: Source host authentication, data integrity, and privacy are provided by the encapsulating security payload (ESP) protocol.

Security in the World Wide Web can be met through the security protocol at the transport level known as transport layer security or TLS. The TLS lies between the application layer and the transport layer (TCP). The handshake protocol and the data exchange or record protocol are the two protocols that make up the TLS.

- Handshake protocol: Apart from defining message exchange between the browser and the server the protocol also defines other communication parameters. It is responsible for security negotiation and authentication of the server to the browser.
- Data exchange protocol: It uses a secret key for encrypting the data for secrecy and encrypting the message digest for integrity. The algorithm details and specification are agreed upon during the handshake phase.

Now coming to the top most application layer, there are several protocols that offer application level security. One of the application layer security (ALS) protocols is PGP. The pretty good privacy (PGP) protocol is used at the application layer to provide security including privacy, integrity, authentication, and nonrepudiation in sending an e-mail. Combination of hashing and public key encryption is called digital signature and provides integrity, authentication, and nonrepudiation.

## 31.7 Implementation Example

This example shows implementation of 3DES or triple DES (data encryption standard) symmetric encryption. 3DES is one of the most secure and unbroken cryptographic algorithm. This implementation example is written using C# programming language. The program is split to help the reader understand the code flow, so that a similar implementation can be done in the reader's choice of programming language.

1. Add the cryptography namespace

```
// Cryptography namespace
using System.Security.Cryptography;
```

2. Next declare the DESCryptoServiceProvider object. This object will contain the symmetric keys required to encrypt and decrypt files.

```
private DESCryptoServiceProvider descriptor;
```

3. The file selected for encryption is termed "fileName" in this example. This can be changed to the specific file to be encrypted. Next we write the function to encrypt. The code has been commented so that the reader can follow the steps involved.

```
/* ---------------------------
Encrypt function
--------------------------- */

void Encrypt_Function( )
{
// Ouput file has same file name with extension ".enc"
string encryptedFile = fileName + ".enc";

descriptor = new DESCryptoServiceProvider();
string stringInput;

// Read file Contents to string
FileStream fileRead = new FileStream(encryptedFile,
FileMode.Create,FileAccess.Write);
StreamReader strRead = new StreamReader(fileName);
stringInput = (strRead).ReadToEnd();
strRead.Close(); // Close stream reader

// Convert string to byte array
byte[] input = Encoding.Default.GetBytes(stringInput);

// Encrypt data
ICryptoTransform desencrypt = descriptor.
CreateEncryptor();
CryptoStream output = new CryptoStream(fileRead,
desencrypt, CryptoStreamMode.Write);
output.Write(input, 0, input.Length);
output.Close();

fileRead.Close(); // Close file reader

}
```

4. Next we can implement decryption.

```
/* ---------------------------
Decrypt function
--------------------------- */

void Decrypt_Function( )
{

// FileStream reads the cipher text from the file
FileStream fileRead = new FileStream(fileName, FileMode.
Open, FileAccess.Read);

// CryptoStream decrypts the data from the stream
ICryptoTransform decrypt = descriptor.CreateDecryptor();
```

```
CryptoStream streamCrypt = new CryptoStream(fileRead,
decrypt, CryptoStreamMode.Read);

// StreamReader uses the ReadToEnd method to pull the
// decrypted data into a string
string decryptedFile = new StreamReader(streamCrypt).
ReadToEnd();

// To remove .enc extension from the filename
FileInfo info = new FileInfo(fileName);
string originalFile = fileName.Substring(0, fileName.
Length - info.Extension.Length);

// StreamWriter dumps the string containing the
// decrypted data to disk
StreamWriter fileWrite = new StreamWriter(originalFile);
fileWrite.Write(decryptedFile);
fileWrite.Close();

}
```

## 31.8 Conclusion

The intent of this chapter is to introduce the reader to the basic concepts of security implementation. Some of the fundamental concepts of symmetric and asymmetric cryptography were discussed. The security services and protocols will give the reader an understanding of security implementation at different layers. The implementation example discussed in this chapter uses 3DES.

## Additional Reading

1. Neil Daswani, Christoph Kern, and Anita Kesavan. *Foundations of Security: What Every Programmer Needs to Know*, New York: Apress, 2007.
2. Vasant Raval and Ashok Fichadia. *Risks, Controls, and Security: Concepts and Applications*. New York: Wiley, 2007.

# Chapter 32

# Application Layer

This chapter is about NETCONF protocol and its implementation. The basic features in NETCONF are discussed along with the XML message for performing NETCONF operations. Considering that most legacy implementations are based on SNMP and the recent technology is centered around XML, a sample program to convert SNMP MIB to XML is also discussed.

## 32.1 Introduction

In the preceeding chapters we have discussed implementation of several application layer protocols like FTP, SMTP, and HTTP. While the application layer protocols discussed so far were suited for a specific management activity like transfer of files, event handling as mails, and so on, these protocols do not fall in the category of a network management protocol for message exchange between the agent and server.

In order to be used as a management protocol, the protocol needs to have functions to perform the following activities between server and agent:

- Function invoked by server to get data from the network element/agent.
- Function invoked by server for setting data in the management database of network element.
- Function invoked by agent to respond back to messages from the server.
- Function invoked by agent to send synchronous and asynchronous event notifications.

There are several protocols like SNMP and NETCONF that satisfies these conditions and are used for network management. Considering that SNMP is mostly phased out with the advent of XML-based protocols, this chapter is focused on implementation with NETCONF, which is an XML-based protocol. It should be noted that most legacy NMS implementations are still SNMP based. A program to convert SNMP event MIB to XML is also discussed. The program uses libsmi library to access SMI MIB modules.

The features of NETCONF that make it the next generation management protocol are:

- It is much easier to manage devices: Compared to the command based implementation of TL1, the object oriented implementation in CORBA and the object identifier based implementation in SNMP, the document based approach in NETCONF makes it much easier to manage network devices.
- Data exchange based on APIs: NETCONF allows devices to expose API that can be used for exchanging configuration data.
- Inherently XML based: – NETCONF is an XML-based protocol, so all the advantages of using XML for structured data exchange is applicable to NETCONF.
- Secure connection: NETCONF uses a secure and connection oriented session for communication between agent and server.
- Ease of adding new features: The functionality of NETCONF can closely mirror the device functionality. This results in quick adoption of new features at much lower implementation costs.
- Multiple database support: The NETCONF system can query one or more databases for network management data.

Most industry leaders in operation support space have embraced XML-based protocols. NETCONF is one of the most popular XML-based protocols and hence has been used in this chapter to discuss management operations.

## 32.2 NETCONF

NetConf, short for network configuration, is based on the client–server model and uses a remote procedure called (RPC) paradigm to facilitate communication. The messages are encoded in XML and are sent over a secure, connection-oriented session. The contents of both the request and the response messages are described using XML. The client is a script or application typically running as part of a network manager and the server is the agent in the network device.

The layers in NETCONF are (see Figure 32.1):

1. Content layer: This is the topmost layer and corresponds to configuration data being handled using NETCONF.
2. Operation layer: Has a set of methods for communication between agent and manager.

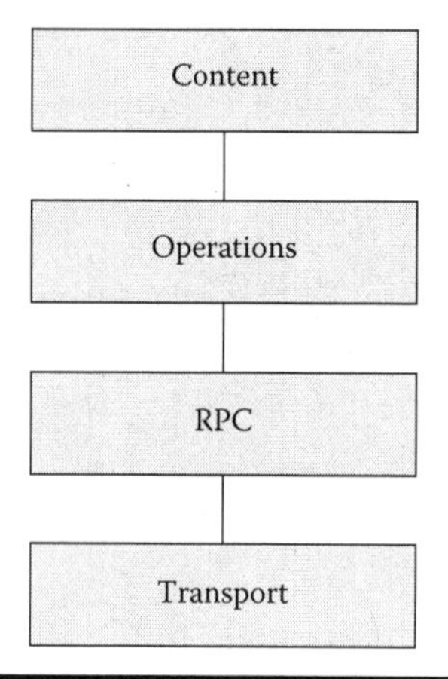


**Figure 32.1 Layers in NETCONF.**

3. RPC layer: Used for encoding the remote procedure call.
4. Transport layer: This layer corresponds to the transport path for communication between the server and client. NETCONF supports transport over SSH, SSL, BEEP, and console.

The configuration data depends on the type of device being managed and varies based on the device implementation. The transport layer can be any basic transport protocol that satisfies some requisites of NETCONF. In this implementation, the XML-based messaging patterns of the RPC model and the NETCONF operations are discussed.

## 32.3 RPC Elements in NETCONF

I. <rpc> element

The <rpc> element has the following characteristics:

- It has a message-id attribute to identify the RPC request from the sender. The receiver can use the same identifier in response, to indicate what request has triggered the specific response.
- This element is used in both management and operation channels.
- The method to be invoked is specified inside the <rpc> tags

```
<rpc message-id = 212>
       < method>
       <!-- This can be any method  -->
       </ method>
</rpc>
```

II. <rpc-reply> element
   The <rpc-reply> element has the following characteristics:
   - This message is sent in response to a <rpc> operation.
   - The value of the message-id attribute is the same as the value of the <rpc> for which the response is sent.
   - The response data is specified inside the <rpc-reply> tags.

```
<rpc-reply message-id = 212>
       < content>
       <!—The response content comes within the rpc-reply
       tags  -->
       </ content>
</rpc-reply>
```

III. <rpc-error> element
   The <rpc-error> element has the following characteristics:
   - This message is sent in <rpc-reply> messages.
   - It is used to capture errors that occur during the processing of an <rpc> request.
   - The error message has the following attributes:
     – <tag> This tag is used to identify the error condition using a string.
     – <error-code> This tag is used to identify the error condition using an integer.
     – <severity> This tag is used to identify the error severity.
     – <statement> This tag gives the cause of the error.
     – <message> This is a string description of the error condition.
   - Additional attributes like <action> can be used to describe the action to be taken when the specific error is encountered and <edit-path> to describe the error.

```
<rpc-error message-id = 215>
       <tag>SAMPLE_CRC_LOG</tag>
       <error-code>512</error-code>
       <severity>error</severity>
       <statement>CRC 250; </statement>
       <message>CRC 250 on Interface Card</message>
</rpc-error>
```

## 32.4 NETCONF Operations

Some of the basic protocol operations in NETCONF are discussed in this section.

I. <get-config> This operation retrieves the configuration data.

```
<!-- Example Request   -->

<rpc message-id="212" xmlns="http://ietf.org/
xmlconf/1.0/base">
       <get-config>
              <source><running/></source>
              <config xmlns="http://example.com/
              schema/1.2/config">
              <users>
                      <user>
                      <name>jithesh</name>
                      </user>
              </users>
              </config>
              <format>xml</format>
       </get-config>
</rpc>
```

```
<!-- Example Response   -->

<rpc-reply message-id="212" xmlns="http://ietf.org/
xmlconf/1.0/base">
         <config xmlns="http://example.com/schema/1.2/
         config">
                 <users>
                       <user>
                       <name>jithesh</name>
                       <type>admin</type>
                       <full-name>Jithesh Sathyan</
                       full-name>
                       </user>
                 </users>
         </config>
</rpc-reply>
```

II. <edit-config> This operation is used to load all or part of a specified configuration to a defined target configuration. The new configuration can be expressed inline or using local/remote file.

```
<!-- Example Request   -->

<rpc message-id="216" xmlns="http://ietf.org/
xmlconf/1.0/base">
      <edit-config>
              <target> <running/> </target>
              <config xmlns="http://example.com/
              schema/1.2/config">
```

```
                        <interface>
                                <name>ExternalRouter 6</
                                name>
                                <router>15</router>
                        </interface>
                </config>
        </edit-config>
</rpc>
```

```
<!-- Example Response  -->

<rpc-reply message-id="216" xmlns="http://ietf.org/
xmlconf/1.0/base">
        <ok/>
</rpc-reply>
```

III. <copy-config> This operation is used to copy a configuration file onto another configuration file. In this process the file to which content is copied is overwritten.

```
<!-- Example Request   -->

<rpc message-id="218" xmlns="http://ietf.org/
netconf/1.0/base">
        <copy-config>
                <source> <running/> </source>
                <target>
                        <url>ftp://crcNMS.com/conf_files/
                        test_Config20.txt</url>
                </target>
                <format>text</format>
        </copy-config>
</rpc>
```

```
<!-- Example Response  -->

<rpc-reply message-id="218" xmlns="http://ietf.org/
netconf/1.0/base">
        <ok/>
</rpc-reply>
```

IV. <delete-config> This operation is used to delete a configuration datastore.

```
<!-- Example Request   -->

<rpc message-id= "220" xmlns="http://ietf.org/
netconf/1.0/base">
```

```
        <delete-config>
                <target>
                        <!-- The current or <running>
                        configuration cannot be deleted -->
                        <startup/>
                </target>
        </delete-config>
</rpc>
```

```
<!-- Example Response  -->

<rpc-reply message-id = "220" xmlns="http://ietf.org/
netconf/1.0/base">
        <ok/>
</rpc-reply>
```

V. <get-state> This operation is used to retrieve the state information of network element.

```
<!-- Example Request   -->

<rpc message-id="222" xmlns="http://ietf.org/
xmlconf/1.0/base">
        <get-state>
                <state >
                        <interface>
                                <name>ExternalRouter 6</name>
                        </interface>
                </state >
        <format >xml</forrat>
        </get-state>
</rpc>
```

```
<!-- Example Response  -->

<rpc-reply message-id="222" xmlns="http://ietf.org/
xmlconf/1.0/base">
        <state>
                <interface>
                        <name>ExternalRouter 6</name>
                        <status>Active</status>
                        <connections> 5 </connections>
                </interface>
        </state>
</rpc-reply>
```

VI. <kill-session> This operation is used to terminate a specific session.

```
<!-- Example Request   -->

<rpc message-id="224" xmlns="http://ietf.org/
xmlconf/1.0/base">
       <kill-session>
              <session-id> 2 </session-id>
       </kill-session>
</rpc>
```

```
<!-- Example Response   -->

<rpc-reply message-id ="224" xmlns="http://ietf.org/
netconf/1.0/base">
       <ok/>
</rpc-reply>
```

VII. <lock> This operation is used to lock the configuration source. The lock will not be granted if a lock is already held by another session on the source or the previous modification has not been committed.

```
<!-- Example Request   -->

<rpc message-id="226" xmlns="http://ietf.org/
xmlconf/1.0/base">
       <lock>
              <target>
                     <running/>
              </target>
       </ lock>
</rpc>
```

```
<!-- Example Response   -->

<rpc-reply message-id ="226" xmlns="http://ietf.org/
netconf/1.0/base">
       <ok/>
</rpc-reply>
```

VIII. <commit> This operation is used to publish a completed configuration on the network element.

```
<!-- Example Request   -->

<rpc message-id="228" xmlns="http://ietf.org/
xmlconf/1.0/base">
      <commit>
            <confirmed/>
            <confirm-timeout>15</confirmed-timeout>
      </commit>
</rpc>
```

```
<!-- Example Response  -->

<rpc-reply message-id ="228" xmlns="http://ietf.org/
netconf/1.0/base">
      <ok/>
</rpc-reply>
```

IX. <unlock> This operation is used to release an existing configuration lock applied using <lock> operation.

```
<!-- Example Request   -->

<rpc message-id="232" xmlns="http://ietf.org/
xmlconf/1.0/base">
      <unlock>
            <target>
                  <running/>
            </target>
      </unlock>
</rpc>
```

```
<!-- Example Response  -->

<rpc-reply message-id ="232" xmlns="http://ietf.org/
netconf/1.0/base">
      <ok/>
</rpc-reply>
```

X. <discard-changes> This operation will revert the configuration to the last committed configuration.

```
<!-- Example Request  -->

<rpc message-id ="234" xmlns="http://ietf.org/
netconf/1.0/base">
      <discard-changes/>
</rpc >
```

```
<!-- Example Response  -->
<rpc-reply message-id ="234" xmlns="http://ietf.org/
netconf/1.0/base">
      <ok/>
</rpc-reply>
```

## 32.5 MIB Event to XML

a. Write the program (mibtrap2xml.c) to convert a MIB file to XML using libsmi library. In this example C programming language has been used.

Step 1: First include the required header files.

```
#include <stdio.h>
#include <smi.h>
#include <unistd.h>
```

Step 2: In the main( ) function:

- Define the SMI MIB access objects
- Initialize SMI
- Identify number of modules
- Call the function to create XML with file details

```
// Define objects for accessing SMI MIB modules
      SmiModule* smiModule;
      SmiModule** modules = NULL;

// .........

// Initialize
smiInit(NULL);

// ................. . .

// Identify number of modules
modules = (SmiModule **) malloc( argc * sizeof(SmiModule
*));
moduleCount = 0;

while( optind < argc ) {

      i = optind++;
      modulename = smiLoadModule(argv[i]);
      smiModule = modulename ? smiGetModule(modulename):
      NULL;
```

```
                if ( ! smiModule ) {
                        fprintf(stderr, "Cannot locate
                        module `%s'\n",
                        argv[i]);
                }
                else
                {
                        // Get total module count
                        modules[moduleCount++] =
                        smiModule;
                }
}
// ...................

// Call convert to XML function - createXML

file = fopen(filename, "w");

        for ( i = 0; i < moduleCount; i++ ) {
                smiModule = modules[i];
                createXml(smiModule, file);
        }

fclose(file);

// ...................
```

Step 3: Convert the data to XML. The output needs to describe each event with the object identifier, event identifier, event label, and description. Read through the comments to understand program flow.

```
void createXml(SmiModule* smiModule, FILE* file) {

SmiNode*     smiNode;
SmiNode*     tmpNode;
SmiElement* smiElem;
int   i;
int   j;
int      len;
char*   logmsg;

smiNode = smiGetFirstNode(smiModule, SMI_NODEKIND_
NOTIFICATION);

fprintf(file, "<!-- Start of xml generated from MIB: %s
-->\n", smiModule->name);

for(; smiNode; smiNode = smiGetNextNode(smiNode, SMI_
NODEKIND_NOTIFICATION) )
```

```
{
        fprintf(file, "<event>\n");
        fprintf(file, "\t<objIdf>");
        len = smiNode->oidlen;

        // Print the Object Identifier
        for ( j = 0; j < len; j++ ) {
          fprintf(file, ".%d", smiNode->oid[j]);
        }

        fprintf(file, "</objIdf>\n");

        // Print the Event Identifier
        fprintf(file, "\t<eventIdf>%s</eventIdf>\n",
        smiNode->name);

        // Print the Event Label
        fprintf(file, "\t<eventLabel>%s defined trap
        event: %s</eventLabel>\n",
        smiModule->name, smiNode->name);

        // Print the Event Description
        fprintf(file, "\t<description> %s", smiNode-
        >description);
        fprintf(file, "</descriptor>\n");

        fprintf(file, "</event>\n");

}

        fprintf(file, "<!-- End of xml generated from MIB:
        %s -->\n", smiModule->name);
}
```

b. Compile the program with libsmi library.

```
# gcc mibtrap2xml.c -lsmi
```

c. Run the executable with the MIB file as input and check the converted XML output.

```
// Example - files usually available in MIB folders
// Usual folder in linux systems: /usr/share/snmp/
mibs

# mibtrap2xml SNMPv2-MIB.txt

# mibtrap2xml IF-MIB.txt
```

d. Sample outputs

Example 1: Output for SNMPv2-MIB

```
<!-- Start of xml generated from MIB: SNMPv2-MIB -->
<event>
      <objIdf>.1.3.6.1.6.3.1.1.5.1</objIdf>
      <eventIdf>coldStart</eventIdf>
      <eventLabel>SNMPv2-MIB defined trap event:
      coldStart</eventLabel>
      <description> A coldStart trap signifies that the
      SNMP entity,
supporting a notification originator application, is
reinitializing itself and that its configuration may
have been altered.</descriptor>
</event>
<event>
      <objIdf>.1.3.6.1.6.3.1.1.5.2</objIdf>
      <eventIdf>warmStart</eventIdf>
      <eventLabel>SNMPv2-MIB defined trap event:
      warmStart</eventLabel>
      <description> A warmStart trap signifies that the
      SNMP entity,
supporting a notification originator application,
is reinitializing itself such that its configuration
is unaltered.</descriptor>
</event>
<!-- End of xml generated from MIB: SNMPv2-MIB -->
```

Example 2: Output for IF-MIB

```
<!-- Start of xml generated from MIB: IF-MIB -->
<event>
      <objIdf>.1.3.6.1.6.3.1.1.5.3</objIdf>
      <eventIdf>linkDown</eventIdf>
      <eventLabel>IF-MIB defined trap event: linkDown</
      eventLabel>
      <description> A linkDown trap signifies that the
      SNMP entity, acting in an agent role, has detected
      that the ifOperStatus object for one of its
      communication links is about to enter the down
      state from some other state (but not from the
      notPresent state). This other state is indicated
      by the included value of ifOperStatus.</
      descriptor>
</event>
```

```
<event>
      <objIdf>.1.3.6.1.6.3.1.1.5.4</objIdf>
      <eventIdf>linkUp</eventIdf>
      <eventLabel>IF-MIB defined trap event: linkUp
      </eventLabel>
      <description> A linkUp trap signifies that the
      SNMP entity, acting in an agent role, has detected
      that the ifOperStatus object for one of its
      communication links left the down state and
      transitioned into some other state (but not
      into the notPresent state). This other state
      is indicated by the included value of
      ifOperStatus.</descriptor>
</event>
<!-- End of xml generated from MIB: IF-MIB -->
```

## 32.6 Conclusion

The intent of this chapter is to introduce the reader to the basic concepts of implementing NETCONF. The chapter started with a basic overview of NETCONF. Then the layers that are relevant for NETCONF application development were discussed. The chapter concludes with an implementation example for converting MIB file to XML with libsmi library. Implementing communication between client and server using a management protocol is the base framework over which business logic of operation support solution is added. The basic operations remain the same when using another management protocol. For example NETCONF <get-config> is similar to SNMP "GET" and SNMP "SET" is similar to NETCONF <set-config> operation.

## Additional Reading


1. Alexander Clemm, Marshall Wilensky, and Scott Bradner. *Network Management Fundamentals*. Indianapolis, IN: Cisco Press, 2006.
2. Robert L. Townsend. *SNMP Application Developer's Guide*. New York: Wiley, 1995.
3. Larry Walsh. *SNMP MIB Handbook*. England: Wyndham Press, 2008.
4. Douglas Mauro and Kevin Schmidt. *Essential SNMP*. 2nd ed. California: O'Reilly Media, Inc., 2005.

# Chapter 33

# OSS Design Patterns

This chapter is about design patterns that are commonly used in programming management applications. There are several design patterns that can be applied in the development of specific modules in management application. The discussion in this chapter is limited to a few of the common design patterns used in management application development.

## 33.1 Introduction

This chapter details the following design patterns in the context of developing management applications:

- The Singleton Design Pattern
- The Chain of Responsibility Design Pattern
- The Observer Design Pattern
- The Factory Design Pattern
- The Adapter Design Pattern
- The Iterator Design Pattern
- The Proxy Design Pattern
- The Mediator Design Pattern

The C++ programming language has been used to implement the patterns. Each pattern description has three parts in this chapter. The first part gives the pattern overview. The second part details a specific management application area where the pattern can be used. The third part gives the implementation of the pattern. It is advised skipping the implementation details if the reader is not familiar with C++.

## 33.2 Singleton Design Pattern

Overview: This pattern is used to restrict the number of instances of a class to one object. Normally when an instance of a class is created with the "new" operator, a new object of the class is created. When a singleton pattern is used, the object of a class is created only once and the same object gets reused on subsequent instantiation requests. This is useful when exactly one object is needed to coordinate actions across the system.

Example of usage in telecom management programming:

- In object oriented programming the socket creation and manipulation is usually encapsulated in a singleton class to ensure that when two processes interact, the same socket object is reused for interaction between the two processes, without unnecessary creation of new sockets for interaction.
- The network will be shown in the NMS solution with a tree diagram where data is collected from multiple element management solutions. The root or base node under which the branches corresponding to different element managers are created can be designed as a singleton. This ensures that all element managers create branches under the same singleton object and a change in attributes of root object is reflected in all branches.

*Implementation:*

```
//-----------------------------------------------------------------
// Singleton Design Pattern
// ver0.0 Jithesh Sathyan Aug 02, 2009
//-----------------------------------------------------------------

#include <iostream.h>

class singleton
{
private:
    static singleton *p;
    singleton()
    {
    };

public:
    static singleton* returnAdd()
```

```
    {
        // Additional check required for multithreaded
        applications
        if(p == NULL)
        {
            p = new singleton;
            cout<<"New object created"<<endl;
            return p;
        }
        else
        {
        cout<<"Old object returned"<<endl;
        return p;
        }
    }
    static cleanUp()
    {
        delete p;
    }
};
singleton* singleton::p = NULL;
void main()
{
    singleton *p1 = singleton::returnAdd();
    singleton *p2 = singleton::returnAdd();
    singleton::cleanUp();
}
```

// Output of the singleton program is shown in Figure 33.1.

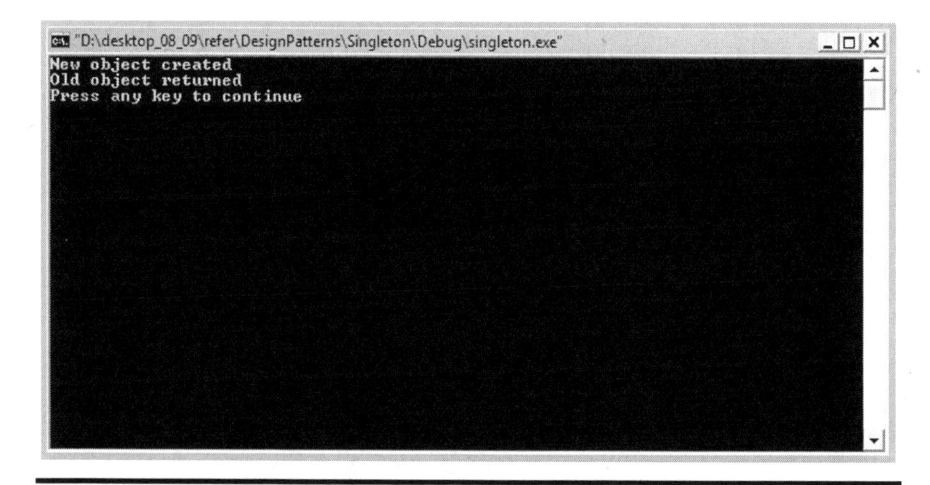


**Figure 33.1 Output obtained on executing singleton program.**

## 33.3 Chain of Responsibility Design Pattern

Overview: This pattern is used when a request sent from a dispatcher would require processing from an object based on the specific context set by the dispatcher. This avoids coupling the sender of a request to its receiver by giving more than one object a chance to handle the request. It is more like launching and leaving a request to a pipeline that contains many handlers for processing the request.

Example of usage in telecom management programming: All alarm management solutions have a help component that gives the user information on a specific alarm. That is when alarm MOD23 is raised; the operator can click help on that alarm to get information on what the cause of the alarm is and how it needs to be fixed. There will be thousands of alarms grouped under a set of modules. Each module maintains an index of alarms that it supports. If the help component is implemented using a chain of responsibility pattern, then a typical help request for MOD23 will result in a search through help modules one by one where each module takes up responsibility to check for MOD23 and leave the responsibility to next module in the chain when the string is not found.

*Implementation:*

```
//-----------------------------------------------------------
// Chain of Responsibility design pattern
// ver0.0 Jithesh Sathyan Aug 06, 2009
//-----------------------------------------------------------


#include <iostream.h>

enum { general, hss, hlr };

class Alarm
{
    // Add constructor :-)
public:
    virtual void needHelp(int x)
    {
        // Implement in child
    }
};

class HSS: public Alarm
{
private:
    Alarm *emp;
```

```
public:
    HSS(Alarm *pass)
    {
        emp = pass;

    }
    void needHelp(int x)
    {
        cout <<"In HSS Alarm module"<<endl;

        if(x != hss)
        {
            cout <<"No help in HSS Alarm module"<<endl;
            emp->needHelp(x);
        }
        else
        {
            cout<<"Help identified in HSS Alarm
            Module"<<endl;
        }
    }
};

class HLR: public Alarm
{
private:
    Alarm *emp;

public:
    HLR(Alarm *pass)
    {
        emp = pass;

    }
    void needHelp(int x)
    {
        cout <<"In HLR Alarm module"<<endl;

        if(x != hlr)
        {
            emp->needHelp(x);
            cout <<"No help in HLR Alarm module"<<endl;
        }
        else
        {
            cout<<"Help identified in HLR Alarm
            Module"<<endl;
        }
```

```
    }
};

class General: public Alarm
{
public:
    General()
    {
        //
    }
    void needHelp(int x)
    {
            cout <<"In General Alarm module"<<endl;
            cout<<"Help identified in General Alarm
            Module"<<endl;
    }
};

void main()
{
    General *gen = new General();
    HLR *ana = new HLR(gen);
    HSS *man = new HSS(ana);
    // Try requesting help for different values
    man->needHelp(2);
}
```

// Output of the responsibility program is shown in Figure 33.2.

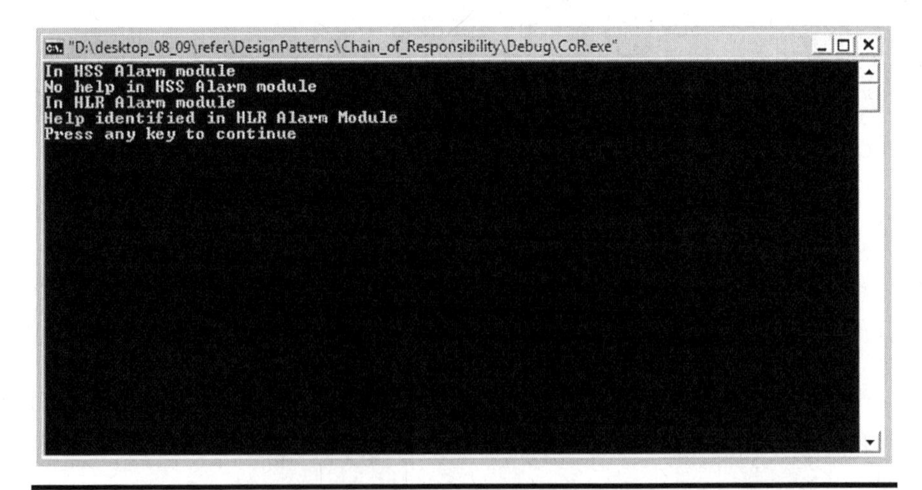


**Figure 33.2 Output obtained on executing chain of responsibility program.**

## 33.4 Observer Design Pattern

Overview: This pattern is used when an event needs to be notified to multiple observers. The typical usages of the observer pattern are:

- Listening for an external event (such as a user action).
- Listening for changes of the value of an object property.
- In a mailing list, where every time an event happens a message is sent to the people subscribed to the list.

This design pattern is used extensively in event-driven programming. The observer pattern is also very often associated with the model-view-controller (MVC) paradigm.

Example of usage in telecom management programming: Events are an important part of OSS applications. When a specific event occurs, multiple OSS modules might be interested in getting a notification to perform a set of predefined actions. This can be easily implemented using observer pattern where the OSS modules can register for a specific event and will get notified when the event occurs.

*Implementation*:

```
//-----------------------------------------------------------
// Observer Design Pattern
// ver0.0 Jithesh Sathyan Aug 01, 2009
//-----------------------------------------------------------

#include <iostream.h>

class Observer
{
    // Add constructor :-)
public:
    int flag_set;
    virtual void update()
    {
    }

}*obsr[3];

// Would suggest to store the observers in a vector
// so that we don't have to keep track of number of
observers

class Subject
{
    // Add constructor :-)
```

```
public:
    void registerObserver(Observer * o);
    void removeObserver(Observer * o);
    void notifyObserver();
};

class Database:public Subject
{
public:
    void registerObserver(Observer* x)
    {
        x->flag_set = 1;
    }
    void removeObserver(Observer* x)
    {
        x->flag_set = 0;
    }
    void notifyObserver()
    {
        for(int j=0; j<3; j++)
        {
            if(obsr[j]->flag_set == 1)
            obsr[j]->update();
        }
    }
    void editRecord()
    {
        notifyObserver();
    }
};

class Administrator:public Observer
{
    void update()
    {
        cout<<"Administrator got notified"<<endl;
    }
};

class Programmer:public Observer
{
    void update()
    {
        cout<<"Programmer got notified"<<endl;
    }
};
```

```
class Engineer:public Observer
{
    void update()
    {
        cout<<"Engineer got notified"<<endl;
    }
};

void main()
{
    Database * database = new Database();
    Administrator administrator;
    Programmer programmer;
    Engineer engineer;

    // Upcasted pointers --------------------
        obsr[0]= &administrator;
        obsr[1]= &programmer;
        obsr[2]= &engineer;
    //------------------------------------

    database->registerObserver(&administrator);
    database->registerObserver(&programmer);
    database->registerObserver(&engineer);
    cout<<"Administrator, Programmer and Engineer
    Registered"<<endl;
    database->editRecord();

    database->removeObserver(&programmer);
    cout<<endl<<"Programmer Removed"<<endl;
    database->editRecord();
}
```

// Output of the observer program is shown in Figure 33.3.

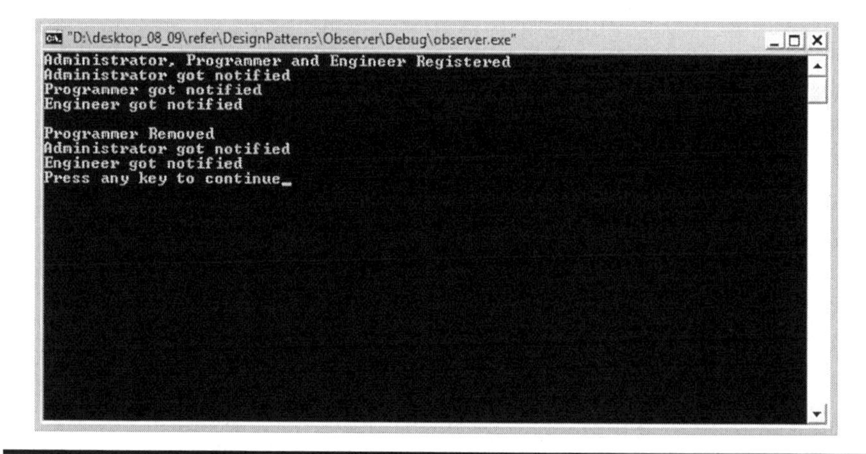


**Figure 33.3 Output obtained on executing observer program.**

## 33.5 Factory Design Pattern

Overview: This pattern is used to eliminate the scenario of frequent change in core code. In this pattern the core code is moved to a factory where decision logic is implemented. The changes specific to an implementation are moved, which ensures that the core code is not complex.

Example of usage in telecom management programming: This pattern is quite common in telecom management application development where ever the programmer finds the core code is becoming complex due to frequent changes being introduced. A typical example of the usage of this pattern is in implementing connection logic with multiple databases. The factory class only decides what database to connect, while the actual implementation of connection logic for a specific database is coded in a separate class.

*Implementation:*

```
//--------------------------------------------------------
// Factory Design Pattern
// ver0.0 Jithesh Sathyan Aug 03, 2009
//--------------------------------------------------------


#include <iostream.h>

static enum { Oracle_type, MySql_type };

class connection
{
public:
    virtual description()
    {
        cout<< "Generic connection"<<endl;
    }
};

class Oracle: public connection
{
public:
    // code for Oracle connection
    description()
    {
        cout<<"Oracle connection established"
        <<endl<<endl;
```

```
    }
};

class MySql: public connection
{
public:
    // code for MySql connection
    description()
    {
        cout<<"MySql connection established"<<endl;
    }
};

class Factory
{
private:
    int connectType;
public:
    Factory(int t)
    {
        connectType = t;
    }

    connection* createConnection()
    {
        // Decision making implemented inside Factory

        switch(connectType){
            case Oracle_type: return new Oracle();
            break;
            case MySql_type: return new MySql(); break;
            default: cout<<"Error. Contact Jithesh"<<endl;
            break;
        }
    }
};

void main()
{
    Factory *fact;
    fact = new Factory(Oracle_type);
    connection *con = fact->createConnection();
    con->description();
}
```

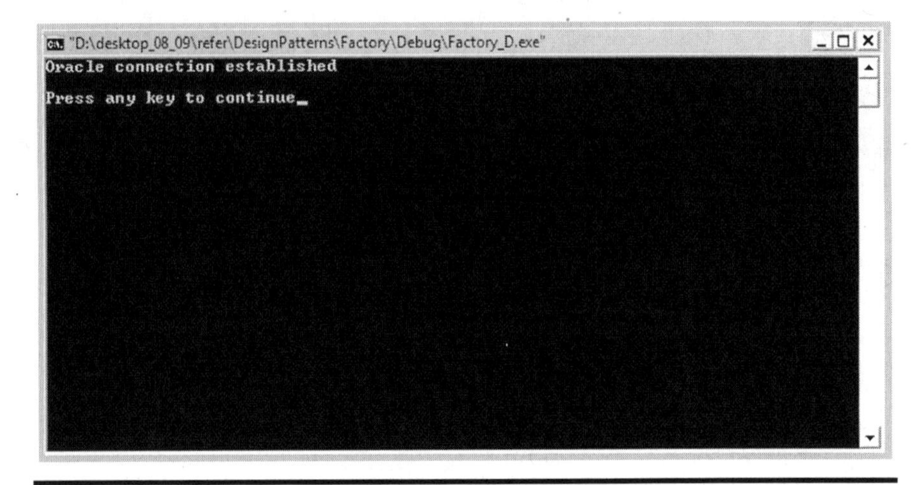


**Figure 33.4 Output obtained on executing factory program.**

// Output of the factory program is shown in Figure 33.4.

## 33.6 Adapter Design Pattern

Overview: This pattern is very useful while handling a legacy code. With this pattern, no change is required when changing the front end or back end of an application. The interaction between two incompactable modules is made possible by adding an adapter to the interface and do necessary conversions to make interaction possible.

Example of usage in telecom management programming: Adapter pattern is quite common in management applications were there will be legacy code that cannot be changed. Consider the scenario where a client application expects an SNMP message and the version information as two separate items while the server sends the version information in the message itself. Rather than changing the client or server, this problem can be fixed by adding an adapter that splits the message from the server to the format required by the client.

*Implementation:*

```
//-----------------------------------------------------------
// Adapter Design Pattern
// ver0.0 Jithesh Sathyan Aug 06, 2009
//-----------------------------------------------------------

#include <iostream>
#include <string>
using namespace std;
```

```
class server_msq
{
public:
    string snmpver_msg;

    setvalue(string msg)
    {
        snmpver_msg = msg;
    }

    getvalue()
    {
        //-----
    }
};

class newClient

{
public:
    string version;
    string msg;

public:
    setvalue()
    {
        //--------
    }
    getvalue()
    {
        cout<<"Version: "<<version<<endl;
        cout<<"Message: "<<msg<<endl;
    }
};

class adapter:public newClient
{
public:
    setvalue(server_msq *obj)
    {
        version = (obj->snmpver_msg).substr(0,2);
        msg = (obj->snmpver_msg).substr(3,(obj->snmpver_
        msg).length());
    }
    getvalue()
    {
        cout<<"Version: "<<version<<endl;
        cout<<"Message: "<<msg<<endl;
    }
};
```

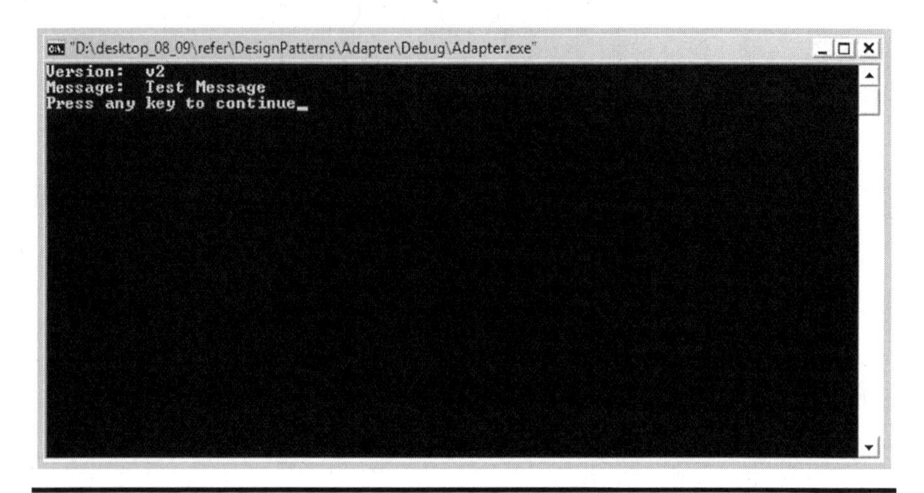


**Figure 33.5 Output obtained on executing adapter program.**

```
void main()
{
     server_msq *serv = new server_msq();
     serv->setvalue("v2 Test Message");
     adapter *adp = new adapter();
     adp->setvalue(serv);
     adp->getvalue();

// ----- now adp output can be used to set newClient object

     delete serv;
     delete adp;
}
```

// Output of the adapter program is shown in Figure 33.5.

## 33.7 Iterator Design Pattern

Overview: This pattern is used to iterate through a list. Most programming languages have standard implementation of the iterator pattern. For example, C++ iterator refines the iterator design pattern into a specific set of behaviors for STL (Standard Template Library). There is also an iterator interface in Java for working with containers like list and vector.

Example of usage in telecom management programming: Data items in management applications are usually stored in containers provided by the specific programming language used in development. To iterate through the items in the container, the iterator design pattern can be used. The data items can be attributes of the elements or the list of elements stored in a hash, stack, or queue.

*Implementation:*

```
//-------------------------------------------------------------
// Iterator Design Pattern
// ver0.0 Jithesh Sathyan Aug 08, 2009
//-------------------------------------------------------------

#include <iostream>
#include <vector>
#include <string>
using namespace std;

class employee
{
public:
    string name;
    string dept;

    employee(string n, string d)
    {
        name = n;
        dept = d;
    }

    employee(const employee& pt)
    {
        name = pt.name;
        dept = pt.dept;
    }

    employee& operator=(const employee& ptx)
    {
        name = ptx.name;
        dept = ptx.dept;
        return *this;
    }
};

class Division
{
private:
    string dept;
    vector<employee> v;
    vector<employee>::iterator itrv;
    Division(string n)
    {
        dept = n;
    }
```

```
     add(string emp)
     {
          v.push_back(employee(emp,dept));
     }
     displayDetails()
     {
          for(itrv = v.begin(); itrv != v.end(); itrv++)
          {
          cout<<"Name: "<<itrv->name<<"   "<<"Section: "<<
          itrv->dept<<endl;
          }
          cout<<"----------------------------"<<endl;
     }
};
void main()
{
     Division *d1 = new Division("Management");
     d1->add("Dan");
     d1->add("Phil");
     d1->add("Steve");
     d1->displayDetails();

     Division *d2 = new Division("Techie");
     d2->add("Mary");
     d2->add("Rick");
     d2->add("Susan");
     d2->displayDetails();

     delete d1,d2;
}
```

// Output of the iterator program is shown in Figure 33.6.

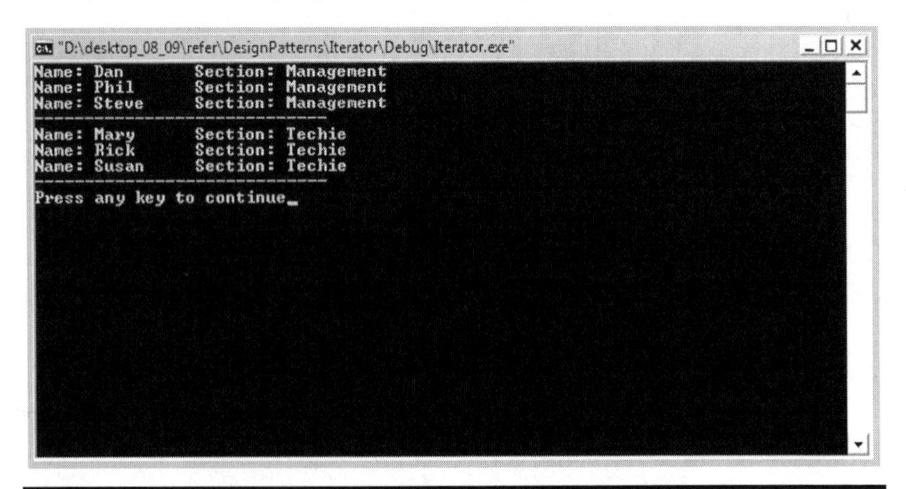


**Figure 33.6 Output obtained on executing iterator program.**

## 33.8 Mediator Design Pattern

Overview: This pattern is used to provide a unified interface to a set of interfaces in a subsystem. It is popular in development of Web-based applications. Instead of implementing navigation logic in an individual Web page, with this pattern the navigation logic is moved to a mediator that makes the link between all the Web pages.

Example of usage in telecom management programming: Most of the OSS are currently Web based. Web-based implementations makes it possible for OSS solutions to be accessed anytime and anywhere. A lot of code reuse is made possible in Web-based application development using mediator design pattern.

*Implementation:*

```
//------------------------------------------------------------
// Mediator Design Pattern
// ver0.0 Jithesh Sathyan Aug 08, 2009
//------------------------------------------------------------

#include <iostream.h>

class MedInf
{
public:
     virtual void navigate() {};
};

class Home
{
private:
     MedInf *med;
public:
     Home(MedInf *x)
     {
          med = x;
     }
     go()
     {
          cout<<"In home page"<<endl;
          med->navigate();
     }
};

class Purchase
{
private:
     MedInf *med;
```

```
public:
    Purchase(MedInf *x)
    {
        med = x;
    }
    go()
    {
        cout<<"In Purchase page"<<endl;
        med->navigate();
    }
};
class Search
{
private:
    MedInf *med;
public:
    Search(MedInf *x)
    {
        med = x;
    }
    go()
    {
        cout<<"In Search page"<<endl;
        med->navigate();
    }
};
class Exit
{
private:
    MedInf *med;
public:
    Exit(MedInf *x)
    {
        med = x;
    }
    go()
    {
        cout<<"Exiting ......"<<endl;
    }
};
class Mediator:public MedInf
{
private:
    Home *home; Purchase *pur;
    Search *sur; Exit *ext;
public:
    Mediator()
```

```
    {
        home = new Home(this);
        pur = new Purchase(this);
        sur = new Search(this);
        ext = new Exit(this);
    }
    void navigate()
    {
        int n;
        cout<<"Make a selection [1]Home [2]Purchase
        [3] Search [4]Exit"<<endl;
        cin>>n;
        switch(n)
        {
            case 1: home->go(); break;
            case 2: pur->go(); break;
            case 3: sur->go(); break;
            case 4: ext->go(); break;
            default: cout<<"Wrong choice"<<endl;
            ext->go(); break;
        }
    }
};

void main()
{
    MedInf *x = new Mediator();
    x->navigate();
}
```

// Output of the mediator program is shown in Figure 33.7.

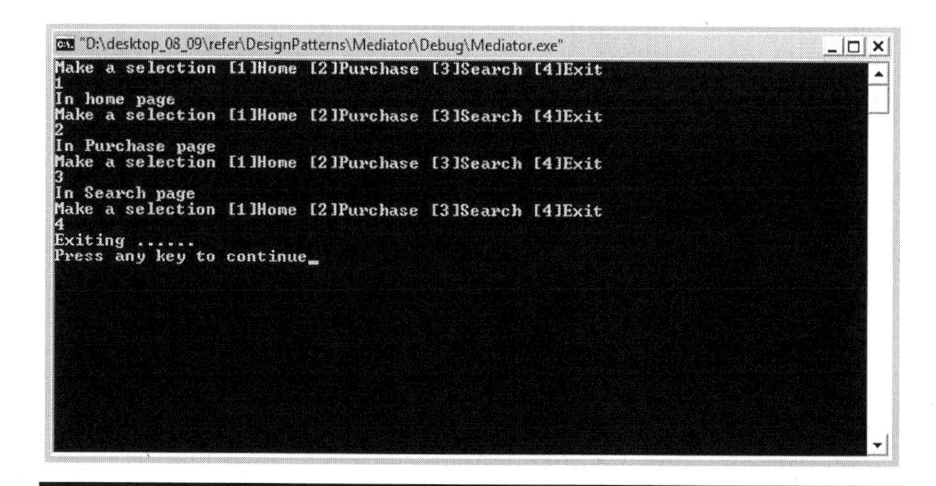


**Figure 33.7 Output obtained on executing mediator program.**

## 33.9 Conclusion

> The important note the author wants to convey from this chapter is that knowledge of the business logic is not enough to develop a good management solution. The service developers need to have a good understanding of the programming language used, the best practices in the programming language, and the design patterns that can be used to implement an effective OSS solution for commercial use.

This chapter only details some of the design patterns that are used in OSS development. There are many more design patterns that are used. For example, when the service developer wants to hide the connection details to a server module a remote proxy design pattern can be used and when state transitions are involved a state method or state object pattern can be used. The choice of design pattern is based on the specific problem that is to be addressed.

## Additional Reading

1. Erich Gamma, Richard Helm, Ralph Johnson, and John M. Vlisside. *Design Patterns: Elements of Reusable Object-Oriented Software*. United Kingdom: Addison-Wesley Professional, 1994.
2. Elisabeth Freeman, Eric Freeman, Bert Bates, and Kathy Sierra. *Head First Design Patterns*. California: O'Reilly Media, Inc., 2004.

# Chapter 34

# Report: NGNM Framework and Adoption Strategy

This chapter is a report analyzing Next Generation Network Management Framework and discussing an adoption strategy for using the same. The report takes into consideration the most important concern of the service provider on how they can effectively utilize the functionality rich legacy solution in adopting a framework that is aimed at next generation management standards.

## 34.1 Report Overview

With the evolution in technology, new and improved networks are coming up in the market and the legacy NMS solution is not scalable to manage these next generation networks or its elements without considerable rework. The capital and operational expenditure in managing these NMS solutions keeps on increasing. The problem most service providers face is either a single legacy NMS that is not scalable and takes considerable effort to modify in managing a new network or the use of multiple NMS solutions for different networks.

There is a lack of experienced professionals who have knowledge of multiple NMS and it is a considerable overhead on the operators to understand and work with multiple NMS. This leads to an increased need for NMS solutions that can cater to management of diverse networks in a multiservice, multitechnology, and multivendor environment. Another important factor influencing NMS architecture is mergers and acquisitions among the key vendors. Ease of integration is a key impediment in the traditional hierarchical NMS architecture. The networks will continue to evolve and the only way to perform network management in a cost effective and operational friendly manner is the migration to a solution that solves the challenges.

Standardizing bodies are coming up with standards and guidelines in developing next generation network management (NGNM) solutions to address current challenges in network management. The major issue here is that the current legacy NMS are functionality rich and highly specialized based on years of development. An adoption strategy of NGNM should involve the transformation from legacy NMS to NGNM in such a manner as to reap the advantages of the NGNM as well as keep the functionalities of legacy NMS.

The important features of this report are:

- The NGNM framework combining standards and popular approaches in the NMS domain.
- The various adoption strategies for moving to the suggested NGNM framework utilizing existing legacy NMS solutions.

There are multiple standards and guidelines given by standardizing bodies in developing multitechnology next generation network management solution. There are also standards on functions and data for the network management solution that are required for a specific network. This report details a NGNM framework that can be easily customized for a specific network; with new functionalities specific to the network or specialized functions added without affecting the generalized functionalities in the network management solution. Also adoption approaches to utilize existing functions in legacy NMS are outlined.

The report is organized as follows. First a brief overview of how the legacy NMS is handled, followed by the attributes and functions required in NGNM solution. Ways of customizing this framework for a specific network like IMS is outlined. Then a few approaches for adopting the framework that utilizes the functionality rich legacy NMS are detailed.

## 34.2 Introduction

With rapid improvements happening in the architecture of core and access networks it is quite clear that legacy NMS (network management system) with proprietary interfaces and protocols will become obsolete, as the cost and effort required in maintaining such systems is very huge and those are not scalable with the requirements. It is difficult to integrate other components to these systems. The solution toward developing an NMS solution for next generation network is to have a scalable NMS suitable for handling legacy and next generation networks with special customizations added specifically to the different networks.

This report presents an NGNM framework incorporating the most popular standards currently available in NMS domain. The discussion on the suggested NGNM framework is organized into four main phases. The first phase details the framework for developing generic network management solution that is based on SOA (service oriented architecture). To the framework, generic functionalities are added based on open standards and interface definitions. Once the architecture and design details of a scalable generic network management solution is ready, the report details the second phase of adding the functionalities of NGNM (next generation network management) as defined by the latest standards on next generation network management. The architecture for building these functionalities is also discussed with the example in the use of EDA (event driven architecture) to satisfy the requirement of minimal manual intervention.

At the end of phase two, a next generation network management solution is ready. The next two phases involve specialization of the solution for a specific network. IP multimedia subsystem (IMS) is considered as an example network in this report. Some of the NMS functionalities specific for IMS are discussed and the architecture for developing the same is detailed. An example of functionality in phase three is HSS (home subscriber server) provisioning agent that sends SOAP (simple object access protocol) triggers to HSS and also updates a common subscriber store based on a generic user profile for a converged network.

The functionalities added are in line with SOA. This ensures that the additions will not require changes in the generic solution. The specialized functions can be managed as separate services. The fourth phase involves customization specific to IMS, like the log format with IMS log parameters and registers for IMS performance data collection based on 3GPP standards. The discussion on NGNM is followed by an analysis of the adoption strategies for working on the NGNM framework.

## 34.3 Legacy NMS

The various modules in a typical legacy NMS are (see Figure 34.1):

a. Platform components: These components include object persistence, logging framework, libraries, and so on that are required by most of the management function implementation modules.
b. Data collection components: Based on the NMS protocol supported on the network element, these components collect data from the network elements. In some management paradigms this component may not be part of the NMS and will be a separate solution called EMS.

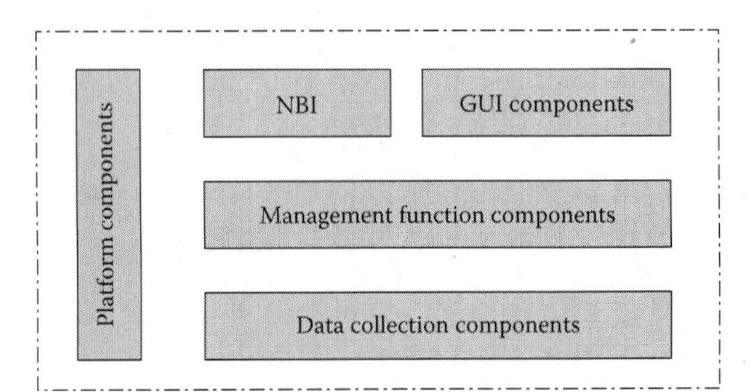


**Figure 34.1 Simplified legacy NMS.**

c. Management function components: These components use the data collected from the network elements to provide FCAPS (fault, configuration, accounting, performance, and security) and other specialized management functionality.
d. Northbound interface components: These components provide management data to the northbound interfaces like another NMS or operations support systems (OSS).
e. GUI components: These provide user interaction screens that enable a user to perform operations on the various modules in the NMS.

## 34.4 Issues with Legacy NMS

From the perspective of the multivendor, heterogeneous network, and convergent service environment, the current legacy NMS has the following issues:

1. No clear demarcation between data collection and functional implementation: For example, the data collection process itself implements fault and configuration management functionality for that network element (NE). The net result is when data needs to be collected from an NE that has the same network management protocol; most of the code needs to be rewritten or modified. If data collection is limited to protocol specific information acquiring, then this problem would not arise.
2. There is much network specific information hard coded in the functional implementation and data collection: For example, when a call trace application is implemented, the table to be queried is specified in the code for functional implementation or data collection, rather than specifying it in a standard xml file and reading from the file.
3. Complex information and process flows that are not based on standards: New services like dynamic event handlers and correlators are being developed.

Legacy NMS cannot easily adopt these services as the fault functionality does not isolate "alarm handling by setting specific threshold" from "forced clearing of alarm" at implementation. The interactions between these modules are closely coupled making the addition of new services difficult without major rework in code.

4. Nonstandard interfaces for interaction: Proprietary interfaces are being used by legacy NMS, to communicate with the NE/EMS in the southbound interface and to OSS in the northbound interface. The communications within the different layers of NMS also use proprietary interfaces. When there is a need to support new NE/EMS/OSS, the complete pipeline in legacy NMS is modified to manage the NE/EMS/OSS. Proprietary interfaces make the NMS very tightly coupled and difficult to maintain.
5. Dependency on management protocols and not based on standard information/data model: The NMS should not be hardwired to the protocol used by the EMS from which it collects data. The functionalities in NMS must be implemented for a standard format unlike the way it is done in legacy NMS.
6. Nonscalable architecture: There are multiple functionalities that are supported in an NMS and in most cases these functionalities may not be implemented by the same vendor. Legacy NMS is not based on scalable architecture like SOA, which makes addition of functionality tedious and causes minimal reuse of components.

# 34.5 NGNM Solution

## *34.5.1 NGNM Framework*

The Phase 1 of the NGNM solution development starts with creation of a generic NGNM framework. This is the NGNM framework shown in Figure 34.2 without any specialized NMS functions. The NGNM framework suggested in this report supports the following:

a. SOA for architecture: The functionalities in the NMS are designed as separate services with standard interfaces for interaction and an XML file for any customization. There is no hard coding of network element specific information like table names. New functionalities can be added as separate services without disturbing or making changes in existing functions. For example, the performance module may interact with the fault module but is independent of the implementation of the fault module. With this technique, when open source components like an independent subscriber provisioning functionality is added to the NMS that interacts with fault or performance module, retesting of existing functions is not required. SOA leads to smooth integration and functionality rich NMS.

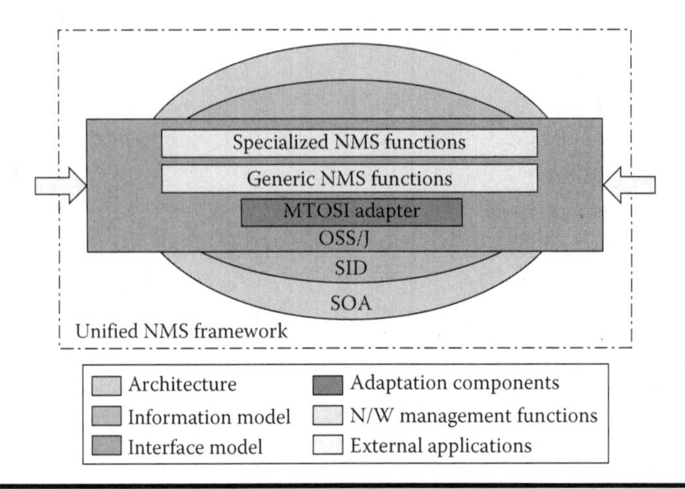


**Figure 34.2 NGNM framework.**

Advantages of using SOA in NGNM framework:

- More reuse
- Reduced effort and risk in implementation with independent services
- Faster development
- Ease of enhancement
- Reduced development and test cost
- Bug fixing and maintenance is easier

b. SID (shared information/data) as the information model: The different modules that make up the NMS and the communication between applications and NMS need to use a common format of data and information for interoperability. SID is currently the most popular information and data model for use in telecom management. The same is adopted for the generic framework development. With SID, first the object in the management framework needs to be identified. This is followed by identification of object attributes and relationship with other objects.

Advantages of using SID in NGNM framework:

- Information in standard format leads to reuse of data.
- More applications that can work on a specific format.
- Published standards of data leads to independent and open source development of product.
- More competition leading to more challenging services that work on the data.

c. Interface model with OSS/J and extended support for MTOSI (multi-technology operations system interface): OSS/J has been adopted for defining the interface language. This is because Java is now extensively used in developing telecom network management modules and most of these components support OSS/J for interface. The queries that are supported with OSS/J are mostly generic and some of the commands on inventory and fault management including certain specific activations cannot be performed with generic OSS/J. This leads to support MTOSI for some specific level queries. If the applications that interact with the NMS have support for MTOSI then there would definitely be a scope to add specialized NMS functions. This can be achieved with an MTOSI adapter.

Advantages of using interface standardization with OSS/J with MTOSI adapter in NGNM framework:

- Standards in interface lead to reuse of application.
- More applications that can request and respond on a specific format.
- Published standards for interface leads to independent and open source development of product.
- More competition leading to more challenging services that work on the specified interface.

d. Generalized NMS functions: This corresponds to modules with generic network management functionalities independent of the network being handled. There are a set of functions that has to be done by an NMS with all kinds of telecom networks whether it is GSM, CDMA, or IMS. Some of these functions are a correlation of fault and performance data, user management, administration, and security.

e. Specialized NMS functions: These are functional modules specific to a network like HSS subscriber provisioning in an IMS network, a charging model for UMTS, and so on. Use of the SOA architecture ensures that the modules in specialized NMS are self-sufficient and does not affect the working of generic NMS.

The addition of specialized functionalities is done only in phase 3 of developing the NGNM solution. Phase 1 of NGNM is limited to the development of a generic NGNM framework.

f. External applications: These are applications that interact with the NMS. It can be an EMS, GUI, OSS, or even a network element that has an agent that can respond to queries from an NMS.

Advantages of having a generalized NMS functions set and isolating it from a specialized set:

- Generalized functions can be reused across different set of networks and isolating it from the specialized set reduces effort to testing both when a specialized function is added and when the generalized function is put to support a new network.
- The specialized functions can be provided as an add-on or plug-ins for a solution suite that comprises of generalized functionalities leading to better revenue and customized NMS.

### 34.5.2 Generic Functionalities for NGNM

The phase 2 of NGNM solution is adding the NGNM functionalities. The company that goes about implementing the NGNM solution has the liberty to choose the set of NGNM functions to be added in the product. The document TR133 of TeleManagement Forum (TMF) can be referred to understand the requirements of next generation network management. The studies in this release of TMF also look into works from different standardizing bodies working on NGNM.

Some of these organizations include ITU-T, 3GPP, OASIS, OIF, MEF, and so on. The supporting document in TR133 of TMF lists the requirements from which the management functional modules to be implemented needs to be identified.

The implementation of these functions should be in line with the existing framework. It should use an SOA based design and have support for common information and interface model. Sample architecture for developing an event engine based on standards is discussed next.

One of the requirements for NGNM is the building of self-learning and survivable networks. This is achieved in the proposed NMS framework by using the event driven architecture. EDA enables the NMS to have dynamic and automatic allocation of network resources. The network management systems built on this framework would be able to do proactive trend monitoring and build survivable networks.

The architecture of such an event engine is shown in Figure 34.3. It can be seen that the architecture builds the event engine as an independent module that can collect data in a standard format or a legacy system. Another generic functionality that can be implemented in NMS is SSO (single-sign-on). TR133 can be referred to for a complete understanding.

The generic NGNM framework in itself leads to satisfying some of the requirements of NGNM. For example M.3060 of ITU-T recommends support for multiple

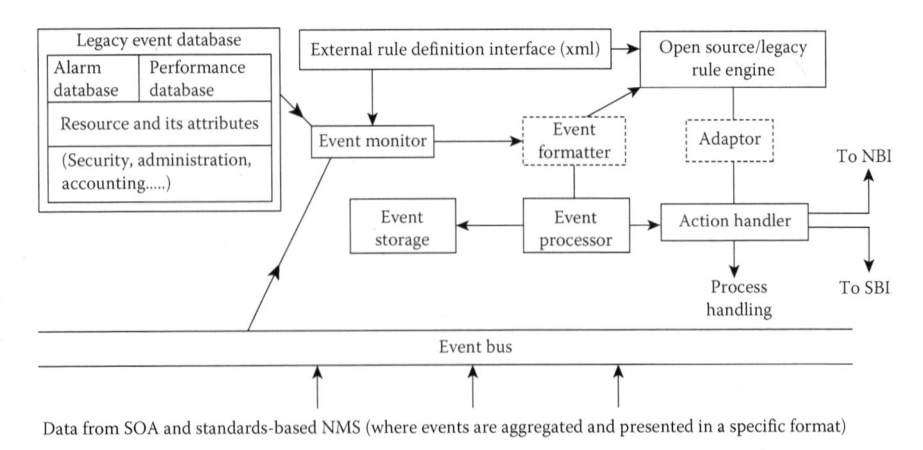


**Figure 34.3 Event engine.**

types of networks. The generic framework and functionalities designed so far can be used with different kinds of networks and not just for IMS.

### 34.5.3 Specialized Functions

In this report a sample application for IMS network is considered to detail a specialized application. Now it is time to add the IMS specific functions as part of phase 3 of the NGNM solution development. One of the functionalities required for an IMS network is provisioning of the subscriber database residing on the HSS. Figure 34.4 in this section on subscriber provisioning shows how the HSS provisioning is implemented in the NMS framework. An HSS provisioning agent sends SOAP triggers to HSS and also updates a common subscriber store based on a generic user profile for a converged network. The interface used is SOAP over HTTP, so it is standardized to use any agent.

Messaging is based on XML schema files for HSS IMS GUP components. This uses the SOAP protocol as defined by 3GPP for subscriber provisioning with http for Web interfacing. The common subscriber store helps in managing multiple HLR and HSS.

"3GPP TS 29.240" Release 7 can be referred to for identifying the IMS GUP components. The use of SOAP in the design of HSS subscriber provisioning is also recommended in this 3GPP document. It can be seen that SOAP is also the suggested protocol for implementation in the MTOSI interface. Another NMS function that can be added is charging management based on 3GPP guidelines. The 3GPP specifications on IMS and OAM&P can be referred to at www.3gpp.org and www.3gpp2.org for more functionality specific to IMS that can be added in the NMS.

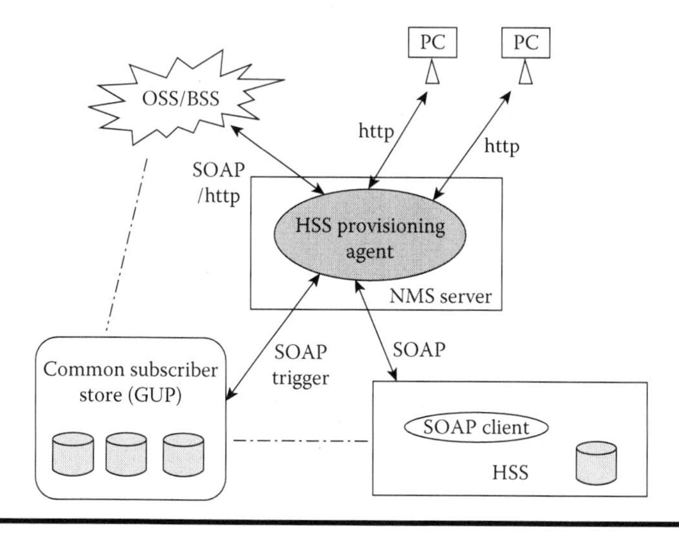


**Figure 34.4 HSS subscriber provisioning.**

### 34.5.4 Customization for a Specific Network

The final phase of the NGNM solution development that involves customization for a specific network whenever required is discussed next. The previous discussion on generic framework and functions specifies the use of xml files for customizing the function module. There is certain level of customization of generic functions required when applying it for a specific network. Again IMS is taken as an example in this report.

The standards specified in 3GPP TS 32.111-2 for the fault management in an IMS network can be used in customizing the parameters in the alarms displayed. All the alarms will be having the attributes as mentioned in Section 5.5.1 of the 3GPP TS 32.111-2 document. It specifies that an alarm needs to have two parts, the notificationHeader and alarmInformation-Body.

Some of the parameters in the notificationHeader are:

- managedObjectClass: Class specific to managed object.
- managedObjectInstance: Instance of the managed object.
- notificationId: Unique id that can differentiate two alarm notifications raised for the same problem scenario on the same managed object.
- eventTime: The time at which the event occurred.

Some of the parameters in the alarmInformationBody are:

- probableCause: It provides information that supplements the eventType on the cause of the event.
- perceivedSeverity: The severity of the alarm.
- proposedRepairActions: How to fix the problem specified in the alarm.
- additionalInformation: Information like the time the alarm was acknowledged.

There is another 3GPP document for performance parameters. The standards specified in 3GPP TS 32.409 can be adopted for defining the performance management parameters to be collected for network elements in IMS network.

## 34.6 Adoption Strategy

Adoption strategy is a discussion of how the NGNM framework can be adopted without doing away with the existing legacy NMS solution.

### 34.6.1 Using Mediation Layer

This approach is suitable when the current architecture uses multiple NMS solutions to manage different networks. In this case, migration of each external application to an NGNM framework is not a cost-effective approach. So it is suggested

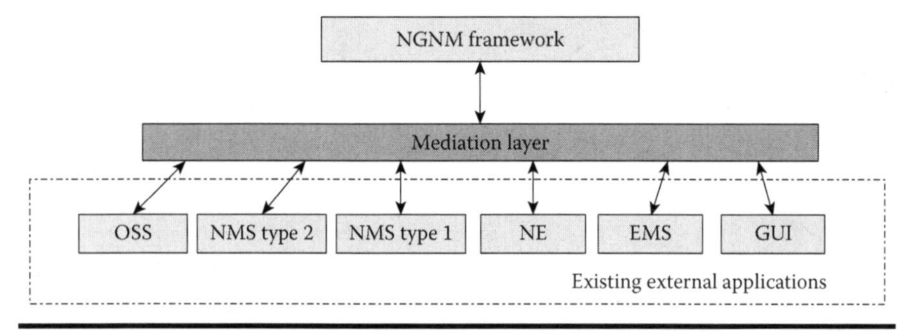


**Figure 34.5 Use of mediation layer.**

creating an NGNM framework as discussed in the previous sections, for future enhancements when adding functions or supporting new networks and integrated the existing legacy NMS solutions to a single GUI of NGNM framework by writing adapters for the existing legacy applications to plug-in to the NGNM framework

In this approach a mediation layer is used by existing applications to interact with the NGNM framework (see Figure 34.5). This mediation layer contains a pool of interface and data/information adaptors. There can be different protocols that collect data from the applications that interact with the NMS. Conversion in data format is required so that the NMS can work on the data, and the interface or language of interaction needs to be based on the interface model supported by NMS. So a conversion might be required if the application does not already support the interface and information model. The adapters need to be written for the external application that interacts with the unified NMS.

This way a single NMS triggers commands and gets results from the different external applications and presents the output on a single GUI. Management of new networks or addition of new functionalities will be supported without using any adapters in the NMS framework.

This adoption approach was used by TMF to demonstrate how legacy/customized NMS solutions can interface with standards based frame in its openOSS catalyst project.

The mediation approach is most popular in the service provider space where there are multiple NMS solutions being put to use to manage the different networks. The service provider would want to integrate the functionalities in the different NMS solutions to a single GUI to make network operations support more convenient and reduce the overhead on the technicians.

### *34.6.2 Staged Migration*

This approach is suitable when a single legacy NMS is used for managing the networks. This is mostly the case with network equipment manufacturers who work on only one or two NMS solutions and has added functions to the NMS solutions with years of development. Years of development have made it a highly specialized NMS, but each time a new network is to be managed or new application needs to be added, there is considerable work involved to make the legacy NMS usable for the new requirement.

In this case there would be many code segments and functionalities in the legacy NMS that can be reused and writing an adapter would mean losing the opportunity to reuse it effectively. A complete rewrite from legacy NMS to NGNM by porting of applications will result in a majority of applications not working properly or ending up with major hard coding and tight coupling in the new framework that still may not be scalable. This makes the staged approach a good solution to migrate to NGNM from legacy NMS in a cost-effective manner.

The staged migration has four main stages:

*Stage 1: Grouping of Functionality*

This stage involves splitting the legacy NMS solution into functional framework and data collector (see Figure 34.6). The data collector section only handles data collection from the network element based on the specific protocol and all functional framework is made independent of data and uses xml or predefined format files to manipulate data. The platform, GUI, and northbound interface components are scalable and does not usually have dependencies. Minor dependencies if any can be easily eliminated. This stage will mainly involve splitting some processes into multiple processes and removing hard coding specific to a particular format of data.

*Stage 2: Make Interactions to Framework Standardized and Data Collection Generic*

In this stage standardization is achieved for any interaction with the functional framework. The information/data model (SID) and interface model (OSS/J and MTOSI) is implemented at this stage. Now data collection and display systems

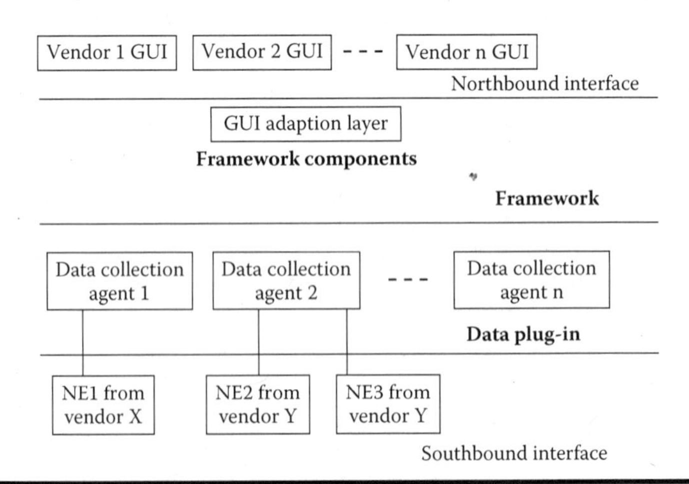


**Figure 34.6 Grouping of functionality in Stage 1.**

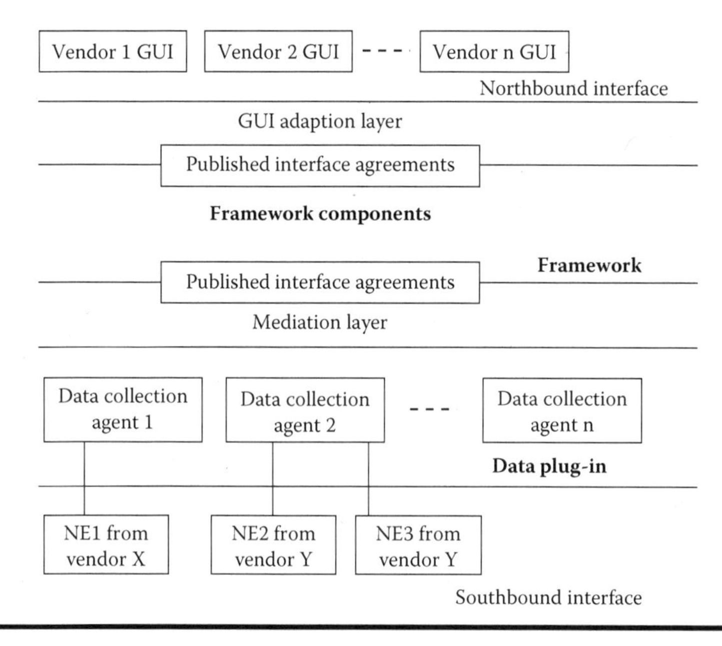


**Figure 34.7 Standardized interaction between framework and data plug-in.**

provided by a different vendor or supplied by open source can be plugged in to the functional framework (see Figure 34.7). The data collection also needs to be made generic, so that the data collector developed for a particular protocol can be used across multiple NEs using the same network management protocol. An example of the change in data collection is to remove any hard coding of tables to read when collecting data from the code to xml files. This stage can be split into interface standard, information standard, and reusable data collectors.

*Stage 3: Split Functional Framework to Services*

The functionalities in the legacy NMS have a high dependency between modules or a major chunk of functionality end up in the same module. We need to isolate functionality into services and make them loosely coupled in this stage. This is similar to the way the functionality is exposed in the user interface where each service is provided by a different option selected in the user interface (see Figures 34.8 and 34.9).

*Stage 4: Standardize Interaction between Services*

The loosely coupled functionalities cannot scale easily for new applications and interoperate with a newly introduced service unless the interfaces for the services are well defined and interactions standardized. Another activity that can be done as part of this stage is to aggregate common services like a report engine for different types of data representation (see Figure 34.10).

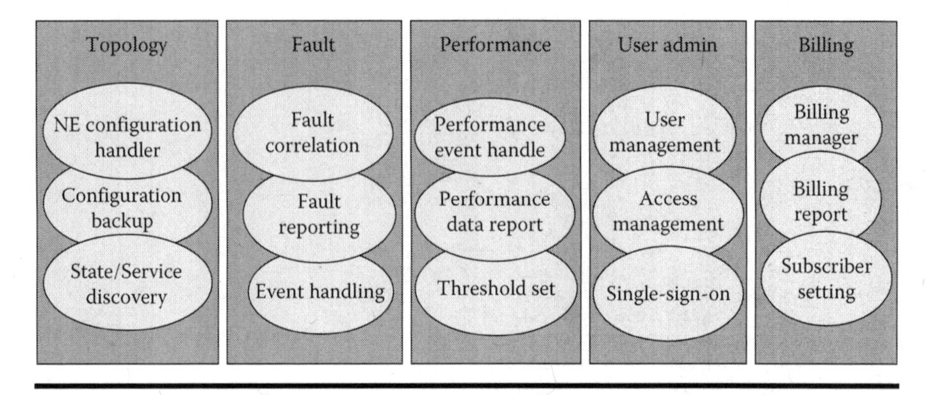


**Figure 34.8 Framework having interdependent processes.**

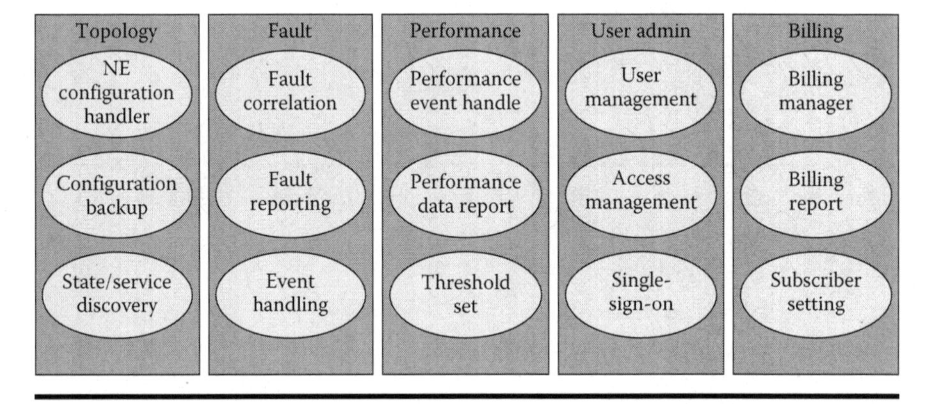


**Figure 34.9 Functions split into services in Stage 3.**

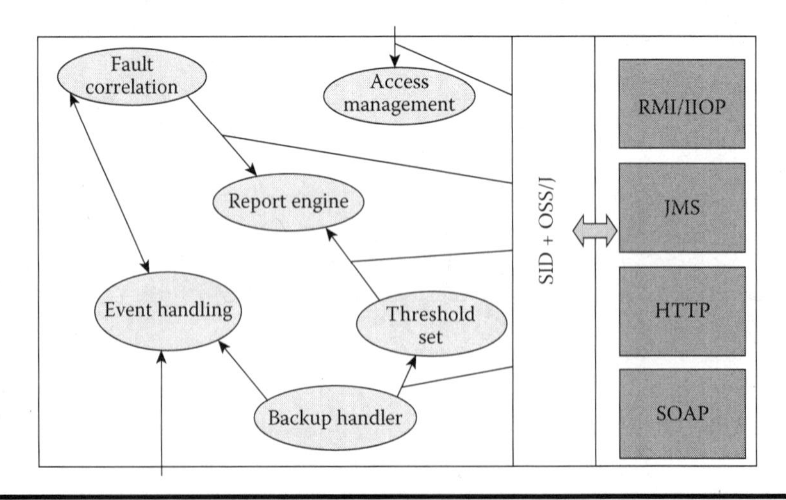


**Figure 34.10 Service assembly in Stage 4.**

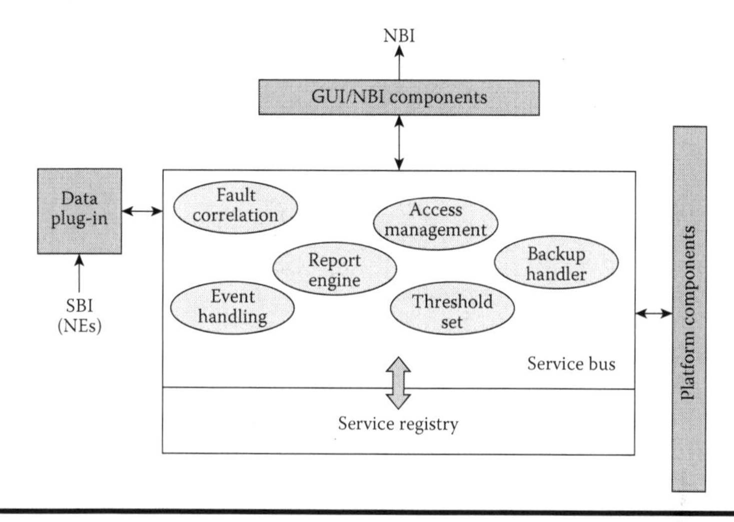


**Figure 34.11 Enhancements on the final stage.**

*Enhancements*

Now add additional components like a service registry and make the functional framework based on the SOA architecture (see Figure 34.11). This way, we can migrate from legacy NMS to unified NMS framework. The advantage of phased approach is that each stage can be tested, further split for more control. This makes the transformation less risky, more manageable, and more cost effective.

| *Stage Description* | *Business Benefits* |
|---|---|
| **Stage 1: Grouping of functionality**<br>This stage involves splitting the legacy NMS solution into framework and data plug-in. The data plug-in only handles data collection from the network element based on the specific protocol. | • Reduced effort and cost to testing |
| **Stage 2: Make interaction to framework standardized**<br>Bring in full standardization for interaction based on SID and OSS/J to have minimal to nil dependency between framework and data collector. Now data collection and display systems provided by a different vendor or supplied by open source can be plugged in to the functional assembly. | • Plug-in can be introduced from the data collection layer and display layer |

*(Continued)*

(Continued)

| *Stage Description* | *Business Benefits* |
|---|---|
| **Stage 3: Split framework components to services**<br>The functionalities in the framework are now highly coupled, hence introducing new functionality or adding an open source module is costly from integration perspective. We need to isolate functionality into services and make them loosely coupled. | • Easier to fix bugs and easier to maintain code<br>• More reuse and return over investment with a service assembly |
| **Stage 4: Standardize interaction between framework components**<br>The loosely coupled functionalities will need to interact based on standards. | • Can introduce new services with minimal change in code<br>• Easily roll in and out of services<br>• Framework can be used for multiple networks |
| **Enhancements**<br>This includes having a centralized registry, repository, and components supporting publish-subscribe architecture. We can enhance it to have a MTOSI adapter. | • A full-fledged NGNM address a multitude of issues in the telecom NMS domain. |

The staged migration can also be performed as a single stage activity where the single stage represents a product release that ensures that the legacy NMS product transforms to NGNM framework.

The migration can be performed in line with existing product development merging the changes with the main code base after changes are made for each stage (see Figure 34.12). This way the product functionalities can be augmented while

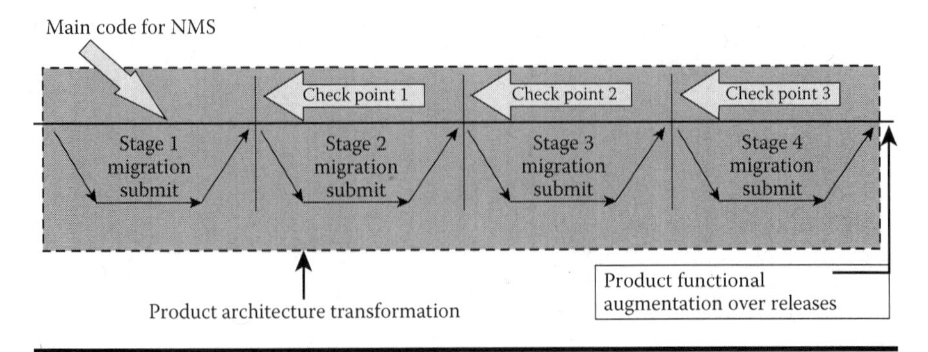


**Figure 34.12 Code migration.**

the transformation of architecture happens. Each stage can be tested and stabilized without effecting delivery to existing clients of NMS product.

### *34.6.3 Combining Mediation and Migration*

Not all clients would require a complete adoption of NGNM framework with mediation or migration approach. Different combinations of mediation and migration can be used based on the system involved.

A few examples include:

- Requires isolation of only a specific functionality: If only a specific functional component in the legacy NMS is planned to be reused, then only migrate the components of that functionality and use mediation for the rest.
- Requires only reuse of data collectors: Detach functionality from data collection as in Stage 1 of migration and get the generic data collectors. Now write mediation adapters for working with functionality block.

## 34.7 Analysis of NGNM Framework

| *Attribute* | *Legacy NMS* | *NGNM Solution* |
|---|---|---|
| Usability | Multiple GUI interfaces and applications make legacy NMS less user friendly, thus increasing the time required to fix a network issue. | Converged solution leads to ease of handling and bug fixing. |
| Modifiability | Due to interdependencies in modules, it is not easy to make changes in a cost-effective manner. | SOA-based NGNM is made up of self-sufficient modules that make addition of functionality or making changes easy. |
| Maintainability | The maintenance cost goes on increasing with minimal reuse and duplication of functionality. | Easier to maintain and fix bugs. |
| Flexibility | The hard coding and dependencies make the application perform only a specific functionality and does not allow flexibility in using parts of the application. | All service components have a published interface and is flexible to use in multiple scenarios. |

*(Continued)*

(Continued)

| *Attribute* | *Legacy NMS* | *NGNM Solution* |
|---|---|---|
| Scalability | There is a separate NMS solution for each type of network. | Simple customizations can make the solution work for a variety of networks. |
| Reusability | Very little reuse possible. | Functionalities can be reused with minimal or no changes. |
| Integrity | Integration with a new functionality requires considerable rework | Integration is easy with service components in SOA framework. |
| Testability | Addition of a new functionality will require retesting of all the modules again due to interdependencies. | Use of service components brings down testing effort and cost considerably. |

## 34.8 Conclusion

It is not cost effective for NMS developers to continue using their legacy solutions and sooner or later adoption of NGNM is inevitable. The major challenge is to adopt NGNM in a manner that best addresses the business needs of the company. This report tries to come up with an NGNM framework that can be considered the best one that can be formulated from existing technology standards in the domain. The adoption methods specified in the report need to be evaluated by different companies and each one should come up with the best strategy that satisfies the company goals. Suggested future work includes white papers that discuss detailed case studies on issues identified in the NGNM frame and its adoption in different companies and how these issues can be avoided. It should be understood that any development on an NGNM solution should ensure that existing scalable architecture is maintained and no new hard codings and dependencies between modules are introduced.

## Supporting Publications

1. Transformation of Legacy Network Management System to Service Oriented Architecture Conference. Next-Generation Communication and Sensor Networks, 2007, USA.
2. Strategies for Developing Next Generation Network Management System Conference. International Seminar and Exhibition on Strategies for Future Defense Networks and Relevance of EMI/EMC in Future Battlefields, 2007, India.

3. Developing Event Driven Intelligent Network Management Systems Conference. 11th World Multiconference on Systemics, Cybernetics, and Informatics (WMSCI, 2007), USA.
4. Unified NMS for Management of Defense Networks Conference: International Seminar and Exhibition on Technology and Anti-technology challenges (DEFCOM, 2008), India.
5. Development of a Scalable NGNM Solution based on TMF Guidelines and its Customization for IMS. Published under whitepapers of TeleManagement Forum (TMF), www.tmforum.org

## References


1. ITU-T SG4: M.3060 (Principles for the Management of Next Generation Networks).
2. 3GPP specification 3gpp TS 32.111-2: Telecommunication management; Fault Management; Part 2: Alarm Integration Reference Point: Information Service (IS).
3. 3GPP specification 3gpp TS 32.409: Telecommunication management; Performance Management (PM); Performance measurements IP Multimedia Subsystem (IMS).
4. TR133-NGN Management Strategy: Policy Paper and associated Addenda, Release 1.0 of TeleManagement Forum.
5. Venkatesan Krishnamoorthy, Naveen Krishnan Unni, and V. Niranjan. Event-Driven Service Oriented Architecture for an Agile and Scalable Network Management System. In Proceedings of the 1st International Conference of Next Generation Web Services Practices. Seoul, Korea, August 2005.
6. Matt Welsh, David Culler, and Eric Brewer. SEDA: Architecture for Well-Conditioned Scalable Internet Services. In Proceedings of the 18th Symposium on Operating Systems Principle, Alberta, Canada, October 2001, pp 230–45.
7. Andreas Hanemann, Martin Sailer, and David Schmitz. 2004. Assured Service Quality by Improved Fault Management. In Proceedings of the 2nd International Conference on Service Oriented Computing, New York, 2004.
8. Boris Gruschke. Integrated Event Management: Event Correlation Using Dependency Graphs. Proceedings of the 9th IFIP/IEEE International Workshop on Distributed Systems: Operations and Management, Newark, DE, October 1998, pp 130–41.
9. Zeng Bin, Hu Tao, Wang Wei, and Li ZiTan. A Model of Scalable Distributed Network Performance Management. In Proceedings of the International Symposium on Parallel Architectures, Algorithms and Networks (ISPAN 2004).
10. Jean Phillippe Martin Flatin, Pierre Alain Doffoel, and Mario Jeckle. Web Services for Integrated Management: A Case Study. In Proceedings of the European Conference on Web Services, September 27–30, 2004.

# Index

## J

## K



## L



## M

## N

## O

## T

## U